Molecular Analysis and Genome Discovery

Second edition

Edited by

Ralph Rapley

University of Hertfordshire, UK

and

Stuart Harbron

The Enzyme Technology Consultancy, UK

A John Wiley & Sons, Ltd., Publication

Library of Congress Cataloguing-in-Publication Data

Molecular analysis and genome discovery / edited by Ralph Rapley and Stuart Harbron. –2nd ed.
   p. ; cm.
Includes bibliographical references and index.
ISBN 978-0-470-75877-9 (cloth)
1. Molecular diagnosis. 2. Genomics. 3. Proteomics. 4. Pharmacogenomics. 5. Polymerase chain reaction. 6. DNA microarrays. I. Rapley, Ralph. II. Harbron, Stuart.
[DNLM: 1. Genomics–methods. 2. Drug Design. 3. Genetic Techniques. 4. Metabolomics–methods. 5. Proteomics–methods. QU 58.5]
RB43.7.M595 2011
615′.7–dc23

2011019770

A catalogue record for this book is available from the British Library.

This book is published in the following electronic formats: ePDF 9781119977445; Wiley Online Library 9781119977438; ePub 9781119978442; Mobi 9781119978459

Set in 10.5/13pt Times by Laserwords Private Limited, Chennai, India
Printed in Singapore by Markono Print Media Pte Ltd.

First Impression 2012

Contents

Preface

In our preface to the first edition of *Molecular Analysis and Genome Discovery* we indicated that the face of diagnostics and drug discovery had changed beyond recognition over the past decade. With the publication of this second edition this statement is even more apposite. There have been numerous advances in the technology and in the discovery of biological systems yielding new areas of analysis such as transcriptomics and metabolomics. There can be no doubt that these continued advances will lead to the ultimate goal of the development and use of personalized and stratified medicines.

This book aims to build upon the discovery and analysis aspects of the first edition by detailing the way in which techniques have been further developed or new methods implemented in the areas of molecular analysis and genome discovery. Following an updated overview of the important areas of genotyping, there are a number of chapters dealing with the methods of DNA analysis. These include the further use of DNA chips and qPCR, two mainstays of the area. Further analysis methods are presented including the use of microfluidic devices, high resolution melt profiling and the ability to analyse DNA on a large scale with parallel sequencing systems. Analysis of nucleic acids using aptamers has also been revisited and updated, providing further exciting analytical approaches for the post-human genome era. A chapter on nanotechnology in cancer biomarker discovery essentially bridges the nucleic acid analysis and discovery aspects and leads into chapters that are more orientated to proteins. Indeed, the emergence of nanotechnology has been spectacular, typifying our opening statement. The advancement of quantum dots, carbon nanotubes and nanoengineering presented in this chapter is a facet which thirty years ago would have been in the realms of science fiction. Chip analysis follows on from the perspective of protein analysis and discovery, after which antibody arrays in proteome profiling and multiplex microbead suspension array based immunoproteomics are addressed. The application of mass spectrometry as applied to metabolomics is detailed in the final chapter.

In compiling this second edition of *Molecular Analysis and Genome Discovery* we have sought again to combine both current and emerging approaches to the analysis of genomes and proteomes. This has been undertaken with an eye on how they may be of benefit for areas such as drug and biomarker discovery. We are again indebted to the panel of expert and distinguished authors who have provided vital insights into these important and exciting areas.

Ralph Rapley
Stuart Harbron

Contributors

Ahmed, Farid E.
GEM Tox Consultants & Labs, Inc., Greenville, NC 27834, USA

Alhamdani, Mohamed Saiel Saeed
Division of Functional Genome Analysis, Deutsches Krebsforschungszentrum (DKFZ), Im Neuenheimer Feld 580, 69120 Heidelberg, Germany

Bailes, Julian
School of Biological Sciences, Royal Holloway, University of London, Egham, Surrey, TW20 0EX, UK

Bayés, Mònica
Centro Nacional de Análisis Genómico, C/Baldiri Reixac 4, 08028 Barcelona, Spain

Bustin, Stephen A.
Academic Surgical Unit, 3rd Floor Alexandra Wing, Royal London Hospital, Whitechapel, London E1 1BB, UK

Dobrowolski, Steven F.
Department of Pathology, University of Utah, School of Medicine, Salt Lake City, Utah, USA

Friedman, Jan M.
Department of Medical Genetics, University of British Columbia, Vancouver, British Columbia, V6H 3N1 Canada and Child & Family Research Institute, Vancouver, British Columbia, V5Z 4H4 Canada

Griffiths, William J.
Institute of Mass Spectrometry, School of Medicine, Room 352 Grove Building, Swansea University, Singleton Park, Swansea SA2 8PP, Wales, UK

Gut, Ivo Glynne
Centro Nacional de Análisis Genómico, C/Baldiri Reixac 4, 08028 Barcelona, Spain

Hoheisel, Jörg D.
Division of Functional Genome Analysis, Deutsches Krebsforschungszentrum (DKFZ), Im Neuenheimer Feld 580, 69120 Heidelberg, Germany

Khan, Imran H.
Center for Comparative Medicine, and Department of Pathology and Laboratory Medicine, University of California, Davis CA 95616, USA

Krishhan, V. V.
Department of Chemistry, California State University, Fresno CA 93740 and Center for Comparative Medicine, and Department of Pathology and Laboratory Medicine, University of California, Davis, CA 95616, USA

Luciw, Paul A.
Center for Comparative Medicine, Department of Pathology and Laboratory Medicine, and California National Primate Research Center, University of California, Davis, CA 95616, USA

Marra, Marco
Department of Medical Genetics, University of British Columbia, Vancouver, British Columbia, V6H 3N1 Canada and BC Cancer Agency Genome Sciences Centre, Vancouver, British Columbia, V5Z 4S6 Canada

Milnthorpe, Andrew
School of Biological Sciences, Royal Holloway, University of London, Egham, Surrey, TW20 0EX, UK

Murphy, Jamie
Academic Surgical Unit, 3rd Floor Alexandra Wing, Royal London Hospital, Whitechapel, London E1 1BB, UK

Nadal, Pedro
Department d'Enginyeria Quimica, Universitat Rovira i Virgili, Avinguda Països Catalans 26, 43007 Tarragona, Spain

Nazar, Ross N.
Department of Molecular and Cellular Biology, University of Guelph, Guelph, Ontario, Canada N1G 2W1

O'Sullivan, Ciara K.
Department d'Enginyeria Quimica, Universitat Rovira i Virgili, Avinguda Països
Catalans 26, 43007 Tarragona, Spain and Institució Catalana de Recerca i Estudis
Avançats, Passeig Lluís Companys, 23, 08010 Barcelona, Spain

Ozdemir, Pinar
Department of Mechanical Engineering, University of Strathclyde, Glasgow,
G1 1XJ, UK

Pinto, Alessandro
Department d'Enginyeria Quimica, Universitat Rovira i Virgili, Avinguda Països
Catalans 26, 43007 Tarragona, Spain

Robb, Jane
Department of Molecular and Cellular Biology, University of Guelph, Guelph,
Ontario, Canada N1G 2W1

Svobodova, Marketa
Department d'Enginyeria Quimica, Universitat Rovira i Virgili, Avinguda Països
Catalans 26, 43007 Tarragona, Spain

Smieszek, Sandra
School of Biological Sciences, Royal Holloway, University of London, Egham,
Surrey, TW20 0EX, UK

Soloviev, Mikhail
School of Biological Sciences, Royal Holloway, University of London, Egham,
Surrey, TW20 0EX, UK

Tucker, Tracy
Department of Medical Genetics, University of British Columbia, Vancouver, British
Columbia, V6H 3N1 Canada

Wang, Yuqin
Institute of Mass Spectrometry, School of Medicine, Room 352 Grove Building,
Swansea University, Singleton Park, Swansea SA2 8PP, Wales, UK

Wittwer, Carl T.
Department of Pathology, University of Utah, School of Medicine, Salt Lake City, Utah, USA

Zhang, Yonghao
Department of Mechanical Engineering, University of Strathclyde, Glasgow, G1 1XJ, UK

1

Overview of Genotyping

Mónica Bayés and Ivo Glynne Gut

Introduction

Several types of variants exist in the human genome: single nucleotide polymorphisms (SNPs), short tandem repeats (STRs) also called microsatellites, small insertions or deletions (InDels), copy number variants (CNVs) and other structural variants (SVs) (Figure 1.1). SNPs are changes in a single base at a specific position in the genome, in most cases with only two alleles (Brookes 1999). By definition the rarer allele should be more abundant than 1% in the general population otherwise referred to as mutations. SNPs are found at a frequency of about one every 100–300 bases in the human genome. Since the completion of the Human Genome Project (HGP) (International Human Genome Sequencing Consortium 2004), SNPs have been discovered at an unprecedented rate and currently there are more than 24 million human reference SNP (rs) entries in the most extensive SNP database (dbSNP Build 132, www.ncbi.nlm.nih.gov/projects/SNP/). SNPs, however, are not randomly distributed across the genome and occur much less frequently in coding sequences than in noncoding regions. SNPs located in regulatory or protein coding regions are more likely to alter the biological function of a gene than those in intergenic regions.

Genotyping is the process of assignment of different variants in an otherwise conserved DNA region. The relative simplicity of methods for SNP genotyping, the abundance of SNPs in the human genome and their low mutation rates have made them very popular in the past decade. SNP genotyping has currently many applications: disease gene localization and identification of disease-causing variants, quantitative trait loci (QTL) mapping, pharmacogenetics, identity testing based on

Molecular Analysis and Genome Discovery, Second Edition. Edited by Ralph Rapley and Stuart Harbron.
© 2012 John Wiley & Sons, Ltd. Published 2012 by John Wiley & Sons, Ltd.

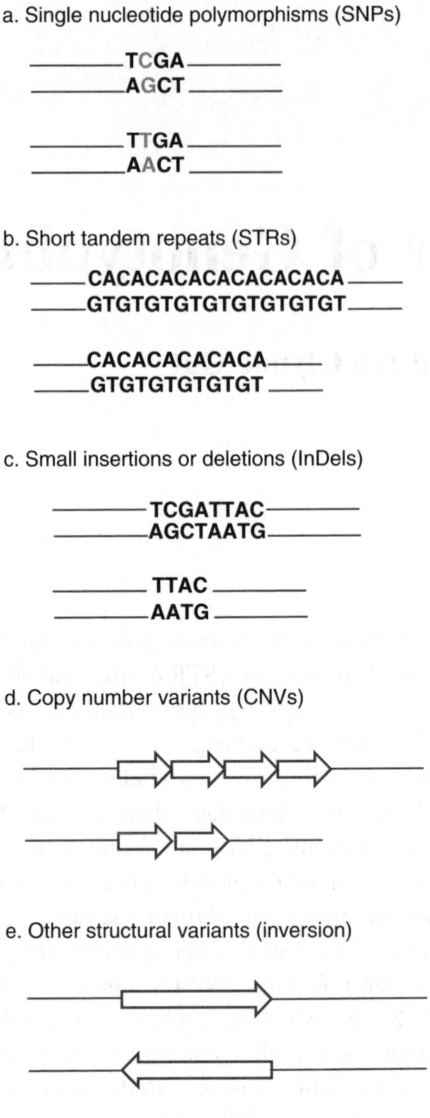

a. Single nucleotide polymorphisms (SNPs)

```
_____TCGA_____
_____AGCT_____

_____TTGA_____
_____AACT_____
```

b. Short tandem repeats (STRs)

```
_____CACACACACACACACACA_____
_____GTGTGTGTGTGTGTGTGT_____

_____CACACACACACA_____
_____GTGTGTGTGTGT_____
```

c. Small insertions or deletions (InDels)

```
_____TCGATTAC_____
_____AGCTAATG_____

_____TTAC_____
_____AATG_____
```

d. Copy number variants (CNVs)

e. Other structural variants (inversion)

Figure 1.1 Types of genetic variants. Each arrow represents a DNA segment of more than 1 kb

genetic fingerprinting, just to mention the major ones. Genotyping applications extend beyond human genetics to animals and plants.

Although some SNP alleles confer susceptibility to complex disorders (asthma, cardiovascular disease, diabetes, etc.), most SNPs are not solely responsible for a disease state. Instead, they serve as biological markers for identifying disease-related variants on the human genome map, based on the fact that alleles of SNPs that are located nearby tend to be inherited together (Jorde 1995). This is termed linkage disequilibrium. For disease gene identification two basic strategies are applied. In the linkage study, related individuals are genotyped with several hundreds to thousands of polymorphisms distributed throughout the genome and attempts are made to identify genetic markers that cosegregate with the disease. Genetic linkage methods have been applied successfully to identify the mutated gene in Mendelian diseases (Risch 1991). If investigating the genetic basis of complex disorders, the association or linkage disequilibrium approach is more powerful (Risch and Merikangas 1996). It involves establishing genotype–phenotype correlations in unrelated individuals that are solely selected on the basis of being affected by a phenotype or not (Clark 2003).

Genetic association studies require a large number of samples to achieve statistically significant results that indicate that a particular allele in a particular region of the genome confers an increased risk of developing the disorder. Many association studies based on the analysis of candidate genes that involve genotyping of tens or hundreds of SNPs in hundreds or thousands of samples have been published. In the past five years, the ability to assay for more than 100 000 SNPs distributed across the genome has enabled the systematic study of complex disorders under a whole genome approach, without any preconceived hypothesis or candidates. Successful genome-wide association studies (GWAS) have been conducted for common diseases such as age-related macular degeneration, rheumatoid arthritis, asthma, Crohn's disease, bipolar disorder, coronary heart disease, type 1 and type 2 diabetes among many others (Klein *et al.* 2005; Wellcome Trust Case Control Consortium 2007; Moffatt *et al.* 2007, 2010; Hindorff *et al.* 2009). A list of all GWAS and associated polymorphisms is kept up to date at www.genome.gov/gwastudies.

In the next sections the most popular methods and platforms for SNP genotyping are discussed, highlighting some practical aspects. Other related applications such as methylation, copy number analysis and second generation sequencing using the same underlying molecular approaches are covered thereafter.

Methods for interrogating SNPs

There are many mature SNP genotyping technologies that have been integrated into large-scale genotyping operations. SNP genotyping methods are still being improved, perfected, integrated and new methods are emerging to satisfy the needs of genomics and epidemiology. No one SNP genotyping method fulfils the requirements of every study that might be undertaken. The choice of a method depends on the scale of the envisioned genotyping project and the resources available. A project might require genotyping of a limited number of SNP markers in a large population or the analysis of a large number of SNP markers in a few samples. Flexibility in choice of SNP markers and DNAs to be genotyped or the possibility to precisely quantify an allele frequency in pooled DNA samples might also be issues.

SNP genotyping methods are very diverse (Syvänen 2001; Kim and Misra 2007). Broadly, each method can be separated into two elements, the biochemical method for discriminating SNP alleles and the actual analysis or measurement of the allele-specific products, which can be an array reader, a plate reader, a mass spectrometer, a gel separator/reader system, or other. In addition, most technologies also require a PCR amplification step to increase the number of target SNP-containing DNA molecules and to reduce the complexity of the template material used for the allele discrimination step.

The most popular methods for allele discrimination are restriction endonuclease digestion, primer extension, hybridization and oligonucleotide ligation (Figure 1.2a).

Restriction endonuclease digestion

Restriction fragment length polymorphisms (RFLPs) are one of the first typing methods described and by far predate the coining of the term SNP (Botstein *et al.* 1980). Restriction endonuclease digestion is still a common format for SNP genotyping in a standard laboratory (Parsons and Heflich 1997). PCR products are digested with restriction endonucleases that are specifically chosen for the base change at the position of the SNP, resulting in a restriction cut for one allele but not the other (Figure 1.2a). In some cases, specific restriction sites can be created during the amplification step by using primers with minor changes in the sequence. Digestion patterns are used for allele assignment after gel electrophoresis. Major limitations of the restriction method are that it is only applicable to a fraction of SNPs and that it does not lend itself to automation.

Primer extension

Primer extension is a stable and reliable way of distinguishing alleles of a SNP. Nucleotides are added by a DNA polymerase generating allele-specific products (Syvänen 1999). Allele-specific primer extension (ASPE) is based on the ability of DNA polymerases to extend with high efficiency those oligonucleotides with 3' perfectly matched ends (Figure 1.2a). It requires two allele-specific primers that have the nucleotide that corresponds to the allelic variant at their 3' ends. In single base primer extension (SBE) an oligonucleotide hybridizes immediately before the SNP nucleotide and the DNA polymerase incorporates a single nucleotide that is complementary to the SNP allele (Figure 1.2a). SBE uses dideoxynucleotides (ddNTP) as terminators.

Hybridization

Alleles differing by one base can be distinguished by hybridizing complementary oligonucleotide sequences to the target DNA (ASO or allele-specific oligonucleotide

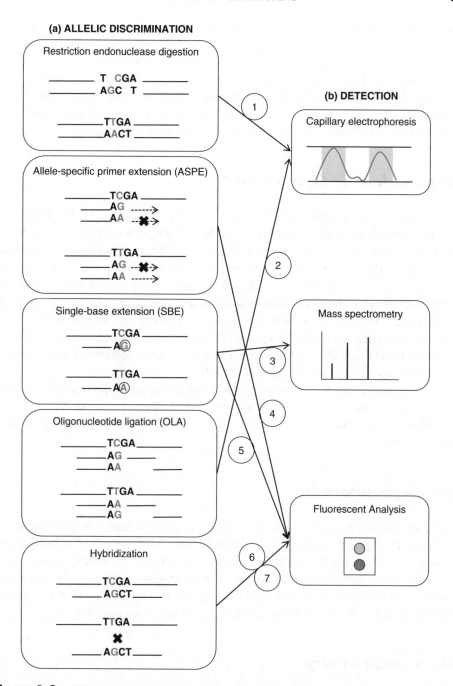

Figure 1.2   SNP genotyping technologies separated into allele discrimination methods (A) and detection of allele-specific products (B). Arrows denote genotyping assays that combine different allele discrimination and detection methods. 1 Restricion endonuclease digestion; 2 SNPlex; 3 iPLEX GOLD assay; 4 GoldenGate assay; 5 Infinium assay; 6 TaqMan assay; 7 GeneChip assay

hybridization), without any enzymatic reaction (Figure 1.2a). As the two alleles of a SNP are very similar in sequence, significant cross-talk can occur. Several approaches have been taken to overcome this problem: the use of multiple probes per SNP, the use of modified oligonucleotides such as peptide nucleic acids (PNAs) (Egholm *et al.* 1993) or locked nucleic acids (LNAs) (Ørum *et al.* 1999) that increase stability of DNA–DNA complexes, the real-time monitoring of the hybridization kinetics or the combination of hybridization and 5′ nuclease activity of polymerases.

Oligonucleotide ligation (OLA)

OLA relies on the specificity of DNA ligases to repair DNA nicks. For OLA, two oligonucleotides adjacent to each other are ligated enzymatically by a DNA ligase when the bases next to the ligation position are fully complementary to the template strand (Barany 1991; Jarvius *et al.* 2003) (Figure 1.2a). The assay requires three probes to be designed: two allele-specific probes that have at their 3′ ends the nucleotide complementary to the SNP variants and one common probe that anneals to the target DNA that is immediately adjacent. Padlock is a variant of OLA that employs two allele-specific oligonucleotides with target complementary sequences separated by a linker. When perfectly annealed to the target sequence, padlock probes are circularized by ligation (Nilsson *et al.* 1994).

Major detection methods include gel electrophoresis, mass spectrometry, fluorescence analysis, and chemiluminescence detection (Figure 1.2b). Nearly all of the above-described methods for allele-distinction have been combined with all of these analysis formats.

Gel electrophoresis

Allele-specific DNA fragments of different sizes can be separated by electrophoretic migration through gels (Szántai and Guttman 2006). Throughput and resolution can be increased if 'gel-filled' capillaries are used. Advantages of capillary systems over slab gel systems include the potential for 24-hour unsupervised operation, the elimination of cumbersome gel pouring and loading, and that no lane tracking is required. Instrumentation with 96- or 384-capillaries is commercially available.

Mass spectrometry

MALDI-TOF MS (matrix-assisted laser desorption/ ionization time-of-flight mass spectrometry) can be used to measure the mass of the allele-specific products. It has been demonstrated as an analysis tool for SNP genotyping (Haff and Smirnov 1997; Tost and Gut 2005). The allele-specific products are deposited onto a matrix on the surface of a chip, ionized with a short laser pulse and accelerated towards the detector (Jurinke *et al.* 2002). The time-of-flight of a product to the detector is directly related to its mass. High resolution and speed are major advantages of the MALDI-TOF MS detection method. Resolution of the current generation of mass spectrometers allows the distinction of base substitutions in the range of 1.000–6.000 Da (this corresponds to product sizes of 3–20 bases, the smallest mass difference for a base change thymine to adenine is 9 Da).

Fluorescent analysis

Allele-specific products can be labelled with different fluorescent dyes and detected using fluorescent readout systems, either microtitre plate or array based (Landegren *et al.* 1998). Most readers use a white light source and optical filters to select specific excitation and emission wavelengths. Some of them can also measure parameters such as fluorescence polarization (FP, measures the increase in polarization of fluorescence caused by the decreased mobility of larger molecules) (P.Y. Kwok 2002) and Förster resonance energy transfer (FRET, measures the changes in fluorescence due the separation of two dyes of a donor/acceptor system) (Tong *et al.* 2001). Most popular fluorescent dyes used in SNP genotyping are Cy3 and Cy5.

Most current genotyping methods are generally based on the combination of one of the allelic discrimination and one of the detection methods described above (Figure 1.2). Often very different methods share elements, for example, reading out a fluorescent tag in a plate reader, or the primer extension method, which can be analysed in many different analysis formats.

Commercial platforms for SNP genotyping

A plethora of SNP genotyping platforms is currently commercially available (Ragoussis 2009). Many of them require purchasing expensive proprietary equipment and expensive laboratory set-up. However, they offer streamlined laboratory and analysis workflows. They range from individual SNP genotyping platforms (Life Technologies TaqMan) to focused content genotyping (Sequenom iPLEX Gold, Illumina GoldenGate) and to platforms for whole genome genotyping (WGG) (Illumina Infinium and Affymetrix GeneChips) (Fan *et al.* 2006). WGG arrays contain from 100 000 to 2.5 million SNPs selected by different approaches and with minor allele frequencies >0.05 in the general population.

TaqMan assay

The TaqMan assay (Life Technologies, www.appliedbiosystems.com) is based on allele-specific hybridization coupled with the 5′ nuclease activity of Taq polymerase during PCR (Holland *et al.* 1991; Livak 2003; Livak *et al.* 1995). The detection is performed by measuring the decrease of FRET from a donor fluorophore to an acceptor-quencher molecule. TaqMan probes are allele-specific probes labelled with a fluorescent reporter at the 5′ end and a common quencher attached to the 3′ end that virtually eliminates the fluorescence in the intact probe. Each assay uses two TaqMan probes that differ at the SNP site, and one pair of PCR primers. During PCR, successful hybridization of the TaqMan probe due to matching with one allele of the SNP results in its degradation by the 5′- to 3′-nuclease activity of the employed DNA polymerase whereby the fluorescent dye and quencher are separated, which promotes fluorescence. TaqMan probes can be designed to detect multiple nucleotide polymorphisms (MNPs) and insertion/deletions (InDels). Because of the simplicity in chemistry, the reaction set-up can be easily automated using liquid handling robots. The 7900HT Fast Real-Time PCR system (Life Technologies) allows up to eighty-four 384-well plates to be processed without manual intervention in less than 4 days. It is a very contamination-safe procedure as plates do not need to be opened after PCR for reading. In contrast, the limiting factors of the technology are the low SNP multiplexing level and the relatively high cost of the dual-labelled probes. Life Technologies has developed a library with 4.5 million genome-wide human TaqMan assays (of which 160 000 are validated assays) for which reagents are commercially available.

In the recent years, a couple of high-throughput real-time PCR instruments have been introduced. The Biomark system (Fluidigm, www.fluidigm.com) contains integrated fluidic circuits or 'dynamic arrays' that allow setting up 9216 genotyping reactions in a single experiment (Wang *et al.* 2009). The user has to simply dispense 96 DNA samples and 96 TaqMan genotyping assays and the dynamic array will then do the work of assembling the samples in all possible combinations. The OpenArray system (www.appliedbiosystems.com) (Morrison *et al.* 2006) can also perform SNP analysis using TaqMan probes. The OpenArray plate contains 3072 reaction through-holes generated by a differential coating process that deposits hydrophilic coatings on the interior of each through-hole and hydrophobic coatings on the exterior. This enables OpenArray plates to hold solutions in the open through-holes via capillary action. The company provides the researcher with OpenArray plates that are preloaded with the selected TaqMan probes (from 16 to 256 different assays per plate depending on the plate format). The main advantages of the Biomark and OpenArray systems compared to conventional thermocyclers are higher throughput, small sample requirement, low reagent consumption and less liquid handling.

iPLEX GOLD assay

The iPLEX Gold reaction (Sequenom, www.sequenom.com) is a method for detecting insertions, deletions, substitutions, and other polymorphisms that combines multiplex PCR followed by a single-base extension and MALDI-TOF MS detection

(Jurinke *et al.* 2002; Oeth *et al.* 2009). After the PCR, remaining nucleotides are deactivated using shrimp alkaline phosphatase (SAP). The SAP cleaves a phosphate from the unincorporated dNTPs, converting them to dNDPs which renders them unavailable to future polymerization reactions. Next, a single base primer extension step is performed incorporating one of the four terminator nucleotides into the SNP site. The extension products are desalted and transferred onto chips containing 384 matrix spots. The allele-specific extension products of different masses are analysed using MALDI-TOF MS. In theory up to 40 different SNPs can be assayed together if the different allele-products have distinct masses; however, generally multiplexes on the order of 24 are more realistic. The whole lab workflow is highly automated and it takes less than 10 hours to process one 384 plate. The MassARRAY Analyzer 4 (Sequenom) can analyse from dozens to over 100,000 genotypes per day, and from tens to thousands of samples. Significant advantages of the method are that it requires standard unmodified oligonucleotides which are cheap and easy to come by. It is a very sensitive method with low input sample requirements and finally generates highly accurate data because it relies on the direct detection of the allele-specific product.

GoldenGate assay

The GoldenGate assay (Illumina, www.illumina.com) (Shen *et al.* 2005) can interrogate 48, 96, 144, 192, 384, 768 or 1536 SNPs simultaneously. The assay combines allele-specific primer extension and ligation for generating allele-specific products followed by PCR amplification with universal primers. Three oligonucleotides are designed for each SNP locus, two of which are allele-specific (ASO) with the SNP allele on their 3′ end, and a locus specific oligonucleotide (LSO) that hybridizes several bases downstream the SNP site. The LSO primer also contains a unique address sequence that allows separating the SNP assay products for individual readout. In the protocol, during the hybridization process, the oligonucleotides hybridize to the genomic DNA that has been first immobilized on a solid support. The complementary ASO is extended and ligated to the LSO, providing high locus specificity. The ligated products are then amplified using universal PCR primers P1, P2 and P3. Primers P1 and P2 are specific for each ASO and carry a fluorescent tag that is used for allele calling.

The separation of the assay products in solution onto a solid format is done using Veracode technology (48, 96, 144, 192 or 384-plex) (Lin *et al.* 2009). It uses cylindrical glass microbeads (240 microns in length) with unique digital holographic codes and coated with capture oligonucleotides that are complementary to one of the addresses present in the PCR products. When excited by a laser, each VeraCode bead emits a unique holographic code image. The BeadXpress reader (Illumina) can identify the individual bead types and in addition detect the results from the two-colour genotyping assay. The Veracode technology contains assay replicates of 20–30 beads per bead type, providing a high level of quality control.

Infinium assay

The Infinium II assay (Illumina, www.illumina.com) uses a two-colour SBE protocol for allelic discrimination coupled with the BeadChip technology for assay detection (Steemers and Gunderson 2007). Whole genome amplified (WGA) samples are hybridized to 50-mer oligonucleotide probes covalently attached to particular microspheres or beads that are randomly assembled in microwells on planar silica slides (BeadChips). After the hybridization, the SNP locus-specific oligonucleotides are extended with the corresponding fluorescently labelled dideoxynucleotides. The intensities of the bead's fluorescence are detected by the iScan Reader (Illumina).

Currently available BeadChips for human allow profiling samples with 300 000 to 2.5 M SNPs distributed throughout the genome. SNP selection in these chips is based on results from the HapMap project (www.hapmap.org) providing high coverage across the genome (see 'SNP databases'). New arrays with up to 5 M common and rare variants from the 1000 Genomes Project (www.1000genomes.org) are in development. The Infinium assay can also be used also to develop BeadChips with customized SNP content (iSelect). Genome-wide genotyping BeadChips are also available for other species such as cattle, pigs and dogs.

GeneChip assay

In the GeneChip assay (Affymetrix, www.affymetrix.com) allelic discrimination is achieved by direct hybridization of labelled DNA to arrays containing allele-specific oligonucleotides. These 25-mer probes are synthesized in an ordered fashion on a solid surface by a light-directed chemical process (photolithography) (Fodor et al. 1991). Oligonucleotides covering the complementary sequence of the two alleles of a SNP are on specific positions of the array. Multiple probes for each SNP are used to increase the genotyping accuracy. The hybridization pattern of all oligonucleotides spanning the SNP is used to evaluate positive and negative signals.

Genomic DNA is digested with a restriction endonuclease and ligated to adaptors that recognize the cohesive 4 bp overhangs. The ligation products are then amplified by PCR using a single universal primer and creating a reduced representation of the genome (Kennedy et al. 2003). Next, PCR amplicons are fragmented, end-labelled and hybridized to the array under stringent conditions. After extensive washing steps, the remaining fluorescence signal is automatically recorded by the GeneChip 3000 scanner (Affymetrix). A specific fluidics station and a hybridization oven are also required to carry out the procedure.

Affymetrix has developed several microarrays designed specifically to interrogate SNPs distributed throughout the human genome. The most comprehensive array, the Genome-Wide Human SNP Array 6.0 has 1.8 million genetic markers, including 906 600 SNPs. The median inter-marker distance over all 1.8 million SNP and copy number markers combined is less than 700 bases. Affymetrix has also launched a new high-throughput genotyping assay, the Axiom Genotyping Solution. It is based on a 96-sample format and can process more than 750 samples per week. The initial Axiom Genome-Wide Human Array contains more than 560 000 SNPs.

Other popular platforms for SNP genotyping are SNPstream (Beckman Coulter) and Pyrosequencing (Qiagen) (Table 1.1) (Syvänen 2001; Sobrino *et al.* 2005; Ragoussis 2009).

Practical recommendations

Different aspects have to be taken into consideration when setting-up a genotyping platform: DNA quality assessment, contamination control, automation and data quality control measures.

In high-throughput laboratories, liquid handling automation is essential both for the SNP allele-discrimination and allele-detection processes (Gut 2001). It not only speeds up the genotyping process but also reduces errors introduced by human handling and pipetting and minimizes the possibility of cross-contamination of samples. Many suppliers of laboratory robotics offer liquid handling robots that can be integrated into high-throughput genotyping workflows. In general, the ease of automation is directly correlated to the complexity of an SNP genotyping protocol. Steps such as gel-filtration and manipulation of magnetic beads can be more problematic to automate. Current liquid handling robots can support both plates and slide microarray formats.

One of the biggest challenges in running SNP genotyping at high-throughput is the management of the production line. A Laboratory Information Management System (LIMS) is a software tool for keeping track of samples, laboratory users, instruments, lab processes, quality standards, and results. Originally, LIMS were developed in-house but currently there are several commercial solutions available such as Biotracker (Ocimum Biosolutions), Geneus (GenoLogics) and StarLIMS (StarLIMS Corporation). Complete systems for the entire high-throughput SNP genotyping process, with automation and LIMS, are marketed as off-the-shelf products. Examples of this are systems from Affymetrix, Sequenom and Illumina. In addition, all platforms discussed in the previous section have developed analysis software for fully automatic scoring of alleles and genotypes and monitoring the performance of all controls (Figure 1.3).

One of the greatest concerns in optimizing a genotyping laboratory is to control for PCR contamination. The high-throughput and repetition of assays with common primer pairs can easily lead to amplification of cross-contamination. The

Table 1.1 Characteristics of commercially available genotyping systems

Platform	Assay Name	Vendor	Allelic Discrimination	Detection	Multiplexing	Format	Throughput (GT per day)[a]
PyroMark MD	Pyrosequencing	Qiagen	Sequencing	Chemiluminescence	1	96 well plate	5000
7900HT	TaqMan	Life Technologies	Hybridization + 5′ nuclease	FRET	1	384 well plate	8064
Biomark	TaqMan	Fluidigm	Hybridization + 5′ nuclease	FRET	1	9216 nanowell dynamic plate	27 648
OpenArray	TaqMan	Life Technologies	Hybridization + 5′ nuclease	FRET	1	3072 nanowell slide	98 304
MassArray Compact Analyzer	iPlex Gold	Sequenom	SBE	Mass spectrometry	2–40	384 position Spectrochip	100 000
SNPstream Imager	SNPstream	Beckman Coulter	SBE	Fluorescence	12–48	384 array plate	3 M
BeadXpress	GoldenGate Veracode	Illumina	ASPE	Fluorescence	48–384	96 well plate	110 000
iScan	Infinium	Illumina	SBE	Fluorescence	3000–2.5 M	12–32 sample Beadchips	185 M
GeneChip 3000	GenChip	Affymetrix	Hybridization	Fluorescence	1.8 M	1 sample array	414 M
GeneTitan	Axiom	Affymetrix	Hybridization	Fluorescence	560 000	96 array plate	430 M

[a] GT: genotypes; assuming a single instrument with autoloader when available and liquid handling automation.

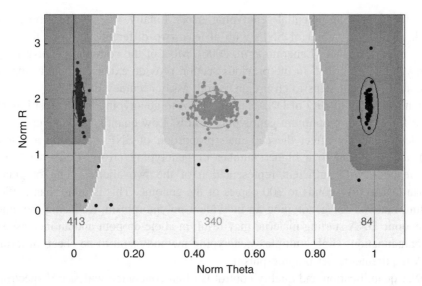

Figure 1.3 Genotype cluster plot for one SNP genotyped across 270 samples using the GoldenGate assay and the Veracode technology. Each data point represents one sample, the y-axis is normalized signal intensity (sum of intensities of the two fluorescent signals) and the x-axis is the theta value that indicates the allelic angle. The software automatically clusters the DNA samples into two homozygous clusters (red and blue) and a heterozygote cluster (yellow). Points depicted in black are unsuccessfully genotyped samples

most important recommendation for preventing contamination is to maintain separate areas, dedicated equipment and supplies for pre-PCR steps (sample preparation and PCR set-up) and post-PCR steps (thermocycling and analysis of PCR products). The rule of thumb should be never to bring amplified PCR products into the PCR set-up area. Uracil-DNA glycosylase (UNG) can also be used to prevent carryover contamination of the PCR products (Longo *et al.* 1990). By using dUTP instead of dTTP in all PCRs, UNG treatment can prevent the reamplification of carryover PCR products by removing any uracil incorporated into the amplicons and then cleaving the DNA at the created abasic sites. Finally, laboratory practices such as the use of disposable filter tips, positive-displacement pipettes, non-contact dispensing options and periodical lab and instrument cleaning also help reduce the risk of carryover contamination (S. Kwok and Higuchi 1989).

Genotyping errors have a deleterious effect on the statistical analysis of the data. To address this issue several quality controls should be carried out in each genotyping experiment: negative controls to monitor cross-contamination, positive controls to check concordance with publicly available data and replicate DNA samples to account for intra- and inter-plate reproducibility (Pompanon *et al.* 2005). Analysis statistics such as deviation from the Hardy–Weinberg equilibrium, Mendelian inconsistencies in pedigrees or the number of inferred recombinants can also be of great value for identifying potential genotyping errors. Finally, it is also recommended to check regularly a subset of SNPs with at least two different platforms to evaluate platform performance (Lahermo *et al.* 2006). Most of the common genotyping platform vendors described in the previous section provide extensive quality measures of several protocol steps to ensure an overall assay accuracy of >99%.

Monitoring the quality of DNA samples prior to genotyping is the most important factor for achieving optimum genotyping results. Low quantity and/or quality DNA samples negatively affect the call rate (proportion of SNPs receiving a genotype call) and also lead to a higher number of genotyping errors. DNA needs to be in a reaction with sufficient representation of the two alleles – 1 ng of genomic human DNA corresponds to 300 copies of the genome. This is more than sufficient starting material for genotyping an individual polymorphism. Reducing the amount of genomic DNA starting material may result in allele-dropout and an increased risk of contamination. High-multiplex genotyping methods tend to be cheap in terms of DNA requirements per polymorphism.

DNA quantification and quality control is often conducted with a UV spectrophotometer at wavelengths of 260 nm and 280 nm. The ratio of absorbance readings at the two wavelengths should be between 1.8 and 2.2, while protein contamination can be assessed by measuring the A260/230 ratio (1.6–2.4). A more precise quantification of the double-stranded DNA target can obtained using a fluorescent nucleic acid stain such as Picogreen (Invitrogen) and a fluorometer (excitation and emission wavelengths of 502 nm and 523 nm, respectively) or by real-time qPCR using a single-copy gene as a copy number reference. Finally, the integrity and molecular weight of DNA are measured by gel electrophoresis using either agarose gels or an instrument such as the Agilent Bioanalyzer.

SNP databases

SNP databases such as dbSNP and HapMap are essential resources for the study of human complex disorders and for evolutionary studies.

The Single Nucleotide Polymorphism database (dbSNP, www.ncbi.nlm.nih.gov/projects/SNP) was launched in 1998 as a public-domain archive of simple genetic polymorphisms. It contains SNP-related information such as SNP flanking DNA sequences, alleles, allele frequencies, validation status and functional relationships to genes (Sherry *et al.* 2001). As of build 132 (September 2010), dbSNP has collected over 244 million submissions corresponding to more than 87 million reference SNP clusters (refSNP) from 100 organisms, including *Homo sapiens*, *Mus musculus*, *Gallus gallus*, *Oryza sativa*, *Zea mays* and many other species. A full list of organisms and the number of reference SNP clusters for each can be found at www.ncbi.nlm.nih.gov/SNP/snp_summary.cgi. The data of dbSNP is also included in repositories such as ENSEMBL (www.ensembl.org) and the UCSC Genome Browser (genome.ucsc.edu).

The international HapMap project (www.hapmap.org) started in 2002 with the aim of cataloguing the vast amount of genetic variation in humans and describing how it is organized in short stretches of strong linkage disequilibrium (haplotype blocks) that coincide with ancient ancestral recombination events (Couzin 2002; Pääbo 2003). Since then more than 3 million SNPs (with an average density of 1 SNP per kb and minor allele frequency >0.05) have been analysed in 270 individuals from populations with African, Asian and European ancestry (International HapMap Consortium 2007). HapMap results provide researchers with a selection of SNP markers that tag haplotype blocks to reduce the number of genotypes that have to be measured for a genome-wide association study (more than 500 000 tag SNPs are required to capture all Phase II SNPs with r2 ≥ 0.8 in a population from Northern Europe (CEU)). In Phase III, 1,184 reference individuals from 11 global populations have been genotyped for 1.6 million SNPs (International HapMap Consortium 2010).

Recent improvements in sequencing technology (see 'Second generation sequencing') fostered the creation of the 1000 Genomes Project (1000 GP, www.1000genomes.org) in 2008. The aim of the project is to obtain a nearly complete catalogue of all human genetic variations with frequencies greater than 1% by sequencing the genomes of 2500 individuals from different populations. Data from three pilot projects is already available: low coverage sequencing of 180 individuals, sequencing at deep coverage of six individuals and sequencing gene regions in 900 individuals. 1000 GP data is further improving the process of identification of disease-associated regions.

Resources such as dbSNP, HapMap and 1000 GP have unquestionably saved medical researchers a lot of time and cost in their projects. All of the information generated by these projects is rapidly released into the public domain. In addition, DNA samples used in the HapMap and 1000 Genomes projects are also publicly available through Coriell Institute (ccr.coriell.org).

Methylation analysis

In mammals, epigenetic modifications are known to play a critical role in the regulation of gene expression across the genome and in maintaining genomic stability (Bernstein *et al.* 2007). Many studies have implicated aberrant methylation in the aetiology of common human diseases, including cancer, multiple sclerosis, diabetes and schizophrenia (Tost 2010). Alterations in DNA methylation can be used as biomarkers for early cancer detection, to discriminate among tumour subtypes or to predict disease outcome (Shames *et al.* 2007).

In animals and plants, methylation occurs preferentially at the 5 position of cytosines within CpG dinucleotides. Reversible methylation of cytosine in the sequence context CpG adds a dynamic component to DNA because it can act as a switch of transcription of a gene if the CpGs are in the promoter region of the gene (Suzuki and Bird 2008). CpGs that can be methylated or not are termed methylation variable positions (MVPs). Methylation information is lost during PCR or primer extension reactions. Nonetheless, measuring the degree of DNA methylation can be done by virtually any SNP genotyping method if a genomic DNA sample is prior treated with bisulfite (L. Shen and Waterland 2007). Bisulfite treatment of DNA results in conversion of non-methylated C into U, while methylated C remains unchanged. After bisulfite conversion and PCR amplification, determination of the degree of methylation at a given MVP in the genomic DNA sample can be achieved by quantifying the degree of C and T at that position. The quantitative resolution of the genotyping method determines the accuracy of measurement that can be achieved. One of the most widely used methods for quantifying MVPs is pyrosequencing (Dupont *et al.* 2004). Pyrosequencing is a sequencing-by-synthesis method based on the detection of pyrophosphate (PPi) which is released during DNA synthesis in a quantity equimolar to the amount of incorporated nucleotide. Assay set-up is straight forward and the accuracy of quantification is better than 2% if using the PyroMark MD instrument (Qiagen, www.pyrosequencing.com). MALDI-TOF mass spectrometry can also be used to detect cytosine methylation using bisulfite conversion biochemistry, followed by PCR and base-specific cleavage process that generate a distinct signal pattern from the methylated and non-methylated template DNA (Sequenom, www.sequenom.com) (Ehrich *et al.* 2005).

There are several array-based methods to identify global patterns of CpG methylation (Beck and Rakyan 2008; Laird 2010). The most popular one is the Human-Methylation27 and HumanMethylation450 BeadChips developed by Illumina (www.illumina.com). It uses the Infium II assay (see 'Commercial platforms for SNP genotyping') to interrogate bisulfite converted DNA for up to 450 000 CpG methylation sites from 99% of RefSeq genes. It reliably detects a difference of 20% in methylation.

The first complete maps of DNA methylation (or methylome) with a one base pair resolution were obtained for *Arabidopsis thaliana* (Lister *et al.* 2008) and two human cell types (Lister *et al.* 2009) using bisulfite treatment and second generation sequencing technologies (see 'Second generation sequencing').

Copy number variation analysis

Copy number variants (CNVs) are a common form of genetic variation in human populations (Database of Genomic Variants, projects.tcag.ca/variation) (Redon *et al.* 2006; McCarroll *et al.* 2008). By analogy to the standard definition of a SNP, a CNV is a copy number polymorphism that ranges from one kilobase to several megabases in size and has a minor allele frequency of 1% or greater.

CNVs may alter gene function by affecting gene dosage, positional effect or by directly interrupting genes. Although most CNVs are neutral polymorphic variants, some of them have been demonstrated to be associated with human diseases such as autism, schizophrenia, mental retardation or psoriasis (Zhang *et al.* 2009). Screening for CNVs, in addition to SNP genotyping, in disease gene identification studies is a major trend in current research projects.

There are several methods to quantify copy number variation across the genome for research and diagnostic purposes. Array comparative genome hybridization (array CGH) is a powerful tool for discovering previously unrecognized submicroscopic aberrations in cancer and genomic disorders just by measuring hybridization intensities (Carter 2007; Gresham *et al.* 2008). Current arrays for CGH contain up to 4.2 million oligonucleotide probes that enable genome-wide detection of CNVs down to 1.5–5 Kb resolution (Nimblegen, www.nimblegen.com) (Agilent, www.agilent.com).

Current WGG SNP arrays, although originally designed for SNP genotyping, can also be used to capture CNVs at a genome-wide scale (Cooper *et al.* 2008; Carter 2007). SNP-CGH, unlike conventional array CGH, can detect copy neutral genetic abnormalities such as uniparental disomy (UPD) and loss of heterozygosity (LOH). HumanOmin2.5 Beadarrays (Illumina) contain nearly 2.5 million genetic markers (median spacing between markers is 1.5 kb), including 60 000 CNV-targeted markers. The human SNP Array 6.0 (Affymetrix) features 1.8 million genetic markers, including more than 946 000 probes for the detection of copy number variation. The Cytogenetics Whole-Genome 2.7 M Array (Affymetrix) provides a comprehensive analysis of structural variation. It contains 2.7 million copy number markers, including 2.3 million of non-polymorphic markers and 400 000 SNPs.

Over the past few years, a number of software packages have been developed to detect changes in copy number by using SNP-CGH data and analysing total signal intensities and allelic intensity ratios (Winchester *et al.* 2009). Different methods based on Hidden Markov Models, circular binary segmentation or mixed models are used to detect CNV segment boundaries. In a second step particular segments that are different in copy number compared with values from a reference individual or group of individuals are identified. For robust detection, a CNV interval requires significant ratio shifts in several consecutive probes.

Because of the relatively low signal-to-noise ratio and high experimental variation that characterizes many of the platforms, candidate CNVs should be validated by alternative low-throughput techniques such as multiplex ligation-dependent probe amplification (MLPA) or real-time PCR.

Second generation sequencing technologies

Second generation sequencing (2ndGS) technologies have dramatically increased the throughput and reduced the costs of DNA sequencing compared with conventional Sanger sequencing methods. Although the cost is still one order of magnitude higher than whole genome genotyping (WGG), the general goal of sequencing a human genome for $1000 in less than one day seems realistic in the near-term. This will open the door for WGG by whole genome sequencing (WGS).

2ndGS platforms allow sequencing of millions of clonally amplified and spatially separated DNA fragments simultaneously. The sequencing process itself is a repetition of cycles of enzymatic reactions (polymerase-based nucleotide incorporations or oligonucleotide ligations, depending on the platform) and imaging-based data collection (Shendure and Ji 2008). The resulting sequence tags or reads are then aligned to a reference genome and genetic polymorphisms (SNPs, InDels, SVs) identified. 2ndGS instruments from Roche/454 (Genome FLX, www.454.com), Illumina (HiSeq2000, www.illumina.com) and Life Technologies (SOLiD 5500xl, solid.appliedbiosystems.com) are commercially available. The best performing of these instruments can generate tens of gigabases of sequence per day.

The main limitations of some of the 2ndGS technologies for WGS are the short read lengths (50 to 500 bp depending on the platform) and the relatively high error rate. The first problem can be partially circumvented by using an approach that generates sequences from both ends of each DNA molecule (paired-end or mate-pair sequencing) thus facilitating the alignment process. Nevertheless, analysis of large and highly repetitive regions is still not feasible. Increasing the depth of coverage (obtaining multiple reads from the same region) is the best option for improving the consensus read accuracy and ensuring high confidence in determination of genetic variants.

WGS enables the cataloguing of all kinds of genetic variation. The sequences of several individual genomes using 2ndGS technologies have been reported recently. Between 3 and 4 million SNPs per genome were identified using different algorithms (Metzker 2010). A large number of InDels that were undetectable with any of the high-throughput genotyping systems described above are starting to emerge in the databases thanks to 2ndGS methods. Finally, paired-end and mate-pair sequencing methods are also able to discern, at one base pair resolution, many CNVs and other SVs such as inversions and translocations in individual genomes. Recently, *de novo* assembly of two human genomes have been reported, allowing the discovery of new SVs in an unbiased manner (Li *et al.* 2010).

A third generation of sequencing technologies based on single-molecule sequencing is under development (Check Hayden 2009). Companies such as Pacific Biosciences (www.pacificbiosciences.com) and Oxford Nanopore Technologies (www.nanoporetech.com) are currently leading this sector. It is likely that routinely DNA sequencing of whole genomes for clinical or research purposes will happen in the near future.

Conclusions

Methods for genotyping and sequencing have come a long way in the past decade. Microsatellites that were the markers of choice a decade ago are no longer used and have been replaced by high-resolution SNP genotyping. Association studies that were very difficult to carry out and for which few positive results had been achieved a decade ago are now used routinely with great success. Several commercial solutions for WGG using hundreds of thousands of SNPs exist. Use of such methods has reached a level where it is now possible to join datasets that were produced in different laboratories by different groups. Meta-analysis has been very successful and added much insight. All of this has revolutionised molecular genetics and resulted in the identification of an unprecedented number of disease-associated genes. However, markers on the commercial genome-wide genotyping arrays have been selected based on the 'common disease–common variant' hypothesis that says that common pathologies are associated with variants of polymorphisms that are present at quite a high level in the general population – have a high minor allele frequency. Results of GWAS have nearly exclusively been variants that confer only marginal additional risk. Very few of the associated variants lead to amino acid changes and could thus be assigned a change of function. Thinking is now shifting that rare variants, that by chance are distributed unevenly onto frequent haplotypes, might be causative. However, including rarer polymorphisms in the genome-wide genotyping system is a game of diminishing returns. It still holds the risk that disease-causing variants might be very rare and private to a few families and thus not represented on the arrays. With the advent of second generation sequencing methods scientists are starting to look towards identifying/genotyping private and rare variants by sequencing. There is a crossover point where sequencing technologies (see 'Second generation sequencing') will become more cost effective than genotyping methods and a point where high-resolution WGG and WGS will have the same price tag and possibly comparable throughput. At this point it will be of interest to consider merging the linkage and association strategy and population genetic strategies in study design.

The ever-expanding toolbox for genetics has added refinement and standardization to the procedures. Methods for genome-wide, custom and quantitative genotyping exist. They can be applied for the reliable genotyping of SNPs, CNVs, InDels, and DNA methylation. Based on this, many interesting results have already been generated and with the continuing improvement of technologies the era of genomics is well underway.

References

Barany, F. (1991) Genetic disease detection and DNA amplification using cloned thermostable ligase. *Proc Natl Acad Sci USA* **88**: 189–193.

Beck, S. and Rakyan, V.K. (2008) The methylome: approaches for global DNA methylation profiling. *Trends Genet* **24**: 231–237.

Bernstein, B.E., Meissner, A. and Lander, E.S. (2007) The mammalian epigenome. *Cell* **128**: 669–681.

Botstein, D. White, R.L., Skolnick, M. and Davis, R.W. (1980) Construction of a genetic linkage map in man using restriction fragment length polymorphisms. *Am J Hum Genet* **32**: 314–331.

Brookes, A.J. (1999) The essence of SNPs. *Gene* **234**: 177–186.

Carter, N.P. (2007) Methods and strategies for analyzing copy number variation using DNA microarrays. *Nat Genet* **39**(7 Suppl): S16–21.

Check Hayden, E. (2009) Genome sequencing: the third generation. *Nature* **457**: 768–769.

Clark, A.G. (2003) Finding genes underlying risk of complex disease by linkage disequilibrium mapping. *Curr Opin Genet Dev* **13**: 296–302.

Cooper, G.M., Zerr, T., Kidd, J.M., Eichler, E.E. and Nickerson, D.A. (2008) Systematic assessment of copy number variant detection via genome-wide SNP genotyping. *Nat Genet* **40**: 1199–203.

Couzin, J. (2002) Human genome. HapMap launched with pledges of 100 M$. *Science* **298**: 941–942.

Dupont, J.M., Tost, J., Jammes, H. and Gut, I.G. (2004) De novo quantitative bisulfite sequencing using the pyrosequencing technology. *Anal Biochem* **333**: 119–127.

Egholm, M., Buchardt, O., Christensen, L. *et al.* (1993) PNA hybridizes to complementary oligonucleotides obeying the Watson–Crick hydrogen bonding rules. *Nature* **365**: 566–568.

Ehrich, M., Nelson, M.R., Stanssens, P. *et al.* (2005) Quantitative high-throughput analysis of DNA methylation patterns by base-specific cleavage and mass spectrometry. *Proc Natl Acad Sci USA* **102**: 15785–15790.

Fan, J.B., Chee, M.S. and Gunderson, K.L. (2006) Highly parallel genomic assays. *Nat Rev Genet* **7**: 632–644.

Fodor, S.P., Read, J.L., Pirrung, M.C. *et al.* (1991) Light-directed, spatially addressable parallel chemical synthesis. *Science* **251**: 767–773.

Gresham, D., Dunham, M.J. and Botstein, D. (2008) Comparing whole genomes using DNA microarrays. *Nat Rev Genet* **9**: 291–302.

Gut, I.G. (2001) Automation in genotyping of single nucleotide polymorphisms. *Human Mut* **17**: 475–492.

Haff, L. and Smirnov, I.P. (1997) Single-nucleotide polymorphism identification assays using a thermostable DNA polymerase and delayed extraction MALDI-TOF mass spectrometry. *Genome Res* **7**: 378–388.

Hindorff, L.A., Sethupathy, P., Junkins, H.A. *et al.* (2009) Potential etiologic and functional implications of genome-wide association loci for human diseases and traits. *Proc Natl Acad Sci USA* **106**: 9362–9367.

Holland, P.M., Abramson, R.D., Watson, R. and Gelfand, D.H. (1991) Detection of specific polymerase chain reaction product by utilizing the 5′–3′ exonuclease activity of *Thermus aquaticus* DNA polymerase. *Proc Natl Acad Sci USA* **88**: 7276–7280.

International HapMap Consortium. (2007) A second generation human haplotype map of over 3.1 million SNPs. *Nature* **449**: 851–861.

International HapMap 3 Consortium. (2010) Integrating common and rare genetic variation in diverse human populations. *Nature* **467**: 52–8.

International Human Genome Sequencing Consortium (2004) Finishing the euchromatic sequence of the human genome. *Nature* **431**: 931–945.

Jarvius, J., Nilsson, M. and Landegren, U. (2003) Oligonucleotide ligation assay. *Methods Mol Biol* **212**: 215–228.

Jorde, L.B. (1995) Linkage disequilibrium as a gene-mapping tool. *Am J Hum Genet* **56**: 11–14.

Jurinke, C., van den Boom, D., Cantor, C.R. and Köster, H. (2002) The use of MassARRAY technology for high throughput genotyping. *Adv Biochem Eng Biotechnol* **77**: 57–74.

Kennedy, G.C., Matsuzaki, H., Dong, S. *et al.* (2003) Large-scale genotyping of complex DNA. *Nat Biotechnol* **21**: 1233–1237.

Kim, S. and Misra, A. (2007) SNP genotyping: technologies and biomedical applications. *Annu Rev Biomed Eng* **9**: 289–320.

Klein, R.J., Zeiss, C., Chew, E.Y. *et al.* (2005) Complement factor H polymorphism in age-related macular degeneration. *Science* **308**: 385–289.

Kwok, P.Y. (2002) SNP genotyping with fluorescence polarization detection. *Hum Mutat* **19**: 315–23.

Kwok, S. and Higuchi, R. (1989) Avoiding false positives with PCR. *Nature* **339**: 237–238.

Lahermo, P., Liljedahl, U., Alnaes, G. *et al.* (2006) A quality assessment survey of SNP genotyping laboratories. *Hum Mutat* **27**: 711–714.

Laird, P.W. (2010) Principles and challenges of genome-wide DNA methylation analysis. *Nat Rev Genet* **11**: 191–203.

Landegren, U., Nilson, M. and Kwok, P.-Y. (1998) Reading bits of genetic information: methods for single-nucleotide polymorphism analysis. *Genome Res* **8**: 769–776.

Li, R., Zhu, H., Ruan, J. *et al.* (2010) De novo assembly of human genomes with massively parallel short read sequencing. *Genome Res* **20**: 265–272.

Lin, C.H., Yeakley, J.M., McDaniel, T.K. and Shen, R. (2009) Medium- to high-throughput SNP genotyping using VeraCode microbeads. *Methods Mol Biol* **496**: 129–142.

Lister, R., O'Malley, R.C., Tonti-Filippini, J. *et al.* (2008) Highly integrated single-base resolution maps of the epigenome in *Arabidopsis*. *Cell* **133**: 523–536.

Lister, R., Pelizzola, M., Dowen, R.H. *et al.* (2009) Human DNA methylomes at base resolution show widespread epigenomic differences. *Nature* **462**: 315–322.

Livak, K., Marmaro, J. and Todd, J.A. (1995) Towards fully automated genome-wide polymorphism screening. *Nat Genet* **9**: 341–342.

Livak, K.J. (2003) SNP genotyping by the 5′-nuclease reaction. *Methods Mol Biol* **212**: 129–147.

Longo, M.C., Berninger, M.S. and Hartley, J.L. (1990) Use of uracil DNA glycosylase to control carry-over contamination in polymerase chain reactions. *Gene* **93**: 125–128.

McCarroll, S.A., Kuruvilla, F.G., Korn, J.M. *et al.* (2008) Integrated detection and population-genetic analysis of SNPs and copy number variation. Nat Genet 40: 1166–74.

Metzker, M.L. (2010) Sequencing technologies – the next generation. *Nat Rev Genet* **11**: 31–46.

Moffatt, M.F., Gut, I.G., Demenais, F. *et al.*; GABRIEL Consortium (2010) A large-scale genomewide association study of asthma. *New Engl J Med* **363**: 1211–1221.

Moffatt, M.F., Kabesch, M., Liang, L. *et al.* (2007) Genetic variants regulating ORMDL3 expression contribute to the risk of childhood asthma. *Nature* **448**: 470–473.

Morrison, T., Hurley, J., Garcia, J. *et al.* (2006) Nanoliter high throughput quantitative PCR. *Nucleic Acids Res* **34**: e123.

Nilsson, M., Malmgren, H., Samiotaki, M. *et al.* (1994) Padlock probes: circularizing oligonucleotides for localized DNA detection. *Science* **265**: 2085–2088.

Oeth, P., del Mistro, G., Marnellos, G. *et al.* (2009) Qualitative and quantitative genotyping using single base primer extension coupled with matrix-assisted laser desorption/ionization time-of-flight mass spectrometry (MassARRAY). *Methods Mol Biol* **578**: 307–343.

Ørum, H., Jakobsen, M.H., Koch, T. *et al.* (1999) Detection of the factor V Leiden mutation by direct allele-specific hybridization of PCR amplicons to photoimmobilized locked nucleic acids. *Clin Chem* **45**: 1898–1905.

Pääbo, S. (2003) The mosaic that is our genome. *Nature* **421**: 409–412.

Parsons, B.L. and Heflich, R.H. (1997) Genotypic selection methods for the direct analysis of point mutations. *Mutat Res* **387**: 97–121.

Pompanon, F., Bonin, A., Bellemain, E. and Taberlet, P. (2005) Genotyping errors: causes, consequences and solutions. *Nat Rev Genet* **6**: 847–859.

Ragoussis, J. (2009) Genotyping technologies for genetic research. *Annu Rev Genom Hum Genet* **10**: 117–133.

Redon, R., Ishikawa, S., Fitch, K.R. *et al.* (2006) Global variation in copy number in the human genome. *Nature* **444**: 444–454.

Risch, N. (1991) Developments in gene mapping with linkage methods. *Curr Opin Genet Dev* **1**: 93–8.

Risch, N. and Merikangas, K. (1996) The future of genetic studies of complex human diseases. *Science* **273**: 1516–1517.

Shames, D.S., Minna, J.D. and Gazdar, A.F. (2007) DNA methylation in health, disease, and cancer. *Curr Mol Med* **7**: 85–102.

Shen, L. and Waterland, R.A. (2007) Methods of DNA methylation analysis. *Curr Opin Clin Nutr Metab Care* **10**: 576–581.

Shen, R., Fan, J.B., Campbell, D. *et al.* (2005) High-throughput SNP genotyping on universal bead arrays. *Mutat Res* **573**: 70–82.

Shendure, J. and Ji, H. (2008) Next-generation DNA sequencing. *Nat Biotechnol* **26**: 1135–1145.

Sherry, S.T., Ward, M.H., Kholodov, M. *et al.* (2001) dbSNP: the NCBI database of genetic variation. *Nucleic Acids Res* **29**: 308–311.

Sobrino, B., Brión, M. and Carracedo, A. (2005) SNPs in forensic genetics: a review on SNP typing methodologies. *Forensic Sci Int* **154**: 181–194.

Steemers, F.J. and Gunderson, K.L. (2007) Whole genome genotyping technologies on the BeadArray platform. *Biotechnol J* **2**: 41–49.

Suzuki, M.M. and Bird, A. (2008) DNA methylation landscapes: provocative insights from epigenomics. *Nat Rev Genet* **9**: 465–476.

Syvänen, A.-C. (1999) From gels to chips: 'minisequencing' primer extension for analysis of point mutations and single nucleotide polymorphisms. *Hum Mutat* **13**: 1–10.

Syvänen, A.-C. (2001) Accessing genetic variation: genotyping single nucleotide polymorphisms. *Nat Rev Genet* **2**: 930–942.

Szántai, E. and Guttman, A. (2006) Genotyping with microfluidic devices. *Electrophoresis* **27**: 4896–4903.

Tong, A.K., Li, Z., Jones, G.S. *et al.* (2001) Combinatorial fluorescence energy transfer tags for multiplex biological assays. *Nat Biotechnol* **19**: 756–759.

Tost, J. (2010) DNA methylation: an introduction to the biology and the disease-associated changes of a promising biomarker. *Mol Biotechnol* **44**: 71–81.

Tost, J. and Gut, I.G. (2005) Genotyping single nucleotide polymorphisms by MALDI mass spectrometry in clinical applications. *Clin Biochem* **38**: 335–350.

Wang, J., Lin, M., Crenshaw, A. *et al.* (2009) High-throughput single nucleotide polymorphism genotyping using nanofluidic Dynamic Arrays. *BMC Genomics* **10**: 561.

Wellcome Trust Case Control Consortium (2007) Genome-wide association study of 14,000 cases of seven common diseases and 3,000 shared controls. *Nature* **447**: 661–678.

Winchester, L., Yau, C. and Ragoussis, J. (2009) Comparing CNV detection methods for SNP arrays. *Brief Funct Genomic Proteomic* **8**: 353–366.

Zhang, F., Gu, W., Hurles, M.E. and Lupski, J.R. (2009) Copy number variation in human health, disease, and evolution. *Annu Rev Genom Hum Genet* **10**: 451–81.

2
DNA Chip Analysis in Genome Discovery

Ross N. Nazar and Jane Robb

Introduction

Since first appearing in the early 1990s (for reviews see Southern 2001; Stoughton 2005) DNA microarrays or 'chips' have become invaluable tools, not only in molecular genetics but also in virtually every area of biology and biotechnology, from medical clinics to ecological studies involving the most remote areas of the planet. At first basically representing a refinement of 'dot blot' technology, DNA microarrays soon became the standard for high throughput or global analyses of mRNA and gene expression profiling. Microarrays can quickly and efficiently be used to examine the effects of virtually any single factor on thousands of genes at a specific time point and under many conditions. The study of thousands of genes simultaneously can provide exciting and important insights into the molecular mechanisms of biological processes. For example, coupled with the Human Genome Project (Collins *et al.* 2003), changes in gene expression in diseases such as cancer, heart disease or mental illness can be readily detected and the progression of such complex diseases can be linked to therapeutic methods (Gerhold *et al.* 2002; Brennen *et al.* 2005).

Microarray expression analysis, in its most common form, involves thousands of DNA fragments derived from mRNAs of known genes and immobilized as microscopic spots (typically less than 200 μm in diameter) on glass microscope slides. cDNA probes prepared from samples of both control and experimental or diseased materials are differentially labelled and hybridized with the glass arrays providing hybridization profiles that can be analysed for intensity differences which reflect

Molecular Analysis and Genome Discovery, Second Edition. Edited by Ralph Rapley and Stuart Harbron.
© 2012 John Wiley & Sons, Ltd. Published 2012 by John Wiley & Sons, Ltd.

actual changes in gene expression. Over the past two decades, such analyses have become standardized and routine and have been commercialized in many forms. Equipment and supplies for microarray preparation or actual microarrays are easily purchased, as are extraction and labelling kits for probe preparation. Highly sensitive microarray scanners as well as image analyses software also are readily obtainable. Since review articles and manuals in support of these methods also are commonly available (e.g. Stoughton 2005; Gresham *et al.* 2008; Miller and Tang 2009), they will not be considered here in further detail.

Expression profiling has continued to represent the dominant DNA microarray usage over the past two decades with comparisons of different tissues from specific organisms, different disease states and genetic backgrounds, as well as differing experimental conditions such as those represented by drug treatments or gene disruption. Baseline levels for mRNAs in specific tissues, sometimes referred to as 'body maps' (Stoughton 2005), have been extensively reported on with many data bases both publicly and commercially available (e.g. Su *et al.* 2004; Dezso *et al.* 2008). Equally, comparisons of two biological conditions (e.g. normal vs. disease state) have been representative of the most common experimental design as well as the bulk of reports in the literature for analyses based on DNA microarrays. More recently, functional response patterns and signalling pathways have become popular subjects for systematic analyses of expression profiles based on DNA chips. Such studies have provided important new information such as major pathway groupings (e.g. Michaut *et al.* 2003; Wolter *et al.* 2009) and evidence for crosstalk between signalling pathways (Roberts *et al.* 2000).

Applications of DNA chip technology with respect to expression profiling will undoubtedly continue and will lead to many new discoveries in these traditional areas of study. In the interim, however, applications of this technology have greatly expanded in other directions (aside from mRNA expression profiling) with equal promise. The present chapter seeks to review developments in such expanding areas of DNA microarray research, which continue to include new clinical applications but also represent new tools in such diverse areas as ecology, population genetics, evolution and the environment. This review focuses on four areas often referred to as interrogation of genomes, cross-species hybridization (CSH), comparative genomic hybridizations (CGH) as well as the classification, identification and DNA barcoding of pathogens and other living organisms.

Interrogating a genome

Apart from the common use of microarray technology, which compares the transcriptome list of two individuals of the same species, newer more specialized microarrays are having a significant impact on our current understanding of the structure and function of eukaryotic genomes. All of these might be thought of as an 'interrogation of a genome'. Traditionally the genome was interrogated with respect to the transcriptome or changes in transcription (and RNA levels). This has been expanded to include alternative splice forms of a gene, the detection of single nucleotide polymorphism (SNP) among alleles within and between populations, the detection of alternative patterns of gene methylation and, when combined with chromatin immunoprecipitation, the detection of DNA sequence bound to a protein.

Traditionally, information about gene structure was obtained using computational gene prediction methods based on known sequence along with large-scale cloning and sequencing of cDNA molecules corresponding to expressed gene transcripts. These approaches tend to underestimate both the number of genes and the products, however, since transcripts are often missed that are rare, non-polyadenylated or environmentally, developmentally or tissue specific. An important advance in microarray platforms that allowed more in-depth studies of genomes was the development of high-density oligonucleotide-based whole genome tiling arrays. These saturate parts of chromosomes, entire chromosomes or even an entire genome and, depending on probe length and spacing, different degrees of resolution can be obtained. For example, Kapranov and co-workers (2002) and Kampa and co-workers (2004) used tiling arrays to investigate exon skipping in genes on human chromosomes 21 and 22. In the plant world, Ner-Gaon and Fluhr (2006) and Campbell and co-workers (2006) were able to study the transcriptional activity and alternative splicing across the entire genome of the plants *Arabidopsis thaliana* and rice, respectively. The studies indicated that the total number of possible transcripts and the frequency of genes undergoing alternative splicing events were much higher than previously supposed.

Most eukaryotic genes have one or more introns and alternative retention or removal of both exons and introns increases the variety of RNA transcripts that can be derived from a single gene. In many cases the alternative transcripts are translated to produce variant or even novel proteins and microarray studies have demonstrated striking variations in putative protein isoforms among various tissues and developmental stages (Johnson *et al.* 2003; Sugnet *et al.* 2006). Also, many pathological conditions have been associated with deficient splicing or the appearance of unusual transcripts resulting in aberrant proteins not found in healthy tissues (Wang and Cooper 2007). Microarray analyses suggest that 70–80% of mammalian genes and approximately 20% of plant genes exhibit alternative splicing. Tiling arrays are particularly advantageous for studying alternative splicing since they can detect altered splice site use as well as intron retention and the use of nonannotated exons; however, these microarray platforms are much more expensive and the computational programs needed for analysis are very large (Cuperlovic-Culf *et al.* 2010). In addition to tiling arrays, there are other microarray alternative splicing platforms which permit more specialized analyses. Splice-junction microarrays use predetermined sets of splicing events and known genes; such arrays include both constitutive exons and controls for transcriptional activity, as well as probes of intron–exon junctions of known genes (e.g. Johnson *et al.* 2003; Sugnet *et al.* 2006; Fagnani *et al.* 2007; Ip *et al.* 2007). Exon arrays, which are available for the human, mouse or rat, comprise multiple oligonucleotide probes within all annotated or predicted exons, such as that used by Xing and co-workers (2006) to detect novel transcripts. Other examples include papers by McKee and co-workers (2007), French and co-workers (2007) and Okoniewski and co-workers (2007). The use of microarrays to study alternative splicing events has been reviewed recently by Moore and Silver (2008) and Hui *et al.* (2009).

Two other related profiling techniques have been applied specifically to study the regulation of gene splicing. These are called chIP-chip and RIP-chip, respectively, and are based on chromatin (ch) or RNA (R) immunoprecipitation (IP) (for review see Moore and Silver 2008). These approaches can identify populations of genes or transcripts regulated by individual protein factors. ChIP protocols cross-link co-transcriptional processing factors to DNA, which is then profiled using tiling arrays (e.g. Tardiff *et al.* 2006; Moore *et al.* 2006; Swinburne *et al.* 2006). RIP protocols require purification of RNPs and profiling of associated transcripts (e.g. Keene *et al.* 2006; Gama-Carvalho *et al.* 2006). The various commercially available microarray platforms and requisite validation studies have been reviewed by Hui and co-workers (2009).

Microarray technologies, together with high-throughput sequencing methods, have also greatly increased our capacity to identify and map SNPs, a variation in a DNA sequence in which a single nucleotide at a particular locus varies among equivalent chromosomes (i.e. different alleles). Short oligonucleotide probes can discriminate single nucleotide differences but the arraying of 15 million SNPs in the human genome would be a truly daunting task. The high correlation between near or adjacent polymorphisms and linked genes (i.e. linkage disequilibrium), however, has allowed the grouping of traits and SNPs into haplotypes. The International HapMap Project (http://www.hapmap.org) has defined a clear set of 300 000 tag SNPs (each tagging a specific haplotype), which define about 70% coverage of the human genome. Such arrays have made whole genome analysis (WGA), otherwise known as microarray genotyping, possible. SNP arraying has three important components:

1. target nucleic acid sequences, immobilized on either glass slides (Affymetrix) or microbeads (Infinium);

2. one or more labelled allele-specific oligonucleotide probes; and

3. a detection system to record and 'translate' the hybridization signal.

Various high-throughput genotyping array platforms based on SNP identification are available and have been reviewed by Beaudet and Belmont (2008).

Epigenetics, the study of heritable modifications without a change in DNA sequence, has widespread implications in many fundamental processes including development, genomic imprinting, gene silencing and chromatin stability. DNA methylation has been recognized as a primary epigenetic modification; in vertebrates modifications are clustered in 'CpG islands' or regions where the frequency of the CG sequence is higher. About 60% of all gene promoters have CpG islands. Modifications to these regions have been associated with gene silencing and activation (see Flintoft 2009) and recently they have even been shown to regulate their expression of many non-protein-coding RNAs (Yazaki *et al.* 2007). Research in this area has and continues to benefit significantly from the use of tiling array technology, which now permits methylation mapping on a whole genome basis (methylome). Because epigenetic changes are also caused by histone methylation, chromatin immunoprecipitation has been combined with whole genome tiling microarrays (ChIP-chip) to identify patterns of protein methylation (e.g. Zhang *et al.* 2007). Since ultra high-throughput DNA sequencing can be viewed as a 'virtual' tiling array, next generation sequencing has played equally important roles in these types of analyses with advantages and disadvantages. In fact the two strategies (ChIP-chip vs. ChIP-PET) have been compared directly in a study of transcription factor binding sites (Euskirchen *et al.* 2007) with strong agreement in many instances and complementary roles resulting in the most comprehensive data. It seems reasonable to expect that both approaches will continue to be applied and expanded for an even greater understanding of genome complexity.

Cross-species hybridization

An important area of potential expansion for DNA chip technology is the use of cross-species hybridization (CSH) in studies of diverse genomes. CSH studies, in general, are defined as studies where arrays developed for one species are used to examine gene expression in another. Issues related to hybrid specificity generally have discouraged such expansion and some researchers have questioned whether CSH studies can generate valid biological results. They point out that array sensitivity decreases in CSH (see Bar-Or *et al.* 2007; Buckley 2007) and this is likely to give rise to more false negatives and a significant reduction in accuracy. Others have pointed out that the advent of much faster, new generations of genome sequencing equipment are permitting a rapid expansion of genome sequencing projects that inevitably will lead to many new species-specific DNA arrays. Nevertheless, given the vast number of biological organisms and the costs of developing a species-specific microarray, most research programs on diverse organisms cannot make use of standard DNA chip technology and are unlikely to benefit from it in the foreseeable future. In the interim, various forms of CSH have or are being applied with considerable promise. Initially, such studies focused on very closely related genomes, but even very distantly related organisms have now been examined raising a broad range of potential applications.

A variety of different strategies have been adopted in the application of CSH methods. Most have been conservative in that RNA samples of one species have been used with microarrays designed for very closely related species. For example, human microarrays have been used to examine chimpanzees, orang-utans and rhesus macaques (Bigger *et al.* 2001; Huff *et al.* 2004). Some strategies have also been applied to more distantly related species such as cats, dogs and pigs (Ji *et al.* 2004; Grigoryev *et al.* 2005). In all these studies, sufficient nucleotide sequence conservation has been assumed to permit detectable signals with equivalent genes. Studies of this nature have usually used one of two general experimental designs based on a heterologous DNA chip that is specific for a closely related organism. In the first instance, samples are prepared from only one species but differ in some variables such as growth temperature or alternative tissues. The array is used to detect differences in gene expression between the two temperatures or tissues as for a homologous chip. In the second example, samples are prepared from an equivalent source (e.g. tissues) in two different species and the array is used to detect differences between the two organisms. More complicated combinations also are possible, including a combination of the two designs, but the basic principle remains the same, namely homology shared with the source of the heterologous chip is exploited to detect differences in the samples.

As noted earlier, questions regarding the high reliability of CSH have concerned many scientists and have curtailed significantly the expansion of its use. The major concern is that in cross-species comparisons, sequence mismatches in equivalent genes can reduce the hybridization signal strength (see Buckley 2007) making the results less reliable and greatly increasing the frequency of false negatives. Duplex stabilities and reassociation kinetics for hybridization are complex and many factors – such as temperature, salt concentration and the nature of each individual probe – are important and can greatly influence the result for each spot on an array (Tijssen 1993). This is somewhat less serious with comparisons of samples from the same species (e.g. from two tissues) but is still a likely complicating factor with all CSH comparisons. Numerous studies have suggested approaches that may improve this problem (e.g. Bar-Or *et al.* 2007; Buckley 2007; Chen *et al.* 2010). For example, the use of assays based on several short oligonucleotide probes (typically 25 base pairs) for each gene (e.g. Affymetrix high-density oligonucleotide GeneChips) rather than longer cDNA clones (hundreds to thousands of base pairs) can make both standard and CSH microarray analyses more reliable (e.g. Enard *et al.* 2002). On the other hand, when only single probes per gene are used, other studies (Kane *et al.* 2000; Walker *et al.* 2006) have concluded that microarrays with longer PCR products or oligomers, respectively, actually identify a greater number of significantly regulated genes at a lower false discovery rate. Some studies have attempted to add masking procedures to filter out poorly hybridizing probes in a data pre-processing step to improve the sensitivity of cross-species analyses (see Chen *et al.* 2010). Filters have been based on available genomic data (Khaitovich *et al.* 2004; Bar-Or *et al.* 2007) or genomic DNA hybridizations (e.g. Ranz and Machado 2006). Although consensus filtration schemes have not been adopted, the trials suggest that such steps can lead to more valuable information. Also, mathematical or statistical approaches or models have been examined as a means of controlling the false discovery rate (see Kammenga *et al.* 2007), particularly in studies of natural variation (Gilad and Borevitz 2006). Again a consensus does not exist but future developments of generalized models are anticipated. In the interim, it seems wise to stress that any conclusions need to be confirmed by independent methods such as RT-PCR.

While concerns about hybridization specificity and signal strength have resulted in many useful studies on the application of heterologous DNA chips for comparison of closely related genomes, they have also strongly discouraged less conservative strategies with more distantly related organisms. Although many such applications in ecology, environmental studies, population genetics or even evolution can be envisaged, generally it has been assumed that the reduced signal strength would make the use of more heterologous chips unreliable. Despite this, some attempts have and are being made to extend the use of unrelated DNA microarrays with promising results.

Initially these attempts were based entirely on high-density oligonucleotide arrays. For example, Ji and co-workers (2004) used an Affymetrix human high-density oligonucleotide array (GeneChip®) looking for sequences conserved within mammals. While most hybridization signals were low, some were equivalent to those observed with same-species hybridization. Using an algorithm to select the reliable signals and to mask the poorly hybridized probes, they found good correlations and suggested a broader application to both animal and plant research. In a similar vein, Belosludtsev and co-workers (2004) attempted to develop a genome sequence-independent universal microarray based on 14 283 unique probes. They reported intensity patterns that reproducibly differentiated various organisms including *Bacillus subtilis, Yetsimia pestis, Streptococcus pneumoniae, Bacillus antheacis* and *Homo sapiens*. Such approaches have continued to be developed and will be considered further with organism identification through DNA barcodes and microarrays.

An even less conservative approach was explored (Nazar *et al.* 2010) when a standardized human cDNA microarray was used to probe transcripts in tomato infected with *Verticillium* wilt pathogens. The results were compared directly with an equivalent study using a standardized tomato cDNA microarray. This study did find similar changes with some known genes expressing protein homologues. More important, the range of changes – as detected by data scatter plots and overall frequency of spots indicating changes in gene expression – were similar with each microarray. Although not normally anticipated, the range of intensity changes seemed more similar with the longer cDNA probes than experienced earlier in the various studies using oligonucleotide arrays. Possibly the much longer sequences provide more opportunities for stable cross hybridization. While this study did not report or even assume any general equivalence in the spots, it showed that very heterologous chips could detect interesting changes in gene expression that might be exploited to provide biomarkers and footprints for studies in biodiversity, ecology or population genetics. It suggested that the relative intensities or ratios of hybridization signals could be treated as 'phenotypes' and analysed using the same mathematical approaches commonly used by population geneticists and breeders even though the actual gene identities are unknown. Such an approach might be used to estimate fundamental parameters of gene expression including dominance and heritability of expression or to test the co-variation of expression among clines or even to access the effects of population structure on gene expression in a variety of species. Potentially it also could be used to compare the magnitude of global responses to changing environmental parameters or to assess toxicological effects or other stresses in the environment. While not a substitute for standard microarray procedures, this study has lent support to the possibility that a heterologous chip could allow opportune studies years before many genomes are sufficiently sequenced and specific arrays become available. Although not applied in this study, electronic masking or filtration as discussed earlier could further enhance the reliability of conclusions drawn this way.

Whether used only in a conservative fashion with closely related organisms as described initially, or in a more adventurous less certain fashion as described above, the application of CSH with heterologous DNA arrays seems to hold considerable potential and undoubtedly deserves further application and testing. The fact that spots are not readily related to a completed genome sequence does initially limit the information that might be derived directly from changes in spot intensities, but, as noted previously (Nazar *et al.* 2010), the results could be used in the reverse direction to clone and ultimately fully identify genes that are discovered in this way. What is perhaps most important is a need to recognize the limitations of weak hybridization signals and to focus on signals that are reliable and, ultimately, can be confirmed by quantitative PCR or knock out experiments.

Comparative genomic hybridization and microarray-based genotyping

Microarray technologies are being increasingly used for the molecular identification and characterization of a specific individual or tissue. There are two different approaches. Microarray-based genotyping or comparative genomic hybridization (CGH) refer to the characterization of an individual or tissue according to the number, kind and distribution of either SNPs (SNPs-array) or copy number variations (i.e. CGH array), respectively, in the genome. These comprise a unique identifiable fingerprint. Applications of such approaches are numerous, including uses in human and veterinary medicine, agriculture and forensics.

There are two common uses of microarray genotyping for medical purposes. In the first, the presence of a specific SNP polymorphism can be correlated with a complex disease, or a particular drug response (i.e. association studies). Where the SNP polymorphism occurs within a protein-coding exon, such a correlation has led to the actual identification of dysfunctional protein pathways that are critical in that disease. However, even if no causative gene can be clearly identified in the haplotype, the presence of the specific SNP can be used for predictive or diagnostic purposes or to design individual-based drug treatment regimes. Specific human diseases, and associated loci identified by SNP whole genome analysis (WGA) have been reviewed by Beaudet and Belmont (2008). The second common medical use of microarray genotyping is to study or detect loss of heterozygosity (LOH). LOH is essentially an allelic imbalance in which one allele of the usual two in humans is either absent or duplicated or, in the case of uniparental disomy (UDP), the allele from one parent is lost but replaced by a second copy of the allele from the other parent. If the duplicated 'replacement' gene is a detrimental mutant allele, UPD can result in a disease situation.

CGH arrays are used for the diagnosis and study of copy number variants (CNVs) in the genome. A CNV is operationally defined as a segment of DNA, larger than 1 kb with a variable copy number compared with a reference genome (Feuk *et al.* 2006; Lee *et al.* 2007). Only through modern microarray approaches have we become aware of the high frequency of CNVs in humans; indeed, the genomes of unrelated individuals can differ from one another at thousands of CNV loci (Iafrate *et al.* 2004; Sebat *et al.* 2004). Clearly, most such differences are benign but some result in disease. Standard cytogenetic techniques, used in the past for diagnosis of abnormal karyotypes, cannot detect genomic imbalances (i.e. duplications and deficiencies) smaller than 5–1.0 kb (Leung *et al.* 2004) and even higher resolution FISH approaches cannot detect differences less than 2 Mb in length (Klinger *et al.* 1992). However, most CNVs are much shorter. Modern DNA microarray technologies, with increased ability to analyse complex data at reasonable cost, have made possible large-scale, high-resolution genome-wide analysis of copy number variation for both research and clinical diagnostics.

In a research context much of the current effort is focused on determining the extent and location of CNV variation in normal healthy humans. A Database of Genomic Variants (http://projects.tcog.ca/variation) has been established specifically to accumulate CNV data obtained from the normal population. These variations provide a host of new cytogenetic markers that can be used in future studies to shed light on disease susceptibility as well as on human evolution and population genetics. For example, one study has already revealed extensive differences among individuals of different races (Redon *et al.* 2006).

DNA microarray identification has clear advantages over traditional cytogenetic techniques to detect genomic mutations in karyotypes; resolution is much superior and its high-throughput capacity at reasonable expense is much more efficient and cost-effective for screening CNVs since one clinical assay using microarray can replace hundreds of FISH tests (Beaudet and Belmont 2008). Unfortunately, it has proven more difficult to use whole genome microarrays to determine the functional significance of new CNVs than one might suppose. The problem lies with the very extensive distribution of genomic mutations in the genomes of healthy individuals and how to sort out which are potentially detrimental and which are not. Because of modifying factors, such as incomplete penetrance, variable expressivity and gene dosage effects, the presence of a particular deletion or duplication in a healthy individual is no guarantee that the mutation is necessarily always benign. Nonetheless considerable progress has been made and several databases are available documenting both CNVs and phenotypes identified during clinical diagnoses. These include the Database of Chromosomal Imbalance and Phenotype in Humans using Ensembl Resources (DECIPHER, http://www.sanger.ac.uk/Post Genomics/decipher/) and the Chromosomal Abnormality Database (CAD; http://www.ukcad.uk/cocoon/ukcad/) among others. Also, a number of specialized arrays have been patented to detect chromosome anomalies as well as monogenic disease-related mutations and SNPs (Yoo *et al.* 2009). Both array CGH and combination SNP/CNV arrays have now been used in large clinical studies to identify genomic mutations in children associated with complex clinical phenotypes including mental retardation, developmental delay and the presence of multiple dysmorphic features (Shaffer *et al.* 2006; Lu *et al.* 2007). Also, autism seems to be linked to a large variety of chromosomal deletions and/or duplications (Vorstman *et al.* 2006; Jacquemont *et al.* 2006; Sebat *et al.* 2007).

Because of the difficulties inherent in interpreting the clinical significance of CNVs, the medical profession has been somewhat reluctant to adopt microarray-based tests for prenatal diagnosis. Nevertheless, this approach has many advantages including higher resolution of small anomalies, cost effectiveness and speed, which is quite important for prenatal assessment. In addition, karyotyping of prenatal cells from amniotic fluid is more difficult and less reliable than from peripheral white blood cells. Arrays are already available to detect CNVs such as trisomies, monosomies and well-known deletions and duplications, as well as mutations for detrimental single gene disorders; available arrays for prenatal diagnostics have been reviewed by Yoo and co-workers (2009). Initial clinical studies using cell-free fetal DNA from amniotic fluid were carried out as early as 2006 (Miura *et al.* 2006) and clinical trials using noninvasive sampling methods (i.e. maternal blood cell samples) coupled with microarray are ongoing (Chu *et al.* 2009). Many reviews discuss the application of array CGH for prenatal diagnosis (Waddell *et al.* 2008).

Both microarray-based genotyping and CGH arrays are used extensively in studies of cancer development as well as in diagnosis. For example, research has shown that high rates of both LOH (e.g. Laiho *et al.* 2003) and CNVs (Pinkel and Albertson 2005; Dear 2009) are characteristic of many malignancies. Also cancerous cells frequently display unusual CNVs that are absent from the patient's normal cells (for review see: Dear 2009). In addition, unusual methylation of CpG islands in tumours has been associated with pathological gene expression (Moskalyov *et al.* 2007) and specialized oligonucleotide microarrays have been developed to test for this (Kamalakaran *et al.* 2009).

While much of the development and use of SNPs and GCH arrays have been in the medical area, many other fields are adapting array strategies for identification and diagnostic purposes. One of the more obvious candidates is the breeding industry. For example, arrays directed towards analyses of microsatellites, SNPs (Van Tassel *et al.* 2008; Matukumalli *et al.* 2009) and, most recently, CNVs (Liu *et al.* 2010a, 2010b) are now being used as measures of genetic variation in cattle. The importance of these new molecular approaches to the cattle industry has been discussed in a recent review by Humblot and co-workers (2010), outlining the development of genomic selection and new reproductive technologies and how these can be improved to meet modern needs of farmers as well as consumers. Similar SNP arrays are being used in breeding programs for other agriculturally important animals such as sheep and chickens (Muir *et al.* 2008; Kijas *et al.* 2009) and Agilent Technologies has recently made available CGH arrays for these, as well. In the plant world, array-based marker systems are being used to similar advantage for the breeding of agriculturally important crops as well as studies of genetic variation, linkage mapping, population studies and map-based gene isolation (for reviews see: Fan *et al.* 2007; Gupta *et al.* 2008; Ganal *et al.* 2009). Another area in which array-based identification and characterization has become increasingly important during the last decade is in forensic sciences. Since 2005, SNP arrays have been in common use in both criminal and historic investigations (Divne and Allen 2005; Bogus *et al.* 2006). Identification systems such as criminal databases, forensic DNA testing and genetic genealogy require reliable, cost-effective genotyping of autosomal, mitochondrial and Y-chromosome markers from different biological materials. Modern forensic chips can assess SNPs from both nuclear and mitochondrial DNA (eg. Krjutskov *et al.* 2009) with high efficiency. While CNVs are also potentially important in analysing this type of material, CGH arrays are often not suitable since specimens are frequently very limited and may be badly degraded. Short fragments are a minor problem in testing for single base SNPs; however, they become a much more serious roadblock when attempting to analyse for CNVs that can be much larger (Dear 2009). Despite this limitation, the range and depth of use to which these new genotyping techniques are being applied are clearly impressive and further ongoing developments continue to hold even greater promise.

Barcodes, DNA microarrays and organism identification

The rapid and accurate identification of organisms is critical to many endeavours including large-scale biodiversity monitoring programs. In addition to immunological methods, the many technical advances in molecular genetics over recent decades have provided numerous assay methods to identify and classify organisms based on hybridization, PCR or DNA sequencing. Each seems to offer specific advantages and they have been widely adopted. Even though assays based on DNA microarrays are probably the most recent entries, they hold much promise and are offering clear advantages in many instances. Furthermore, barcodes linked to short DNA sequences that uniquely identify certain features such as genes, mutations or even whole organisms also have become widely adopted in many areas of research. In these circumstances, the DNA microarray can also serve an invaluable role, representing a fast and cost-effective method of reading the barcodes. The most elaborate example of these kinds of application is the Yeast Knock Out (YKO) collection, used for screens of pooled yeast (*Saccharomyces cerevisiae*) deletion mutants (Giaever *et al.* 2002; Ammar *et al.* 2009), a project of the Yeast Deletion Consortium. In this case two unique 20-mer oligonucleotide tags were added to each knockout gene and a 12 400 spot microarray is used to assess contributions from each of the 6200 yeast genes. As described in this section, most applications are less elaborate but, whether used as standalone assays or linked to DNA barcoding, the DNA microarray offers clear advantages, which are being adopted widely.

Microbial surveillance or discovery assays, based on DNA in uncharacterized samples, often use sequence analyses when in-depth information is sought, or qPCR when cost, speed and sensitivity are an issue. Both approaches have a limited capacity for multiplexing and are intolerant of primer-target mismatches. This is an advantage when the microbe sequence is precisely known but represents a disadvantage for the discovery of unknown species or variant strains. DNA microarrays often represent a middle ground, offering high probe densities and the possibility of unexpected target discovery, frequently at much lower cost with processing times of a day or less. For some or all of these reasons there has been a rapidly expanding use of DNA microarrays as a powerful tool in DNA-based molecular diagnostics for the classification and identification of organisms, including pathogens. While originally designed to target a limited range of organisms (e.g. Sintchenko *et al.* 2007; Miller and Tang 2009), more recent efforts have attempted to develop broad-spectrum microarrays. For example, Pasquer and co-workers (2010) have attempted to develop a DNA chip for the reliable identification of a wide range of bacterial taxa at the subspecies level by using hybridization fingerprints generated by widely distributed anonymous markers. The array carries 95 000 unique 13-mer probes and is able to discriminate bacteria at the species, subspecies and strain levels. Mismatch probe hybridization was observed but seemed to have no effect on the discriminatory capacity. To permit the development of similar arrays with fewer spots, Dai and co-workers (2009) have been trying to develop arrays that operate using group testing and compressive sensing principles in which probes respond to a group of targets rather than a single target organism. Small-scale design and lab experimental calibration of the model to date remain promising. Equally, some generalized microarrays have been redesigned for application in specialized environments. For example, Ehrenfeld and co-workers (2009) have used results from a high-density microarray with 32 392 50-mer oligonucleotides to manufacture a new slide (BMS 3.0) with 560 specific oligonucleotrides for the identification of biomining microorganisms. Finally, in an effort to identify the suites of viruses and bacteria in complex samples, a pan-Microbial Detection Array (MDA) was recently developed and reported to detect all known viruses (including pathogens), bacteria and plasmids based on a novel statistical analysis method to identify mixtures of organisms from complex samples hybridized to the array (Gardner *et al.* 2010).

The application of DNA microarrays for species identification is not limited to bacteria or microbes and its application also is expanding rapidly in studies of biodiversity and even environmental regulation. For example, issues pertaining to the release of genetically modified organisms, including patent applications and related commercial consequences, have provided opportunities for the application of microarray assay in the identification of genetically modified organisms (GMOs) and their presence in processed foods. Kim and co-workers (2010) recently developed a microarray system that could reliably detect 19 genetically modified organisms (soybeans, maizes, canolas and cotton) in processed foods. Equally, the very high value of fish and fishery products throughout the world has spurred many scientists and/or countries to develop assays that would protect them from commercial fraud by mislabelling, which can even threaten consumer health if toxic species (e.g. pufferfish) enter the market. Current management of fisheries, fishing quotas and future stocks also requires accurate identification. Fish DNA barcodes, based on the sequencing of three mitochondrial genes, the 16S rRNA (16S), cytochrome b (*cytb*) and cytochrome oxidase subunit I (COI) have represented a very popular approach, as has also been the case in many other barcoding projects (Hajibabaei *et al.* 2007) aimed at studies of biodiversity or taxonomy. Next generation sequencing methods are enabling the analysis of mixed samples (for references see Kochzius *et al.* 2010) but require highly sophisticated and expensive equipment. Since DNA microarrays can differentiate hundreds of specimens simultaneously at lower cost, studies have been undertaken to compare both technologies (Kochzius *et al.* 2010) with 50 European marine fish species. This study showed that the sequence markers performed differently in DNA barcoding and microarray analyses; where *cytb* and COI were equally suited for sequence analyses, 16S-probes performed appreciably better in DNA hybridizations. This study has developed an effective way for the identification of 30 fish species, which represents a significant step towards an automated and easy-to-handle assay for ichthyoplankton surveys.

As in the other newer areas of application, many aspects from probe design to standardized supporting software are still under development and important questions remain. Pleas have been issued for greater collaboration and standardization (e.g. Radulovici *et al.* 2010). Considerable uncertainty also remains with respect to current competing technologies (barcoding vs. DNA microarrays) or perhaps a combination of both (see Kochzius *et al.* 2010). Nevertheless, there is little dispute that DNA microarrays have a very important future with respect to organism identification, both in the environment and clinically, and that applications will continue and likely expand.

Concluding remarks

Probably first used by Roy Britten as a way of analysing the composition of genomes in the 1960s (e.g. Britten and Kohne 1968), DNA/RNA hybridization is possibly the most understated molecular technique in biology. Fundamentals of hybridizations underlie many of the breakthrough technologies in molecular genetics from 'sticky ends' in DNA cloning to gel blots to the priming of reverse transcription, DNA sequencing, targeted mutagenesis and the polymerase chain reaction. As noted earlier, hybridization has reappeared once more as the basis of the first 'global' genetic analysis technique when DNA microarrays were adopted for gene expression profiling. As outlined in this review, 20 years after their introduction DNA microarrays have 'come of age' with applications as diverse as anyone can imagine. From interrogating genomes, to cross-species analyses to genome comparisons and finally to the detection and identification of a living organisms in diverse environmental niches, the technology represents powerful tools that continue to develop into ever more complex but promising approaches to push back the frontiers of biological and medical research.

References

Ammar, R., Smith, A.M., Heisler, L.E. *et al.* (2009) A comparative analysis of DNA barcode microarray feature size. *BMC Genomics* **10**: 471.

Bar-Or, C., Novikov, E., Reiner, A. *et al.* (2007) Utilizing microarray spot characteristics to improve cross-species hybridization results. *Genomics* **90**: 636–45.

Beaudet, A.L. and Belmont, J.W. (2008) Array-based DNA diagnostics: let the revolution begin. *Annu Rev Med* **59**: 113–129.

Belosludtsev, Y.Y., Bowerman, D., Weil, R. *et al.* (2004) Organism identification using a genome sequence-independent universal microarray probe set. *Biotechniques* **37**: 654–658, 660.

Bigger, C.B., Brasky, K.M. and Lanford, R.E. (2001) DNA microarray analysis of chimpanzee liver during acute resolving hepatitis C virus infection. *J Virol* **75**: 7059–7066.

Bogus, M., Sobrino, B., Bender, K. A. *et al.* (2006) Rapid microarray-based typing of forensic SNPs *International Congress Series* **1288**: 37–39.

Brennan, D.J., O'Brien, S.L., Fagan, A. *et al.* (2005) Application of DNA microarray technology in determining breast cancer prognosis and therapeutic response. *Expert Opin Biol Ther* **5**: 1069–1083.

Britten, R.J. and Kohne, D.E. (1968) Repeated sequences in DNA. Science 161: 529–540.

Buckley, B.A, (2007) Comparative environmental genomics in non-model species: using heterologous hybridization to DNA-based microarrays. *J Exp Biol* **210**: 1602–1606.

Campbell, M.A., Haas, B.J., Hamilton, J.P. *et al.* (2006) Comprehensive analysis of alternative splicing in rice and comparative analyses with *Arabidopsis*. *BMC Genomics* **7**: 327.

Chen, Y., Wu, R., Felton, J. *et al.* (2010) A method to detect differential gene expression in cross-species hybridization experiments at gene and probe level. *Biomed Inform Insights* **2010**: 1–10.

Chen, Y.A., Chou, C.C., Lu, X. *et al.* (2006) A multivariate prediction model for microarray cross-hybridization. *BMC Bioinformatics* **7**: 101.

Chu, T., Burke, B., Bunce, K. *et al.* (2009) A microarray-based approach for the identification of epigenetic biomarkers for the noninvasive diagnosis of fetal disease. *Prenat Diagn* **29**: 1020–1030.

Collins, F.S., Morgan, M. and Patrinos, A. (2003) The human genome project: lessons from large-scale biology. *Science* **300**: 286–290.

Cuperlovic-Culf, M., Barnett, D.A., Culf, A.S. and Chute, I. (2010) Cell culture metabolomics: applications and future directions. *Drug Discov Today* **15**: 610–621.

Dai, W., Sheikh, M.A., Milenkovic, O. and Baraniuk, R.G. (2009), Compressive sensing DNA microarrays. *J Bioinform Syst Biol* **2009**: 1–12.

Dear, P.H. (2009) Copy-number variation: the end of the human genome? *Trends Biotechnol* **27**: 448–454.

Dezso, Z., Nikolsky, Y., Sviridov, E. *et al.* (2008) A comprehensive functional analysis of tissue specificity of human gene expression. *BMC Biol* **6**: 49.

Divne, A.M. and Allen, M.A. (2005) DNA microarray system for forensic SNP analysis. *Forensic Sci Int* **154**: 111–121.

Ehrenfeld, N., Aravena, A., Reyes-Jara, A. *et al.* (2009) Design and use of oligonucleotide microarrays for identification of biomining microorganisms. *Adv Mater Res* **71–73**: 155–158.

Enard, W., Khaitovich, P., Klose, J. *et al.* (2002) Intra- and interspecific variation in primate gene expression patterns. *Science* **296**: 340–343.

Euskirchen, G.M., Rozowsky, J.S., Wei, C.L. *et al.* (2007) Mapping of transcription factor binding regions in mammalian cells by ChIP: comparison of array- and sequencing-based technologies. *Genome Res* **17**: 898–909.

Fagnani, M., Barash, Y., Ip, J.Y. *et al.* (2007) Functional coordination of alternative splicing in the mammalian central nervous system. *Genome Biol* **8**: R108.

Fan, C., Vibranovski, M.D., Chen, Y. and Long, M. (2007) A microarray-based genomic hybridization method for identification of new genes in plants: case analyses of *Arabidopsis* and rice. *J Integr Plant Biol* **49**: 915–926.

Feuk, L., Carson, A.R. and Scherer, S.W. (2006) Structural variation in the human genome. *Nat Rev Genet* **7**: 85–97.

Flintoft, L. (2009) Gene regulation: sequence, chromatin, action! *Nat Rev Genet* **10**, 512–513.

French, P.J., Peeters, J., Horsman, S. *et al.* (2007) Identification of differentially regulated splice variants and novel exons in glial brain tumors using exon expression arrays. *Cancer Res* **67**: 5635–5642.

Gama-Carvalho, M., Barbosa-Morais, N.L., Brodsky, A.S. *et al.* (2006) Genome-wide identification of functionally distinct subsets of cellular mRNAs associated with two nucleocytoplasmic-shuttling mammalian splicing factors. *Genome Biol* **7**: R113.

Ganal, M.W., Altmann, T. and Röder, M.S. (2009) SNP identification in crop plants. *Curr Opin Plant Biol* **12**: 211–217.

Gardner, S.N., Jaing, C.J., McLoughlin, K.S. and Slezak, T.R. (2010) A microbial detection array (MDA) for viral and bacterial detection. *BMC Genomics* **11**: 668.

Gerhold, D.L., Jensen, R.V. and Gullans, S.R. (2002) Better therapeutics through microarrays. *Nature Genetics* **32**: 547–552.

Giaever, G., Chu, A.M., Ni, L. *et al.* (2002) Functional profiling of the *Saccharomyces cerevisiae* genome. *Nature* **418**: 387–391.

Gilad, Y. and Borevitz, J. (2006) Using DNA microarrays to study natural variation. *Curr Opin Genet Dev* **16**: 553–558.

Gresham, D., Dunham, M.J. and Botstein, D. (2008) Comparing whole genomes using DNA microarrays. *Nat Rev Genet* **9**: 291–302.

Grigoryev, D.N., Ma, S.F., Simon, B.A. *et al.* (2005) In vitro identification and in silico utilization of interspecies sequence similarities using GeneChip technology. *BMC Genomics* **6**: 62.

Gupta, P.K., Rustgi, S. and Mir, R.R. (2008) Array-based high-throughput DNA markers for crop improvement. *Heredity* **101**: 5–18.

Hajibabaei, M., Singer, G.A., Hebert, P.D. and Hickey, D.A. (2007) DNA barcoding: how it complements taxonomy, molecular phylogenetics and population genetics. *Trends Genet* **23**: 167–172.

Huff, J.L., Hansen, L.M. and Solnick, J.V. (2004) Gastric transcription profile of Helicobacter pylori infection in the rhesus macaque. *Infect Immun* **72**: 5216–5226.

Hui, J., Kishore, S., Khanna, A. and Stamm, S. (2009) Analysis of alternative splicing with microarrays. *Bioinformatics for Systems Biology*, Part III, 267–279.

Humblot, P., Le Bourhis, D., Fritz, S. *et al.* (2010) Reproductive technologies and genomic selection in cattle. *Vet Med Int* **2010**: 192787.

Iafrate, A.J., Feuk, L., Rivera, M.N. *et al.* (2004) Detection of large-scale variation in the human genome. *Nat Genet* **36**: 949–951.

Ip, J.Y., Tong, A., Pan, Q. *et al.* (2007) Global analysis of alternative splicing during T-cell activation. *RNA* **13**: 563–572.

Jacquemont, M.L., Sanlaville, D., Redon, R. *et al.* (2006) Array-based comparative genomic hybridisation identifies high frequency of cryptic chromosomal rearrangements in patients with syndromic autism spectrum disorders. *J Med Genet* **43**: 843–849.

Ji, W., Zhou, W., Gregg, K. *et al.* (2004) A method for cross-species gene expression analysis with high-density oligonucleotide arrays. *Nucleic Acids Res* **32**: e93.

Johnson, J.M., Castle, J., Garrett-Engele, P. *et al.* (2003) Genome-wide survey of human alternative pre-mRNA splicing with exon junction microarrays. *Science* **302**: 2141–2144.

Kamalakaran, S., Kendall, J., Zhao, X. *et al.* (2009) Methylation detection oligonucleotide microarray analysis: a high-resolution method for detection of CpG island methylation. *Nucleic Acids Res* **37**: e89.

Kammenga, J.E., Herman, M.A., Ouborg, N.J. *et al.* (2007) Microarray challenges in ecology. *Trends Ecol Evol* **22**: 273–279.

Kampa, D., Cheng, J., Kapranov, P. *et al.* (2004) Novel RNAs identified from an in-depth analysis of the transcriptome of human chromosomes 21 and 22. *Genome Res* **14**: 331–342.

Kane, M.D., Jatkoe, T.A., Stumpf, C.R. *et al.* (2000) Assessment of the sensitivity and specificity of oligonucleotide (50mer) microarrays. *Nucleic Acids Res* **28**: 4552–4557.

Kapranov, P., Cawley, S.E., Drenkow, J. *et al.* (2002) Large-scale transcriptional activity in chromosomes 21 and 22. *Science* **296**: 916–919.

Keene, J.D., Komisarow, J.M. and Friedersdorf, M.B. (2006) RIP-Chip: the isolation and identification of mRNAs, microRNAs and protein components of ribonucleoprotein complexes from cell extracts. *Nat Protoc* **1**: 302–307.

Khaitovich, P., Weiss, G., Lachmann, M. *et al.* (2004) A neutral model of transcriptome evolution. *PLoS Biol* **2**: e132.

Kijas, J.W., Townley, D., Dalrymple, B.P. *et al.* (2009) A genome wide survey of SNP variation reveals the genetic structure of sheep breeds. *PLoS One* **4**: e4668.

Kim, J.H., Kim, S.Y., Lee, H. *et al.* (2010) An event-specific DNA microarray to identify genetically modified organisms in processed foods. *J Agric Food Chem* **58**: 6018–6026.

Klinger, K., Landes, G., Shook, D. *et al.* (1992) Rapid detection of chromosome aneuploidies in uncultured amniocytes by using fluorescence in situ hybridization (FISH). *Am J Hum Genet* **51**: 55–65.

Kochzius, M., Seidel, C., Antoniou, A. *et al.* (2010) Identifying fishes through DNA barcodes and microarrays. *PLoS One* **5**: e12620.

Krjutskov, K., Viltrop, T., Palta, P. *et al.* (2009) Evaluation of the 124-plex SNP typing microarray for forensic testing. *Forensic Sci Int Genet* **4**: 43–48.

Laiho, P., Hienonen, T., Mecklin, J.P. *et al.* (2003) Mutation and LOH analysis of ACO2 in colorectal cancer: no evidence of biallelic genetic inactivation. *J Med Genet* **40**: e73.

Lee, C., Iafrate, A.J. and Brothman, A.R. (2007) Copy number variations and clinical cytogenetic diagnosis of constitutional disorders. *Nat Genet* **39**: S48–54.

Leung, W.C., Waters, J.J, and Chitty, L. (2004) Prenatal diagnosis by rapid aneuploidy detection and karyotyping: a prospective study of the role of ultrasound in 1589 second-trimester amniocenteses. *Prenat Diagn* **24**: 790–795.

Liu, G.E., Hou, Y., Zhu, B. *et al.* (2010) Analysis of copy number variations among diverse cattle breeds. *Genome Res* **20**: 693–703.

Liu, G.E., Van Tassel, C.P., Sonstegard, T.S. *et al.* (2010) Detection of germline and somatic copy number variations in cattle. *Dev Biol (Basel)* **132**: 231–237.

Lu, X., Shaw, C.A., Patel, A. *et al.* (2007) Clinical implementation of chromosomal microarray analysis: summary of 2513 postnatal cases. *PLoS One* **2**: e327.

Matukumalli, L.K., Lawley, C.T., Schnabel, R.D. *et al.* (2009) Development and characterization of a high density SNP genotyping assay for cattle. *PLoS One* **4**: e5350.

McKee, A.E., Neretti, N., Carvalho, L.E. *et al.* (2007) Exon expression profiling reveals stimulus-mediated exon use in neural cells. *Genome Biol* **8**: R159.

Michaut, L., Flister, S., Neeb, M. *et al.* (2003) Analysis of the eye developmental pathway in *Drosophila* using DNA microarrays. *Proc Natl Acad Sci USA* **100**: 4024–4029.

Miller, M.B. and Tang, Y.W. (2009) Basic concepts of microarrays and potential applications in clinical microbiology. *Clin Microbiol Rev* **22**: 611–633.

Miura, S., Miura, K., Masuzaki, H. *et al.* (2006) Microarray comparative genomic hybridization (CGH)-based prenatal diagnosis for chromosome abnormalities using cell-free fetal DNA in amniotic fluid. *J Hum Genet* **51**: 412–417.

Moore, M.J., Schwartzfarb, E.M., Silver, P.A. and Yu, M.C. (2006) Differential recruitment of the splicing machinery during transcription predicts genome-wide patterns of mRNA splicing. *Mol Cell* **24**: 903–915.

Moore, M.J. and Silver, P.A. (2008) Global analysis of mRNA splicing. *RNA* **14**: 197–203.

Moskalyov, E.A., Eprintsev, A.T. and Hoheisel, J.D. (2007) DNA methylation profiling in cancer: from single nucleotides towards methylome. *Mol Biol (Mosk)* **41**: 793–807.

Muir, W.M., Wong, G.K., Zhang, Y. *et al.* (2008) Genome-wide assessment of worldwide chicken SNP genetic diversity indicates significant absence of rare alleles in commercial breeds. *Proc Natl Acad Sci USA* **105**: 17312–17317.

Nazar, R.N., Chen, P., Dean, D. and Robb, J. (2010) DNA chip analysis in diverse organisms with unsequenced genomes. *Mol Biotechnol* **44**: 8–13.

Ner-Gaon, H. and Fluhr, R. (2006) Whole-genome microarray in Arabidopsis facilitates global analysis of retained introns. *DNA Res* **13**: 111–121.

Okoniewski, M.J., Hey, Y., Pepper, S.D. and Miller, C.J. (2007) High correspondence between Affymetrix exon and standard expression arrays. *Biotechniques* **42**: 181–185.

Pasquer, F., Pelludat, C., Duffy, B. and Frey, J.E. (2010) Broad spectrum microarray for fingerprint-based bacterial species identification. *BMC Biotechnol* **10**: 13.

Pinkel, D. and Albertson, D.G. (2005) Comparative genomic hybridization. *Annu Rev Genom Hum Genet* **6**: 331–354.

Radulovici, A.E., Archambault, P. and Dufresne, F. (2010) DNA barcodes for marine biodiversity: moving fast forward? *Diversity* **2**: 450–472.

Ranz, J.M. and Machado, C.A. (2006) Uncovering evolutionary patterns of gene expression using microarrays. *Trends Ecol Evol* **21**: 29–37.

Redon, R., Ishikawa, S., Fitch, K.R. *et al.* (2006) Global variation in copy number in the human genome. *Nature* **444**: 444–454.

Roberts, C.J, Nelson, B., Marton, M.J. *et al.* (2000) Signaling and circuitry of multiple MAPK pathways revealed by a matrix of global gene expression profiles. *Science* **287**: 873–880.

Sebat, J., Lakshmi, B., Malhotra, D. *et al.* (2007) Strong association of de novo copy number mutations with autism. *Science* **316**: 445–449.

Sebat, J., Lakshmi, B., Troge, J. *et al.* (2004) Large-scale copy number polymorphism in the human genome. *Science* **305**: 525–528.

Shaffer, L.G., Kashork, C.D., Saleki, R. *et al.* (2006) Targeted genomic microarray analysis for identification of chromosome abnormalities in 1500 consecutive clinical cases. *J Pediatr* **149**: 98–102.

Sintchenko, V., Iredell, J.R. and Gilbert, G.L. (2007) Pathogen profiling for disease management and surveillance. *Nat Rev Microbiol* **5**: 464–470.

Southern, E.M. (2001) DNA microarrays. History and overview. *Methods Mol Biol* **170**: 1–15.

Stoughton, R.B. (2005) Applications of DNA microarrays in biology. *Annu Rev Biochem* **74**: 53–82.

Su, A.I., Wiltshire, T., Batalov, S. *et al.* (2004) A gene atlas of the mouse and human protein-encoding transcriptomes. *Proc Natl Acad Sci USA* **101**: 6062–6067.

Sugnet, C.W., Srinivasan, K., Clark, T.A. *et al.* (2006) Unusual intron conservation near tissue-regulated exons found by splicing microarrays. *PLoS Comput Biol* **2**: e4.

Swinburne, I.A., Meyer, C.A., Liu, X.S. *et al.* (2006) Genomic localization of RNA binding proteins reveals links between pre-mRNA processing and transcription. *Genome Res* **16**: 912–921.

Tardiff, D.F., Lacadie, S.A. and Rosbash, M. (2006) A genome-wide analysis indicates that yeast pre-mRNA splicing is predominantly posttranscriptional. *Mol Cell* **24**: 917–929.

Tijssen, P. (1993) Overview of principles of hybridization and the strategy of nucleic acid probe assays. *Laboratory Techniques in Biochemistry and Molecular Biology: Hybridization with Nucleic Acid Probes Part I: Theory and Nucleic Acid Preparation* Vol 24: pp. 19–78. Elsevier, Amsterdam.

Van Tassell, C.P., Smith, T.P., Matukumalli, L.K. *et al.* (2008) SNP discovery and allele frequency estimation by deep sequencing of reduced representation libraries. *Nat Methods* **5**: 247–252.

Vorstman, J.A., Staal, W.G., van Daalen, E., van Engeland, H. *et al.* (2006) Identification of novel autism candidate regions through analysis of reported cytogenetic abnormalities associated with autism. *Mol Psychiatry* **11**: 18–28.

Waddell, N. (2008) Microarray-based DNA profiling to study genomic aberrations. *IUBMB Life* **60**: 437–440.

Walker, S.J., Wang, Y., Grant, K.A. *et al.* (2006) Long versus short oligonucleotide microarrays for the study of gene expression in nonhuman primates. *J Neurosci Methods* **152**: 179–189.

Wang, G.S. and Cooper, T.A. (2007) Splicing in disease: disruption of the splicing code and the decoding machinery. *Nat Rev Genet* **8**: 749–761.

Wolter, T.R., Wong, R., Sarkar S.A. and Zipris, D. (2009) DNA microarray analysis for the identification of innate immune pathways implicated in virus-induced autoimmune diabetes. *Clin Immunol* **132**: 103–115.

Xing, Y., Kapur, K. and Wong, W.H. (2006) Probe selection and expression index computation of Affymetrix Exon Arrays. *PLoS One* **1**: e88.

Yazaki, J., Gregory, B.D. and Ecker, J.R. (2007) Mapping the genome landscape using tiling array technology. *Curr Opin Plant Biol* **10**: 534–542.

Yoo, S.M., Choi, J.H., Lee, S.Y. and Yoo N.C. (2009) Applications of DNA microarray in disease diagnostics. *J Microbiol Biotechnol* **19**: 635–646.

Zhang, X., Clarenz, O., Cokus, S. *et al.* (2007) Whole-genome analysis of histone H3 lysine 27 trimethylation in Arabidopsis. *PLoS Biol* **5**: e129.

3

qPCR, Theory, Reliability and Use in Molecular Analysis

Jamie Murphy and Stephen A. Bustin

The real-time reverse-transcription polymerase chain reaction (RT-qPCR) is the most convenient and popular method for quantitating steady-state RNA levels (Bustin 2000). It is rapid, has the potential for high throughput and, being a homogeneous assay, minimizes contamination problems. Another key advantage is the wide dynamic range of RT-qPCR assays, allowing the researcher to quantify RNA from samples containing hugely different levels of nucleic acid. Nevertheless, it has become clear that while its use ameliorates some of the problems associated with legacy, gel-based RT-PCR assays, it also introduces new complications that must be addressed if this technology is to produce biologically relevant results (Bustin 2002). Critical areas that require attention include sample preparation, RNA quality, reagents (Wolffs *et al.* 2004), assay design, normalization, data analysis (Ginzinger 2002) and transparency of data analysis. This latter point, which influences all key steps of a RT-qPCR assay, is of particular importance when the aim is the development of clinically relevant diagnostic or prognostic assays. Unfortunately, in these circumstances RT-qPCR data are frequently used in an inappropriate manner to support conclusions that are not reliably derived from the actual results obtained. Therefore, until standardized protocols, appropriate assay design and data analysis are universally implemented, RT-qPCR will not be able to realize its potential and progress beyond its current role as a research tool.

Molecular Analysis and Genome Discovery, Second Edition. Edited by Ralph Rapley and Stuart Harbron.
© 2012 John Wiley & Sons, Ltd. Published 2012 by John Wiley & Sons, Ltd.

Sample preparation

Conventionally, RNA extractions from human/animal tissues and the subsequent analyses have been carried out using whole tissue biopsies with no attempt to differentiate the separate cell lineages contained within any one particular sample. This practice inevitably results in the detection of average RNA levels across a whole range of cell types; thus expression levels of a cell type-specific RNA may be masked or incorrectly ascribed to that of a different cell lineage (Chelly *et al.* 1989). Therefore, it is not surprising that considerable differences have been reported in the gene expression profiles of microdissected and bulk tissue preparations taken from the same sample (Fink *et al.* 2002; Sugiyama *et al.* 2002).

This finding is of special significance when assessing certain pathological conditions, for example comparison between gene expression profiles for adjacent normal tissue and cancer biopsies, profiling of intravasated tumours or even profiling of different areas within the same tumour (Figure 3.1). Adjacent normal tissue may appear phenotypically unremarkable; however, it frequently exhibits altered gene expression or miRNA profiles ('field-change') or contains premalignant genotypic abnormalities (Deng *et al.* 1996). This is further complicated by the fact that immune cell infiltrates vary between tumours derived from the same tissue of origin (Bustin *et al.* 2001). Therefore, quantitative assessment of mRNA targets from bulk tissue preparations must be interpreted with considerable caution.

The introduction of technologies such as laser capture microdissection (LCM) allows for experimental designs that deal with this problem (Emmert-Buck *et al.* 1996), since it permits the extraction of pure subpopulations of cells from the background of heterogeneous tissue samples for detailed molecular analysis (Figure 3.2; Walch *et al.* 2001). For example, LCM samples subjected to RT-qPCR analysis might consist of single, tens, hundreds or thousands of cells. Conventionally, isolation of total RNA from very small samples has been both difficult and inefficient. However, extraction kits are now available which permit total RNA isolation from single cells and although this claim may seem rather extraordinary, it is supported by studies

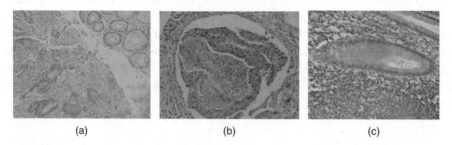

(a)   (b)   (c)

Figure 3.1 Tissue biopsies consist of different cell types: (a) a cancer biopsy contains significant amounts of normal tissue; (b) a colorectal cancer intravasated into a blood vessel surrounded by bulk tumour; (c) well and poorly differentiated sections of the same tumour

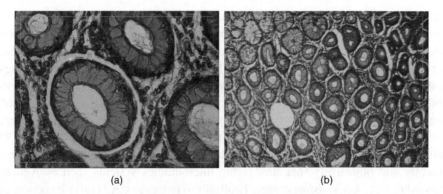

(a)   (b)

Figure 3.2 Laser capture microdissection: (a) a single colonic crypt is circled using an electronic pen (green circle labelled 1); (b) the single crypt has been catapulted into a microfuge tube for gene expression profiling

detailing extraction of total RNA from tens of cells with subsequent RT-PCR analysis (Bohle *et al.* 2000; Dolter and Braman 2001).

LCM particularly lends itself to the extraction of total RNA from archival formalin-fixed, paraffin-embedded tissue specimens, especially given the ready availability of tissue samples prepared in this manner and that associated long-term clinical data are often available for such specimens, facilitating retrospective investigation of disease mechanisms or prognostic indicators (Lehmann *et al.* 2000). It must be appreciated that special considerations apply in this setting, however, since considerable mRNA degradation can occur before (Mizuno *et al.* 1998) or during (Klimecki *et al.* 1994) the formalin fixation process, with covalent modification of RNA making extraction, reverse transcription (RT), amplification and subsequent quantification challenging (Masuda *et al.* 1999). Furthermore, different tissue preparation methods will invariably lead to different results from different laboratories (Goldsworthy *et al.* 1999). Nevertheless, reliable commercial extraction kits are now available for total RNA extraction from formalin-fixed, paraffin-embedded tissues and general use of such kits should aid standardization of quantitative protocols.

RNA quality

The quality of total RNA extract is perhaps the most important determinant of the reproducibility and biological relevance of RT-qPCR analyses (Bomjen *et al.* 1996). Unlike DNA, RNA is extremely prone to degradation once it is extracted from its natural cellular environment. Consequently, RNA purification is considerably more demanding than that of DNA. Template suitable for inclusion in an RT-qPCR assay must be: (1) of the highest integrity if quantitative results are to be biologically relevant; (2) free of DNA, especially if the target is an intronless gene; (3) free from inhibitors of the RT reaction; and (4) free of nucleases to allow prolonged storage.

RNA is most easily quantified using Thermo Scientific's (Wilmington, DE, USA) NanoDrop or Implen's (Munich, Germany) NanoPhotometer spectrophotometers, which use as little as 1 µl of sample to calculate accurate RNA concentrations. Clearly every RNA preparation must be rigorously assessed for quality and, while assessment of RNA integrity by inspecting 28S and 18S ribosomal RNA gel electrophoresis bands for each sample was acceptable for qualitative analysis, it is no longer sufficient for accurate quantitative expression profiling. Furthermore, gel electrophoresis is a low-throughput procedure that requires considerable quantities of RNA. The routine analysis of large numbers of total RNA preparations is much more conveniently and objectively measured using microfluidics systems such as Agilent's (Palo Alto, CA, USA) Bioanalyzer, BioRad's (Hercules, CA, USA) Experion and Lab901's (Loanhead, UK) TapeStation that can also quantitate RNA concentration.

The second determinant of template suitability is the absence of RT inhibitors. There are numerous components within mammalian tissues that can inhibit RT-qPCR assays, for example the haeme compound (Akane *et al.* 1994; Al Soud, Jonsson and Radstrom 2000; Al Soud and Radstrom 2001) and calcium (Bickley *et al.* 1996). Further confounders include high levels of copurified RNA (Pikaart and Villeponteau 1993), culture media or even nucleic extraction reagents (Rossen *et al.* 1992). Confusingly, some evidence suggests that inhibitors may selectively interfere with one specific RT polymerase, leaving others totally unaffected (Belec *et al.* 1998). Therefore, each RNA preparation must be tested for the presence of inhibitors. This is most easily accomplished by amplifying a known amount of an amplicon that is not present within the sequence of the target RNA and comparing the copy numbers obtained in the presence of RNA to those obtained in the control. Any decrease in apparent copy number is indicative of inhibition of the RNA preparation. A technical description outlining how to perform inhibition assays can be found elsewhere (Nolan *et al.* 2006).

Reagents

The variability of RT-qPCR results obtained from identical samples assayed in different laboratories continues to be a significant problem (Bolufer *et al.* 2001). As there are so many steps involved in converting a tissue sample to a 'quantitative' result, the single most likely source of variation is the person undertaking the experiment. Consequently, this provides a convincing argument for the use of robots to aliquot reagents/mRNA preparations when reliability and robustness are the main concerns. Similarly, the use of commercially available kits rather than laboratory-specific 'master mixes' makes assay consistency within specific lot numbers more likely, further improving reproducibility of results. However, it is important to realize that different batches of commercial assay kits or primer/probe pairings synthesized by the same manufacturer at different times may result in significantly different calculated mRNA copy numbers. Therefore, it is important that target-specific standard curves are obtained from reference oligonucleotides or standard RNA preparations are included with every RT-qPCR run, especially when fresh batches of RT-qPCR kit or primer and probe sets are used.

Assay design

Priming strategy

The ostensibly small task of converting RNA into a PCR-amplifiable DNA template is a substantial contributor to the variability and lack of reproducibility characteristic of published RT-PCR results. Several explanations have been proposed: (1) the dynamic nature of the intracellular environment mandates inevitable variation in the quantity and composition of total RNA extracts prepared from biological samples; (2) purified mRNA is inherently unstable after extraction and may be of variable quality; (3) RNA–cDNA conversion efficiency is dependent upon template abundance, having been demonstrated to be significantly lower when the target is scarce (Karrer et al. 1995) and negatively affected by the presence of bystander nucleic acids (Curry, McHale and Smith 2002); and (4) different priming strategies used to synthesize cDNA may result in widely disparate calculated mRNA copy numbers.

There are four different ways to prime cDNA synthesis, each of which can generate different results for both specificity and cDNA yield (Zhang and Byrne 1999; Lekanne Deprez et al. 2002): (1) random primers; (2) oligo-dT; (3) a combination of random and oligo-dT priming; and (4) target-specific priming. Non-specific endogenous priming can occur regardless of which primers are used to prime the RT reaction, as specific PCR products can be obtained without adding any primers to the RT reaction (Ambion 2001). This is significant as non-specific priming during cDNA synthesis can lead to a lowered or variable signal in the subsequent PCR assay, although the practical extent of this problem remains unclear. Random priming initiates the RT reaction at multiple points along the length of any RNA, including rRNA, and produces multiple cDNA transcripts from each original target. This method is particularly useful for transcripts with complex secondary structures, or if the target is purified mRNA rather than total RNA. However, this priming strategy may overestimate mRNA copy numbers by >15-fold compared with sequence-specific priming (Zhang and Byrne 1999) and may be less sensitive and less reliable than other methods of priming cDNA (Lekanne Deprez et al. 2002). cDNA synthesis by oligo-dT priming is more specific to mRNA than random priming, as it will not reverse transcribe rRNA. However, oligo-dT priming is entirely dependent upon the presence of a polyA tail as well as high-quality RNA, especially if the amplicon is towards the 5'-end of the mRNA. Furthermore, significant difficulties may also be encountered when attempting to reverse transcribe targets with complex secondary structures, since the RT may bypass entire sequences or fall of the template and not reinitiate. A combination of random priming and oligo-dT addresses some of these issues and is the best choice when the main criterion is to target numerous mRNA targets and store the cDNA ready for future analysis. Target-specific priming generates the most specific cDNA and therefore probably represents the most sensitive option for quantification (Lekanne Deprez *et al.* 2002). It provides superior results to assays using random primers or oligo-dT, although there does seem to be gene-specific variation. Therefore, it is important to validate individual assays using standard curve dilutions before coming to conclusions about results obtained from actual samples. The main disadvantage of this method is that it is not possible to use the same preparation to amplify additional targets at a later date.

Transcript detection

There are two RT-qPCR reporting chemistries: detection of fluorescence from double-stranded DNA-specific but not target-specific DNA binding dyes; or, from target-specific probes,

Non-specific detection

Non-specific detection uses fluorescent reporter dyes (e.g. SYBR Green) that bind to double-stranded DNA generated during PCR (Ishiguro *et al.* 1995). While these reporter dyes are non-specific, they are able to yield template specific data when DNA melt curves are used to identify specific amplicons (Ririe, Rasmussen and Wittwer 1997). These assays may be more flexible than their probe-based equivalent, as they can be easily incorporated into previously optimized and long-established protocols. Furthermore, DNA-binding dye assays are significantly cheaper, since they are not associated with any probe-specific costs. Assays using fluorescent reporter dyes need not be less reliable than probe-based assays, despite the non-specific system for amplicon detection. In actual fact, there is at least one report suggesting that SYBR Green I chemistry is more accurate and produces a more linear decay plot than TaqMan probe-based assays (Schmittgen, Zakrajsek and Mills 2000). However, fluorescent reporter dyes bind indiscriminately to any double-stranded DNA, resulting in signal detection from a 'no template control' (NTC) sample caused by dye molecules binding to primer–dimer complexes, although this can be minimized by the use of separate RT and PCR steps. Data interpretation may be more complex with DNA-binding dye assays than probe-based experiments. Fluorescent reporter dye chemistry is no more specific than conventional PCR and thus concomitant melt curve analyses are mandatory. Furthermore, multiple dye molecules bind to a single amplicon. Therefore, if one assumes that two experiments have the same amplification efficiencies, a longer amplicon will generate more signal than a shorter one; furthermore, if amplification efficiencies are different, quantification will be even more inaccurate.

Specific detection

Template-specific detection analysis requires the design and synthesis of custom-made fluorescent probes for each individual assay. Detection systems generally utilize fluorescent resonance energy transfer (FRET) between a donor and a quencher molecule as the basis for target detection. There is a wide range of fluorescent dyes available for template-specific detection, since they are used in other molecular biology procedures such as DNA sequencing. The main advantage associated with specific chemistries is that the use of a probe introduces an additional level of specificity, obviating the need to sequence the target or perform melt curve analyses to confirm the identity of the amplicon. In addition, multiplex PCR can be performed using probes labelled with different reporter dyes, allowing the detection of amplicons from several distinct sequences in a single PCR reaction tube. However, artefacts (e.g. primer) must be used to optimize reaction conditions prior to any probe-based quantification experiments. The considerable expense of template-specific chemistries must also be borne in mind, since costs are incurred rapidly when quantitating multiple targets. Another limiting factor is that although there are numerous dyes to choose from, their excitation and emission spectra must be sufficiently distinct to allow the instruments to detect fluorescence reliably.

Data analysis

The threshold cycle (Cq) has become the parameter most frequently cited when reporting RT-qPCR data, but is a Cq value in isolation sufficient to allow critical appraisal of conclusions drawn from an RT-qPCR assay? Cq is defined as the cycle number when signal intensity from a sample exceeds calculated background fluorescence. However, it must be appreciated that background fluorescence is not a constant value and is influenced by the dynamic conditions of a specific reaction, with a given Cq value varying with fluctuations in background signal intensity. Although a positive Cq (defined as a fluorescence reading of less than the final cycle number) can obviously represent genuine amplification, some Cq values are not due to specific amplification. Moreover, genuine amplification events may fail to register a positive Cq. One important reason for the latter scenario is a drifting baseline secondary to an incorrectly calibrated background Cq range, whereupon an erroneously specified number of early PCR cycles are used to incorrectly calculate threshold fluorescence levels. This background fluorescence is used to determine baseline signal intensity simultaneously across all assays on the reaction plate, although this generalized background signal is not necessarily appropriate for any given individual well.

An analogy may be useful to explain this concept further: If two 'rogue-states' fire missiles into the sky, it is important to confirm that both started their ascendency from sea-level prior to assuming that the missile from one state is more powerful than the other, even though the former entered the stratosphere and the latter did not. This highlights the concept of baseline correction, since an adjustment of the baseline cycles plot may correct erroneous baseline drift, ultimately demonstrating equivalent Cq values, or even missile trajectories. This is especially important when analysing the amplification profile of negative controls, as should baseline correction demonstrate clear evidence of amplification, this result questions the validity of data derived from the samples of interest. This effect may be minimized by selecting adaptive baseline enhancement that is now provided as an option by several qPCR instruments, as this software setting automatically calculates individual baselines for each sample, thereby providing the most accurate Cq.

Normalization

The theory underlying quantification is very straightforward: the more copies of target there are at the start of the assay, the fewer cycles of amplification are necessary to generate a Cq, the fluorescence that can be detected above the background fluorescence. Inevitably, in practice the relationship between amplicon copy number and detection is more problematic. The first hurdle that must be overcome is that reliable quantification of any rare target within a complex nucleic acid mixture is difficult (Karrer *et al.* 1995). Secondly, it is absolutely essential that a normalization strategy is used to control for the concentration of template, variation within amplification efficiencies and biological differences between samples. Despite numerous proposed strategies, published real-time quantification data are plagued by inappropriate normalization methods (Thellin *et al.* 1999).

Normalization against high-quality, accurately measured total RNA has been shown to produce quantification results that are biologically relevant (Bustin 2002; Tricarico *et al.* 2002). However, this approach is crucially dependent on accurate quantification and quality assessment of the RNA using technologies discussed previously. Furthermore, normalization against total RNA does not overcome the problem of variable subpopulations leading to inappropriate quantification and conclusions. Total RNA levels may be elevated in highly proliferating cells and this will also affect the accuracy of any comparison of amplicon numbers, for example between normal and tumour cells. Finally, it is not always possible to quantify transcript against total RNA, especially when very limited amounts of clinical samples are available.

Normalization against ribosomal RNAs has also been widely reported (Bhatia *et al.* 1994; Zhong and Simons 1999), with rRNA transcript number thought to be somewhat less susceptible to variation than mRNA expression (Schmittgen and Zakrajsek 2000) resulting in the suggestion that rRNA may be more reliable than some reference genes (de Leeuw, Slagboom and Vijg 1989; Mansur *et al.* 1993). 18S rRNA has been proposed as the most stable reference target transcript (Bas *et al.* 2004), although concern has been raised since rRNA transcription may be undertaken by a different RNA polymerase from that of the mRNA fraction (Solanas, Moral and

Escrich 2001). Another potential pitfall is that the amplification kinetics are likely to be different between the vastly abundant rRNA and most target mRNAs, thus generating misleading quantification data. Finally, rRNA normalization obviously cannot be used when quantifying targets from polyA-enriched samples.

The use of internal reference genes has been proposed as the most appropriate normalization procedure. There is no consensus on what reference genes to use and, unfortunately, there is a large body of often contradictory studies advocating the use of a whole host of individual reference sequences. It is now clear that no single transcript fulfils the criteria of a universal reference gene, as none are constitutively expressed in all cell types, in all conditions and independent of all experimental designs. To mitigate some of the limitations associated with single reference gene normalization, models using multiple internal control genes have been proposed. One such strategy ranks reference genes by similarity of expression profile, on the assumption that gene clusters demonstrating stable expression patterns relative to each other are appropriate control genes, before using their geometric mean as a normalization factor (Vandesompele *et al.* 2002). Criticisms of this model include the observation that it preferentially selects co-expressed genes, in addition to the considerable number of assays required to validate a group of reference genes suitable for each individual experiment. Several other normalization models exist (Akilesh, Shaffer and Roopenian 2003; Andersen, Jensen and Ørntoft 2004; Szabo *et al.* 2004), but these are not without their limitations and do not represent straightforward 'one-size-fits all' solutions for universal use. Consequently, the multiple reference gene normalization model proposed by Vandesompele remains the gold standard for normalization.

Transparency of published data

As previously suggested, a lack of consensus exists on how best to perform and interpret RT-qPCR assays. The problem is perpetuated by a lack of sufficient experimental detail in most publications, which in turn impedes the readers' ability to evaluate critically the quality of the results presented, repeat the reported experiments or integrate methodological advances into their own studies. Consequently, guidelines tackling this issue have recently been published which aim to promote consistency between laboratories and increase experimental transparency (Bustin *et al.* 2009). The Minimum Information for publication of Quantitative real-time PCR Experiments (MIQE) are a set of guidelines outlining the minimum data set necessary to effectively evaluate RT-qPCR assays and are designed to be used as a checklist to both accompany manuscript submission and to be available alongside the published manuscript to enhance critical appraisal. The full impact of these guidelines is yet to be realized due to their comparatively recent publication; however, it is hoped they will encourage better experimental practice, allowing more reliable and unequivocal interpretation of RT-qPCR results.

Further considerations

The previous sections have outlined key areas where inappropriate experimental design, execution or analysis may result in RT-qPCR data which is not biologically relevant. However, considerable efforts have been made in recent years by both our group and others to establish robust protocols covering the major steps of the RT-qPCR pathway and thus improve the quality of reported data. Those readers who wish an equally broad but more protocol-driven description of sample preparation, RNA quality assessment, reagent selection, assay design, normalization procedures and data analysis strategies are directed to the protocol document previously published by our unit (Nolan, Hands and Bustin 2006).

Conclusion

RT-qPCR assays have the potential to be extremely powerful clinical diagnostic, prognostic and investigative tools. Despite almost a decade of warnings, however, there remains little appreciation of exactly how subjective reported RT-qPCR results may be. Furthermore, considerable doubts remain about the biological validity of quantitative data. This brief perspective has highlighted the central problems associated with RT-qPCR technology. Despite attempts to define set protocols there remains an urgent need for universal agreement on basic issues such as quality and quantity control of RNA, guidelines for analysis and reporting of results. There is a particular requirement to implement recently proposed rules about the information that should be made publicly available on experimental and analytical procedures for any publication involving this technology. Until such proposals are universally implemented, RT-qPCR will not be able to make the most of its potential beyond its current role as a research tool.

References

Akane, A., Matsubara, K., Nakamura, H. *et al.* (1994). Identification of the heme compound copurified with deoxyribonucleic acid (DNA) from bloodstains, a major inhibitor of polymerase chain reaction (PCR) amplification. *J Forensic Sci* **39**: 362–372.

Akilesh, S., Shaffer, D.J. and Roopenian, D. (2003). Customized molecular phenotyping by quantitative gene expression and pattern recognition analysis. *Genome Res* **13**: 1719–1727.

Al Soud, W.A., Jonsson, L.J. and Radstrom, P. (2000). Identification and characterization of immunoglobulin G in blood as a major inhibitor of diagnostic PCR. *J Clin Microbiol* **38**: 345–350.

Al Soud, W.A. and Radstrom, P. (2001). Purification and characterization of PCR-inhibitory components in blood cells. *J Clin Microbiol.* **39**: 485–493.

Ambion (2001) *EndoFree RT*. November 2001. http://www.ambion.com/jp/techlib/prot/bp_1740.pdf (accessed 23 January 2011).

Andersen, C.L., Jensen, J.L. and Ørntoft, T.F. (2004) Normalization of real-time quantitative reverse transcription-PCR data: a model-based variance estimation approach to identify genes suited for normalization, applied to bladder and colon cancer data sets. *Cancer Res* **64**: 5245–5250.

Bas, A., Forsberg, G., Hammarstrom, S. and Hammarstrom, M.L. (2004) Utility of the housekeeping genes 18S rRNA, beta-actin and glyceraldehyde-3-phosphate-dehydrogenase for normalisation in real-time quantitative reverse transcriptase-polymerase chain reaction analysis of gene expression in human T lymphocytes. *Scand J Immunol* **59**, 566–573.

Belec, L., Authier, J. and Eliezer-Vanerot, M.C. (1998). Myoglobin as a polymerase chain reaction (PCR) inhibitor: a limitation for PCR from skeletal muscle tissue avoided by the use of *Thermus thermophilus* polymerase. *Muscle Nerve* **21**: 1064–1067.

Bhatia, P., Taylor, W.R., Greenberg, A.H. and Wright, J.A. (1994) Comparison of glyceraldehyde-3-phosphate dehydrogenase and 28S-ribosomal RNA gene expression as RNA loading controls for northern blot analysis of cell lines of varying malignant potential. *Anal Biochem* **216**: 223–226.

Bickley, J., Short, J.K., McDowell, D.G. and Parkes, H.C. (1996) Polymerase chain reaction (PCR) detection of *Listeria* monocytogenes in diluted milk and reversal of PCR inhibition caused by calcium ions. *Lett Appl Microbiol* **22**: 153–158.

Bohle, R.M., Hartmann, E., Kinfe, T. *et al.* (2000) Cell type-specific mRNA quantitation in non-neoplastic tissues after laser-assisted cell picking. *Pathobiology* **68**, 191–195.

Bolufer, P., Lo, C.F., Grimwade, D. *et al.* (2001) Variability in the levels of PML-RARa fusion transcripts detected by the laboratories participating in an external quality control program using several reverse transcription polymerase chain reaction protocols. *Haematologica* **86**: 570–576.

Bomjen, G., Raina, A., Sulaiman, I.M. *et al.* (1996) Effect of storage of blood samples on DNA yield, quality and fingerprinting: a forensic approach. *Indian J Exp Biol* **34**, 384–386.

Bustin, S.A. (2000). Absolute quantification of mRNA using real-time reverse transcription polymerase chain reaction assays. *J Mol Endocrinol* **25**: 169.

Bustin, S.A. (2002) Quantification of mRNA using real-time reverse transcription PCR (RT-PCR): trends and problems. *J Mol Endocrinol* **29**: 23–39.

Bustin, S.A., Benes, V., Garson, J.A. *et al.* (2009) The MIQE guidelines: minimum information for publication of quantitative real-time PCR experiments. *Clin Chem* **55**, 611–622.

Bustin, S.A., Li, S.R., Phillips, S. and Dorudi, S. (2001) Expression of HLA class II in colorectal cancer: evidence for enhanced immunogenicity of microsatellite-instability-positive tumours. *Tumor Biol* **22**: 294–298.

Chelly, J., Concordet, J.P., Kaplan, J.C. and Kahn, A. (1989) Illegitimate transcription: transcription of any gene in any cell type. *Proc Natl Acad Sci USA* **86**: 2617–2621.

Curry, J., McHale, C. and Smith, M.T. (2002) Low efficiency of the Moloney murine leukemia virus reverse transcriptase during reverse transcription of rare t(8;21) fusion gene transcripts. *Biotechniques* **32**: 768–770.

de Leeuw, W.J., Slagboom, P.E. and Vijg, J. (1989) Quantitative comparison of mRNA levels in mammalian tissues: 28S ribosomal RNA level as an accurate internal control. *Nucleic Acids Res* **17**: 10137–10138.

Deng, G., Lu, Y., Zlotnikov, G. *et al.* (1996) Loss of heterozygosity in normal tissue adjacent to breast carcinomas. *Science* **274**: 2057–2059.

Dolter, K.E. and Braman, J.C. (2001) Small-sample total RNA purification: laser capture microdissection and cultured cell applications. *Biotechniques* **30**: 1358–1361.

Emmert-Buck, M.R., Bonner, R.F., Smith, P.D. *et al.* (1996) Laser capture microdissection. *Science* **274**: 998–1001.

Fink, L., Kohlhoff, S., Stein, M.M. *et al.* (2002) cDNA array hybridization after laser-assisted microdissection from nonneoplastic tissue. *Am J Pathol* **160**: 81–90.

Ginzinger, D.G. (2002) Gene quantification using real-time quantitative PCR: an emerging technology hits the mainstream. *Exp Hematol* **30**: 503–512.

Goldsworthy, S.M., Stockton, P.S., Trempus, C.S. *et al.* (1999) Effects of fixation on RNA extraction and amplification from laser capture microdissected tissue. *Mol Carcinog* **25**: 86–91.

Ishiguro, T., Saitoh, J., Yawata, H. *et al.* (1995) Homogeneous quantitative assay of hepatitis C virus RNA by polymerase chain reaction in the presence of a fluorescent intercalater. *Anal Biochem* **229**: 207–213.

Karrer, E.E., Lincoln, J.E., Hogenhout, S. *et al.* (1995) In situ isolation of mRNA from individual plant cells: creation of cell-specific cDNA libraries. *Proc Natl Acad Sci USA* **92**: 3814–3818.

Klimecki, W.T., Futscher, B.W. and Dalton, W.S. (1994) Effects of ethanol and paraformaldehyde on RNA yield and quality. *Biotechniques* **16**: 1021–1023.

Lehmann, U., Bock, O., Glockner, S. and Kreipe, H. (2000) Quantitative molecular analysis of laser-microdissected paraffin-embedded human tissues. *Pathobiology* **68**: 202–208.

Lekanne Deprez, R.H., Fijnvandraat, A.C., Ruijter, J.M. and Moorman, A.F. (2002) Sensitivity and accuracy of quantitative real-time polymerase chain reaction using SYBR green I depends on cDNA synthesis conditions. *Anal Biochem* **307**: 63.

Mansur, N.R., Meyer-Siegler, K., Wurzer, J.C. and Sirover, M.A. (1993) Cell cycle regulation of the glyceraldehyde-3-phosphate dehydrogenase/uracil DNA glycosylase gene in normal human cells. *Nucleic Acids Res* **21**: 993–998.

Masuda, N., Ohnishi, T., Kawamoto, S. *et al.* (1999) Analysis of chemical modification of RNA from formalin-fixed samples and optimization of molecular biology applications for such samples. *Nucleic Acids Res* **27**: 4436–4443.

Mizuno, T., Nagamura, H., Iwamoto, K.S. *et al.* (1998) RNA from decades-old archival tissue blocks for retrospective studies. *Diagn Mol Pathol* **7**: 202–208.

Nolan, T., Hands, R.E. and Bustin, S.A. (2006) Quantification of mRNA using real-time RT-PCR. *Nat Protocols* **1**: 1559–1582.

Nolan, T., Hands, R.E., Ogunkolade, W. and Bustin, S.A. (2006) SPUD: a quantitative PCR assay for the detection of inhibitors in nucleic acid preparations. *Anal Biochem* **351**: 308–310.

Pikaart, M.J. and Villeponteau, B. (1993) Suppression of PCR amplification by high levels of RNA. *Biotechniques* **14**: 24–25.

Ririe, K.M., Rasmussen, R.P. and Wittwer, C.Q. (1997) Product differentiation by analysis of DNA melting curves during the polymerase chain reaction. *Anal Biochem* **245**: 154–160.

Rossen, L., Norskov, P., Holmstrom, K. and Rasmussen, O.F. (1992) Inhibition of PCR by components of food samples, microbial diagnostic assays and DNA-extraction solutions. *Int J Food Microbiol* **17**: 37–45.

Schmittgen, T.D. and Zakrajsek, B.A. (2000) Effect of experimental treatment on housekeeping gene expression: validation by real-time, quantitative RT-PCR. *J Biochem Biophys Methods* **46**: 69–81.

Schmittgen, T.D., Zakrajsek, B.A. and Mills, A.G. (2000) Quantitative reverse transcription-polymerase chain reaction to study mRNA decay: comparison of endpoint and real-time methods. *Anal Biochem* **285**: 194–204.

Solanas, M., Moral, R. and Escrich, E. (2001) Unsuitability of using ribosomal RNA as loading control for Northern blot analyses related to the imbalance between messenger and ribosomal RNA content in rat mammary tumors. *Anal Biochem* **288**: 99–102.

Sugiyama, Y., Sugiyama, K., Hirai, Y. *et al.* (2002) Microdissection is essential for gene expression profiling of clinically resected cancer tissues. *Am J Clin Pathol* **117**: 109–116.

Szabo, A., Perou, C.M., Karaca, M. *et al.* (2004) Statistical modeling for selecting housekeeper genes. *Genome Biol* **5**: R59.

Thellin, O., Zorzi, W., Lakaye, B. *et al.* (1999) Housekeeping genes as internal standards: use and limits. *J Biotechnol* **75**: 291–295.

Tricarico, C., Pinzani, P., Bianchi, S. *et al.* (2002) Quantitative real-time reverse transcription polymerase chain reaction: normalisation to rRNA or single housekeeping genes is inappropriate for human tissue biopsies. *Anal Biochem* **309**: 293–300.

Vandesompele, J., De Preter, K., Pattyn, F. *et al.* (2002) Accurate normalisation of real-time quantitative RT-PCR data by geometric averaging of multiple internal control genes. *Genome Biol* **3**: 0034.1–0034.11.

Walch, A., Specht, K., Smida, J. *et al.* (2001) Tissue microdissection techniques in quantitative genome and gene expression analyses. *Histochem Cell Biol* **115**: 269–276.

Wolffs, P., Grage, H., Hagberg, O. and Radstrom, P. (2004) Impact of DNA polymerases and their buffer systems on quantitative real-time PCR. *J Clin Microbiol* **42**: 408–411.

Zhang, J. and Byrne, C.D. (1999) Differential priming of RNA templates during cDNA synthesis markedly affects both accuracy and reproducibility of quantitative competitive reverse-transcriptase PCR. *Biochem J* **337**: 231–241.

Zhong, H. and Simons, J.W. (1999) Direct comparison of GAPDH, beta-actin, cyclophilin and 28S rRNA as internal standards for quantifying RNA levels under hypoxia. *Biochem Biophys Res Commun* **259**: 523–526.

4

DNA Analysis in Droplet-Based Microfluidic Devices[1]

Pinar Ozdemir and Yonghao Zhang

Introduction

Modern microfluidic technologies and their applications in biomedical sciences have recently been under rapid development. Microfluidic devices, particularly in conjunction with microdroplet technology, have dramatically changed the polymerase chain reaction (PCR) process for amplifying DNA. Microfluidic technology, micro-total-analysis systems (μTAS) or Lab-on-a-Chip are the methods used for integration of analysis systems in a micro-scaled device. This miniaturization speeds up analysis time and reduces the consumption of sample and reagents. Developing microfluidic technology generates creative solutions for chemical, pharmaceutical, healthcare and food industries.

The earliest appearance of a microfluidic device was in the 1970s as an application for gas chromatography (Terry 1975). The earliest DNA amplification was carried out by Kary Mullis in 1984. With the PCR process, an individual DNA segment can be replicated several million times in over 20 cycles (Bartlett and Stirling 2003). The simplicity of the process has revolutionized many life science and related areas, including clinical diagnoses, and medical, biological and forensic analyses (Auroux *et al.* 2002, 2004; Vilkner, Janasek and Manz 2004; Chen, Manz and Day 2007). The adaptation of PCR to a micro-scaled device came after the concepts of μTAS and Lab-on-a-Chip for biological and chemical analyses emerged (Northrup *et al.* 1993).

---

[1] This chapter is modified and updated from our recent review paper 'Microfluidic DNA amplification – a review', Analytica Chimica Acta 638 (2009) 115–125.

Molecular Analysis and Genome Discovery, Second Edition. Edited by Ralph Rapley and Stuart Harbron.

Since DNA has been studied intensively and we benefit from commercial equipment with reasonable prices, DNA analysis is very appealing. The PCR process is used to amplify specific regions of a DNA strand (i.e. a single gene, a part of a gene or a non-coding sequence) based on chosen specific primers. The PCR process typically comprises 20–40 thermal cycles and each cycle has two to three steps with optimal temperatures. The most commonly used PCR process has three steps for denaturation, annealing and extension with corresponding specific temperatures. In the first denaturation step, a high temperature (94–98 °C) is required to break the hydrogen bonding, so that the double-stranded DNA template becomes two complementary single strands. Once a single-stranded DNA template is generated in the denaturation step, the temperature will be lowered to allow the primers to specifically bind to the complementary sequences of the DNA template. The temperature is dependent on the used primers, typically between 50–65 °C. The final extension step is to replicate a thermostable DNA polymerase. The temperature depends on the DNA polymerase and is often close to 72 °C. The extension time is also determined by both the DNA polymerase and the length of the DNA fragment to be amplified. As a close approximation, the polymerization rate of the DNA polymerase is about 1000 bases per minute, at the optimum temperature. At each extension step, the amount of the DNA fragment is doubled, which leads to exponential amplification of the DNA target.

A conventional PCR device can be made using a test tube, a few simple reagents and a heat source; the whole chamber with the PCR mixture is heated up and cooled down in repeated cycles. Because of this large thermal mass change procedure the reactions typically take 1–2 hours. In 1999, a report showed that the time required for denaturation and annealing steps for the sample to achieve equilibrium temperature can be significantly reduced (Wittwer and Hermann 1999). Therefore, the conventional PCR process is too lengthy. In addition, it has a high consumption of expensive reagents, with a preference for amplifying short fragments and producing short chimeric molecules. The preparations for pre- and post-PCR have to be done offline.

On the other hand, microfluidic devices offer creative solutions for all these problems. For example, microfluidic PCR can achieve rapid heat transfer, due to a large surface-to-volume ratio and fast mixing can be achieved by diffusion, due to its small dimension. At the same time, several functions such as sample handling, detection, mixing and separation can also be integrated into a single chip. Moreover, the thermal cycling time will be significantly reduced, because of the swift thermal responsiveness of the sample to the surrounding environment and the PCR mixture will be exposed to more uniform temperatures during the PCR process, thereby enhancing the yield. A further attractive feature of miniaturized PCR is its portability, making it useful for detection and analysis in the field. There are problems with microfluidic PCR because of the large surface-to-volume ratio between the device surfaces and the sample, however, which will lead to PCR inhibition and carryover contamination. Since the PCR inhibition and carryover contamination are caused by PCR mixture adsorption at device surfaces and interactions between different mixtures, we need to perform PCR reactions in an isolated environment. Microdroplet technology can provide an ideal solution to these problems, so that PCR will now occur inside droplets, which can eliminate sample/reagent surface adsorption and carryover contamination. PCR in droplets can also prevent recombination between homologous gene fragments during the PCR process and eliminate the synthesis of short chimeric products. Therefore, PCR in droplets provides a convenient way for single-molecule and single-cell amplification.

Microfluidic DNA analysis is a diverse multidisciplinary field, so we will select only some important developments to provide a snapshot of the technologies being developed. We will focus on the most important and exciting recent developments in continuous-flow microPCR and microfluidic PCR in droplets. Other aspects of this field – including materials, fabrication techniques, system optimization, integration and applications – can be found in many recent review articles (e.g. Auroux *et al.* 2004; Chen, Manz and Day 2007; Y. Sun, Yien and Kwok 2006; C.S. Zhang *et al.* 2006; C. Zhang, Xing and Li 2007; C. Zhang and Xing 2007; Ong *et al.* 2008; Y.H. Zhang and Ozdemir 2009).

Continuous-flow microPCR chips

Chip-based microPCR devices are commonly classified in two categories: well-based PCR chips (e.g. Northrup *et al.* 1993; Wilding, Shoffner and Kricka 1994; Belgrader *et al.* 1999; Gulliksen *et al.* 2004; Ohashi *et al.* 2007) and continuous-flow PCR chips (e.g. H. Nakano *et al.* 1994; Kopp, de Mello and Manz 1998; Chiou *et al.* 2001; Z.Y. Chen *et al.* 2004; H. Wang *et al.* 2006; Mohr *et al.* 2007). As the name indicates, in well-based PCR the solution is first injected into a well then the whole chip is heated and cooled. This means a long thermal cycling time, leading to unwanted inertial effects caused by large total thermal mass. In contrast, because of their dynamic design, continuous-flow microPCR devices move only the PCR sample through fixed temperature zones to achieve the required thermal-cycling, which reduces the thermal inertia of the system. In doing so, it also reduces the cycling time, consumes less energy and makes the system more amenable to portable applications and integration into μTAS. Continuous-flow has been achieved in microPCR chip in various designs: oscillatory devices (e.g. Bu *et al.* 2003; Auroux, Day and Manz 2005; W. Wang *et al.* 2005), closed-loop devices (e.g. Liu, Enzelberger and Quake 2002; West *et al.* 2002; Sadler *et al.* 2003), and fixed-loop devices (e.g. H. Nakano *et al.* 1994; Kopp, de Mello and Manz 1998; H. Wang *et al.* 2006; Mohr *et al.* 2007; Park, Kim and Hahn 2003; Crews, Wittwer and Gale 2008). The oscillatory devices collect the sample in a chamber which circulates the sample back and forth between the chambers at denaturation and annealing temperatures. However, closed-loop devices use the large thermal mass generated by the PCR process in the device to circulate the sample around a fixed circuit. This is known as the thermo-siphonic affect (Chung, Park and Choi 2010), that is the heated sample becomes buoyant and natural convection causes the movement. The advantage of both designs is that the number of thermal cycles can be flexible, whereas the number of cycles in a fixed-loop system must be determined at the fabrication stage.

The earliest continuous-flow PCR device was capillary-based as demonstrated by H. Nakano et al. (1994). This device was made of coiled Teflon capillary with an internal diameter of 500 μm with 30 cycles. Each cycle was 150 mm long and separated in to three zones: the PCR mixture first went through a 20-mm-long denaturation zone with a temperature of 94 °C, followed by a 30-mm-long annealing zone with a temperature of 50 °C and at the end a 100-mm-long extension zone at 72 °C. In comparison with the commercial thermocycler, a 50% amplification yield was achieved in only 10% of the processing time, that is 12–18 min. It is also reported that some capillary-based PCR devices were coated with an indium-tin oxide (ITO) thin film at the outer surface (Friedman and Meldrum 1998). ITO thin film can function as a heater and, due to its transparency (K. Sun et al. 2002), it can be used for real-time fluorescent monitoring. Another similar device design with the droplet approach was demonstrated by Chiou et al. (2001). First, the capillary was filled with oil. The PCR mixture was then injected as a 1 μl droplet, the PCR completed in 23 min with 30 cycles. It did manage to amplify a 500-base pair product with 78% amplification efficiency. An improved design of this type of device could complete 30 thermal cycles in 2.5 min. One of the important modifications for capillary-based devices is to make them reusable; this has been achieved by adding a washing step between sequential injection of samples to eliminate carryover contamination between sequential runs (Bruckner-Lea et al. 2002; Belgrader et al. 2003). Although these continuous-flow devices demonstrated a high-speed and high-yield PCR with a flexible cycling number, they are not compatible for integration into a microfluidic system.

In 1998, an adaptation of MEMS technology was demonstrated by Kopp, de Mello and Manz (1998). A micro-dimensioned channel was etched into a Corning 0211 glass with a total length of 2.2 m for 20 cycles (Figure 4.1). This device demonstrated amplification of a 176-base pair DNA fragment in 20 cycles. The total amplification time is between 18.7 and 1.5 min with flow rates from 5.8 to 72.9 nl/s. Several others followed this work and developed a broad range of chip-based microfluidic PCR devices (e.g. Bu et al. 2003; Crews, Wittwer and Gale 2008; Schneegaß and Köhler 2001; Schneegaß, Bräutigam and Köhler 2001; Felbel et al. 2002; Fukuba et al. 2004; Obeid et al. 2003; Obeid and Christopoulos 2003). Silicon, one of the commonly used materials for *integrated circuits*, was used to fabricate a 25-cycle silicon-glass PCR chip; the chip had integrated heaters and temperature sensors (Schneegaß, Bräutigam and Köhler 2001). An interesting design feature is rotationary PCR chips; the one presented by West et al. (2002) provides a flexible number of controlled cycles. Another chip which can perform both spatial and temporal cycling was presented by Liu, Enzelberger and Quake (2002). In addition to rotational force-driven devices, magnetic force-driven devices were demonstrated by Y. Sun, Kwoka and Nguyen (2007), where the PCR mixture was modified by the conductive nature of the electrolyte and the sample in an annular microchannel was driven by an

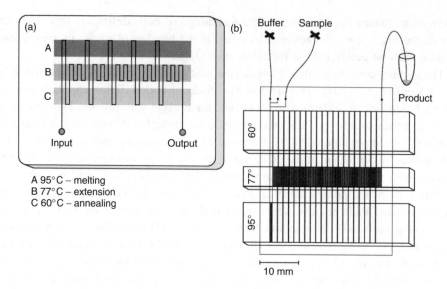

Figure 4.1   Schematic diagram of continuous-flow PCR chip: (a) three temperature zones for PCR thermal cycles are maintained at 95, 77 and 60 °C – a pressure pumping method is used to inject the sample; (b) device layout: three inlets on the left side of the device and one outlet on the right – only two inlets are used to inject both the sample and buffer (Reproduced, with permission, from Kopp, M.U., de Mello, A. J. and Manz, A. (1998) Chemical amplification: continuous-flow PCR on a chip. *Science* **280**: 1046–1048. © 1998 The American Association for the Advancement of Science)

AC magnetic field. Obeid and Christopoulos (2003) demonstrated a device with various outlets that provides a dynamic approach for the number of cycles to be adjusted depending on the demand. This device can amplify RNA through reverse transcription into cDNA.

While continuous-flow chip designs continue to improve, detection systems are also developing fast. Obeid and Christopoulos (2003) developed a laser-induced fluorescence (LIF) detection system which speeds up the quantification of the amplified products and works with smaller volume samples. Fukuba *et al.* (2004) reported a 30-cycle continuous-flow microPCR device with miniature pumps and valves, heating provided by embedded ITO heaters and the temperature controlled via an integrated platinum sensor. In addition, the researchers were able to demonstrate amplification with plug flow, thus reducing the amplification volume to 2 μl per amplified sample. In some examples, laser was used to generate a temperature gradient (e.g. Braun, Goddard and Libchaber 2003), which was to create an effect similar to the Rayleigh–Benard cell for PCR (Krishnan, Ugaz and Burns 2002). The main idea is to induce density variation to drive the sample through different temperature zones for PCR processing. Another rotary PCR chip (Z.Y. Chen *et al.* 2004) managed to achieve an average velocity of 2.5 mm/s for a temperature gradient induced flow. Crews, Wittwer and Gale (2008) proposed a continuous-flow thermal gradient PCR,

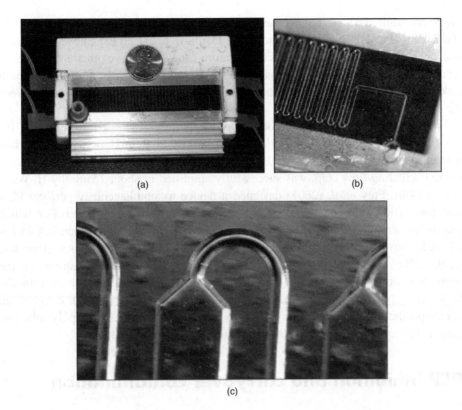

(a)　　　　　　　　　　　　(b)

(c)

Figure 4.2 (a) The continuous-flow thermal gradient PCR chip. (b) The serpentine channel, which has a linear temperature gradient, so that rapid temperature change is achieved in
a narrow channel while slow ramp rates are obtained in a wide channel. (c) The smooth and
curved glass channel surface. (Reproduced, with permission, from Crews, N., Wittwer, C. and
Gale, B. (2008) Continuous-flow thermal gradient PCR. *Biomed Microdevices* **10**: 187–195.
© 2008 Springer Verlag)

with each cycle consisting of temperature spikes to denature and anneal by passing
through the narrow channel, then a moderate thermal ramp through the extension
temperature in the wide channel. The authors demonstrated to achieve a high yield
and specificity amplification for a 40-cycle PCR in less than 9 min (Figure 4.2).

Creating various numbers of the temperature zones is another approach to performing Sanger thermal cycle sequencing reactions. Hashimoto *et al.* (2004) developed
a device in which four different temperature zones were separated into the four
quadrants of a rectangular substrate. This device has a 20-loop spiral microchannel
through each temperature zone repeatedly. Likewise, H. Wang *et al.* (2006) developed
a quadrant heating/spiral channel continuous-flow microPCR.

Microchannels will be affected by thermal 'cross-talk' (S. Li *et al.* 2006), especially when multiple temperature zones are used. Isolation between temperature zones
complicates the design of the continuous-flow PCR (Crews, Wittwer and Gale 2008).

There are some devices (Schneegaß, Bräutigam and Köhler 2001; Hashimoto *et al.* 2004; Yang, Pal and Burns 2005), however, that have managed to achieve better thermal separation between the various temperature zones. There are other approaches to reducing the inter-temperature transition time. For example, S. Li *et al.* (2006) built a device with a 20-cycle serpentine microchannel, narrowing the regions between the three temperature zones.

TaqMan, a fluorophore-based detection technology was used to improve the real-time PCR by Nakayama *et al.* (2006). A 1 μl PCR mixture with a standard human papilloma virus-DNA (HPV-DNA) sample, which was injected into a micro-oscillating-flow chip, was successfully amplified in about 15 min by H. Wang *et al.* (2006). Frey *et al.* (2007) designed a device to simultaneously perform PCR reactions. The device has parallel multiple channels with a single actuator which pumps the sample through three temperature zones. Moreover, 72 parallel 450-pl RT-PCRs have also been performed in a microfluidic chip (Marcus, Anderson and Quake 2006). In Table 4.1, we have summarized the interesting features of the continuous-flow PCR chips. So far, a broad range of applications of microfluidic PCR have been reported. For example, Hashimoto *et al.* (2007) coupled an allele-specific ligation detection reaction to continuous-flow PCR in a polycarbonate chip and successfully detected low-abundant DNA point mutations.

PCR inhibition and carryover contamination

PCR inhibition is one of the major challenges during the amplification process due to temperature gradient and adsorption of the PCR components to the channel surface because of large surface-to-volume ratio. Surface adsorption of sample and reagents will also cause carryover contamination. To understand the adsorption, Gonzalez *et al.* (2007) investigated various polymeric surfaces in the capillary surfaces. When the PCR mixture moves along a long channel, the channel surface area may be sufficiently large to adsorb all the PCR components. The researchers started to focus on using various materials to reduce PCR inhibition. Kolari *et al.* (2008) observed that fluorocarbon-coated silicon had strong inhibition, while native silicon and an SF6 etched surface without oxidation caused significantly little inhibition. On the other hand, the TaqMan real-time PCR label can be adsorbed by the native silicon surface (W. Wang *et al.* 2006). A number of investigations have been conducted to understand the inhibition on silicon and silicon nitride (SiN) surfaces (Wilding, Shoffner and Kricka 1994; Wilding *et al.* 1995; Shoffner *et al.* 1996; Taylor *et al.* 1997; Kricka and Wilding 2003; Erill *et al.* 2003; Krishnan, Burke and Burns 2004; Felbel *et al.* 2004; Panaro *et al.* 2004). The biocompatibility of microfluidic materials is still an interesting research subject for choosing appropriate materials.

To prevent PCR inhibition and carryover contamination, various measures have been taken to minimize surface adsorption of DNA and PCR reagents. One common approach is to treat the channel surfaces with a silanizing agent. This pre-treatment

Table 4.1   Summary of the continuous-flow PCR chips

Device	Layout	Demonstration	Other Features
Kopp, de Mello and Manz (1998)	Fixed-loop with 20 cycles	Amplification of 176-base pair DNA fragment as low as 90 seconds for 20 cycles	First on-chip continuous-flow PCR
Schneegaß, Bräutigam and Köhler (2001)	Fixed-loop with 25 cycles	Amplification of 700-base pair fragment in less than 30 min for 25 cycles	Proposed a liquid /liquid two phase PCR
West et al. (2002)	Closed-loop design	Magnetohydrodynamic force used to pump fluid	
Liu, Enzelberger and Quake (2002)	Closed-loop design	Amplification of 199-base pair DNA fragment in 40 min	Online detection of fluorescence level
Sun et al. (2002)	Fixed-loop with 30 cycles	450-base pair fragment amplified in 19 min for 30 cycles	Online fluorescence monitoring system
Bu et al. (2003)	Oscillatory chip	Numerical simulation for device optimal design and operation	
Obeid et al. (2003); Obeid and Christopoulos (2003)	Fixed-loop with multiple cycles 20, 25, 30, 35, and 40	Amplification of 230-base pair fragments for 30 cycles in only 6 min	DNA and RNA application; laser-induced fluorescence detection system
Fukuba, Yamamato and Nahanuma (2004)	Fixed-loop with 30 cycles	1460-base pair fragments successfully amplified in 60 min for 30 cycles	
Hashimoto et al. (2004)	Fixed-loop with 20 cycles	500 and 997 base pair fragments amplified in 1.7 and 3.2 min for 20 cycles respectively.	Numerical simulation for determining device optimal flow rates

(continued overleaf)

Table 4.1 *(continued)*

Device	Layout	Demonstration	Other Features
Hashimoto et al. (2007)	Fixed-loop with 30 cycles	290-base pair fragment amplified in 18.7 min for 30 cycles.	Coupled with 13 cycles for an allele-specific ligation detection reaction
Nakayama et al. (2006)	Fixed-loop with 50 cycles	113-base pair fragment amplified in 40 min for 50 cycles	A method proposed to prevent bubble generation; laser-induced fluorescence detection system
Wang et al. (2006)	Fixed-loop with 20 cycles	Able to amplify up to 632-base pair fragments in 14.6 min for 20 cycles.	Coupled with a solid-phase reversible immobilization chip
Li et al. (2006)	Fixed-loop with 20 cycles	Successful amplification of 90-base pair fragment	Numerical simulation for device optimal design
Sun, Kwoka and Nguyen (2007)	Closed-loop design	Amplification of 500-base pair fragment in 13.5 min for 30 cycles.	Magnetohydrodynamic force used to pump fluid
Frey et al. (2007)	Oscillatory PCR chip	Demonstrated to determine the threshold cycle number for their device	Proposed a parallel design to increase throughput; Online optical detection system to monitor fluorescence level
Crews, Wittwer and Gale (2008)	Fixed-loop with 30 cycles	For 40 cycles, less than 9 min required to amplify 108/181-base pair fragments with high yield and specificity	Narrowing/widening channels to manipulate temperature gradient
Chung, Park and Choi (2010)	Closed-loop	127 basepair-long DNA amplified in 10 min	Polymeric chip using thermosiphon effect for cycles

step is done before the injection of the PCR mixture. Schneegaß, Bräutigam and Köhler (2001) proposed enhancing the surface biocompatibility of the silicon/glass chip using hexamethyldisilazane (HMDS). Kim *et al.* (2006) modified the hydrophobic PDMS chip surfaces by adding polyvinylpyrrolidone (PVP) to the PCR mixture. Since PDMS is one of the common materials for making microfluidic devices, there have been many attempts to modify its surface properties. Fukuba *et al.* (2004) proposed coating it using 2-methacryloyloxyethyl phosphorylcholine (MPC) with a silane coupler. Prakash, Amrein and Kaler (2008) applied an SU8-Teflon coating to modify the contact angle of a sessile Taq droplet to reduce adsorption. Schneegaß and Kohler (2001) and Xia *et al.* (2007) practised several passivation strategies; a washing step between samples was needed. A stationary reactor generating a bi-directional thermocycling system was also used to minimize surface adsorption of DNA and reagents (L. Chen *et al.* 2007). Here we will focus on droplet technology, which provides one of the most effective means of eliminating surface adsorption and carryover contamination because PCR reactions are contained in droplets.

PCR in droplets

The continuous-flow microPCR chips effectively shortens the process time and reduces the thermal inertia dramatically, but as mentioned in previously, surface adsorption of the sample and reagents due to a large surface-to-volume ratio creates a new major challenge. The PCR performance is also significantly affected by the parabolic velocity profile of the channel flow-field. The sample located in the middle of the stream will be processed in a shorter time then the sample located closer to the channel walls. As mentioned earlier, single-phase continuous-flow PCR has similar problems to conventional PCR; the preference for amplifying short fragments and the production of short chimerical molecules (Taly, Kelly and Griffiths 2007). These problems are amenable to microdroplet technology.

Continuous-flow PCR chips using multi-phase flows introduced microdroplet technology to PCR devices. These devices not only lead to an efficient PCR but also eliminate PCR inhibition and sample carryover contamination. In comparison with conventional continuous-flow microPCR devices working with a single aqueous phase, microdroplet technology has additional benefits such as reducing thermal mass and shortening thermal cycling process time. Any temperature change in each sample and the reagents confined in a microdroplet will be achieved uniformly because of their small volume. In addition, PCR mixture confined in a droplet will achieve a uniform temperature field for the same time intervals.

PCR in droplets has the potential application to amplify a single DNA molecule or single cells, with the nucleic acids from one cell being amplified inside one droplet. This technology will improve our understanding of disease progression and will be helpful in diagnostic stage. As the samples are now contained in droplets, they could

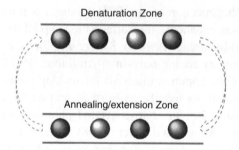

Figure 4.3 An illustration of microfluidic PCR in droplets: the PCR mixture is contained in individual droplets, moving through different temperature zones in the microchannels for DNA denaturation, annealing/extension. The microfluidic PCR in droplets has many advantages, e.g. elimination of carryover contamination between successive samples, adsorption at the surface and diffusional dilution of samples; prevention of the synthesis of short, chimeric products and other artifacts; rapid thermal response for a fast PCR process; low consumption of reagents; easy integration as a function of a μTAS. In addition, individual droplets can contain different PCR samples, so that it is particularly suitable for single cell and single molecule amplification. (Reproduced, with permission, from Zhang, Y. H., and Ozdemir, P. (2009) Microfluidic DNA amplification – a review, *Anal. Chimica Acta*, **638**: 115–125. © 2009 Elsevier)

be detected and subsequently sorted on the chip itself. So PCR function can be conveniently integrated into a lab-on-a-chip system. Review articles (Günther and Jensen, 2006; Kelly *et al.* 2007; Teh *et al.* 2008; Huebner *et al.* 2008) show the rapid development of microdroplet technology including its applications. Figure 4.3 shows a schematic diagram of microfluidic PCR in droplets.

Conventional PCR has already been used to explore genomic and cDNA libraries. When the high-throughput screening of transcription factor targets needs to be achieved, however, droplet-based PCR provides a better approach than conventional PCR (Margulies *et al.* 2005; Shendure *et al.* 2005; M. Li *et al.* 2006; K. Zhang *et al.* 2006; Wetmur *et al.* 2005; Dressmann *et al.* 2003; Kojima *et al.* 2005). Traditional PCR has a limited capacity for amplifying larger DNA fragments and the recombination between homologous regions of DNA leads to artefactual fragments. Continuous-flow droplet PCR started to become more appealing to researchers (Taly, Kelly and Griffiths 2007; Williams *et al.* 2006), as it alleviates these problems and enables the use of small amounts of template DNA and high numbers of PCR cycles. During this process, the template fragments are confined in the small aqueous droplets, which encapsulate the target genes in a water droplet immersed in an immiscible oil carrier phase. These droplets generally contain one or at most a few molecules of template DNA due to their small volume. Therefore, this prevents recombination between homologous or partially homologous gene fragments during the PCR process. Consequently, we can eliminate the synthesis of short, chimeric products and other artefacts. Reducing the number of fragments in the PCR solution also increases the possibility of amplifying smaller fragments (Taly, Kelly and Griffiths 2007). Water droplets in an oil medium are stable at the temperatures used for the PCR process, which creates an additional advantage of performing parallel amplification of single DNA or RNA molecules. Droplets have been used with conventional PCR devices, which provides a high throughput method for DNA sequencing; interested readers can refer to recent review papers (Taly, Kelly and Griffiths 2007; Kelly *et al.* 2007) for more information. As mentioned earlier in this chapter, single-phase flow PCR fills in the whole channel, which can lead to carryover contamination between successive samples, adsorption at the surface and diffusional dilution of samples (e.g. Kopp, de Mello and Manz 1998). Using immiscible liquids to isolate the sample slugs from each other and the device channels can partially solve these problems (Hardt *et al.* 2004). The high-throughput microfluidic PCR device demonstrated by Curcio and Roeraade (2003) injects the samples/reagents in separate aqueous segments that have an immiscible organic liquid as the continuous flow phase. A 15-m-long Teflon coiled tube is used to cycle through three temperature zones. An intermediate cleaning process between two consecutive samples is required to reduce carryover between samples.

In some literature (Park, Kim and Hahn 2003; Obeid *et al.* 2003; Obeid and Christopoulos 2003), a second phase flow (to prevent sample contamination plugs of air) is used to separate the aqueous sample flows. This does not, however, eliminate the adsorption of the samples and reagents or the subsequent transfer of the samples. Even with a surface treatment, carryover contamination cannot be eliminated totally (Park, Kim and Hahn 2003). To avoid PCR inhibition, the key factor is to minimize or stop the interaction between the PCR mixture and the device surfaces. Therefore, microdroplet technology can offer a satisfying solution.

Nisisako, Torii and Higuchi (2002) dispersed the aqueous phase into an immiscible oil phase. Each droplet represents a transportable individual reaction volume which does not interact with its surrounding medium. The droplets contact the surface at specific locations only when they are manipulated to do so. Droplet-based systems can therefore avoid the problems of adsorption, cross-contamination and diffusional dilution associated with single-phase microfluidic systems.

The continuous-flow PCR device of Dorfman *et al.* (2005) encapsulates the PCR mixture in 1 μl droplets. This device is formed by a 4.5-m-long PEA capillary (i.d. 800 μm), which was coiled around a cylinder, to achieve 35 thermal cycles. A sophisticated optical detection and sample preparation implemented device (Chabert *et al.* 2006) was successfully used to amplify a 572-base pair DNA fragment of Litmus 28i. Hartung *et al.* (2009) generated droplets within a Teflon FEP tube and T-connectors. With this low energy consuming device with an integrated continuous reverse-transcription process, they managed to detect viruses even when the sample concentration was very low. An interesting approach has been reported by Ohashi *et al.* (2007) where aqueous droplets contained hydrophilic magnetic beads and an applied magnetic field was used to manipulate the transportation of droplets through different temperature zones in a flat-bottomed tray-type reaction chamber. Teflon capillary tube devices are popular as their assembly is easier than making chip devices. Another approach manipulated droplets containing magnetic beads with a magnet (Tsuchiya *et al.* 2008), which was convenient for the separation and fusion of droplets. Similarly, a recent publication by Okochi *et al.* (2010) performing RT-PCR within five parallel fabricated lanes demonstrated the movement of droplets controlled by magnetic beads.

Gonzalez *et al.* (2007) demonstrated robust detection of the low-copy transcript CLIC5 from 18 cells per microlitre in cultured lymphoblasts with a rotary PCR device. This kind of device creates a potential for development of an integrated system for continuous gene expression directly from cell suspensions. C. Zhang and Xing (2009) presented a closed-loop parallel-flow device. Markey, Mohr and Day (2010) developed a device for high-throughput PCR. The droplets were generated in a T-junction device attached to PTFE tubing. Amplification of the DNA occurs at the PTFE tubing which is coiled around the aluminium cylinder heaters.

Although system integration of droplet-based PCR is difficult to achieve, Mohr *et al.* (2006, 2007) successfully coupled microdroplet technology with a continuous-flow microPCR chip (Figure 4.4a). Figure 4.4b shows the schematic diagram of a PCR chip. This chip was mill machined on a polycarbonate sheet and sealed with a 100 μm thick acetate foil. The whole process was completed in 32 thermocycles and each cycle was 63 mm long. The dimensions of the overall device was $75 \times 74 \times 4 \, mm^3$; the droplet generation occurred at the channel with dimensions of 200 μm wide and 200 μm deep which then expands to 500 μm wide and 400 μm deep. The droplets were reported to be approximately 5 nl volume with diameter of 100–155 μm. With an optical monitoring system, this device can perform quantitative real-time PCR measurements. The successful amplification of a 60-base pair fragment from the RNase P gene was achieved.

A high-throughput chip-based continuous-flow PCR has been reported by Kiss *et al.* (2008). This chip managed to amplify a 245-base pair adenovirus product within 35 min. The authors developed a chip capabile of generating millions of uniform droplets with the aim of amplifying single-base nucleic acids in a complex environment. A fluorescence monitoring system was used to collect data for real-time measurements. Chip layout and droplet generation can be seen in Figure 4.5. Recently, Schaerli *et al.* (2009) developed a novel radial design of continuous-flow PCR chip (Figure 4.6), which amplified 85-base pairs of DNA in 17 min within 34 thermal cycles. Again, the temperature gradient of the droplets were monitored and measured by adding fluorescence (rhodamine B) lifetime imaging to the aqueous PCR mixture.

Mazutis *et al.* (2009) investigated the behaviour of DNA molecules trapped in droplets, which exhibit the Poisson distribution. This research focused on taking measurements and minimizing contamination. Their chip managed to achieve

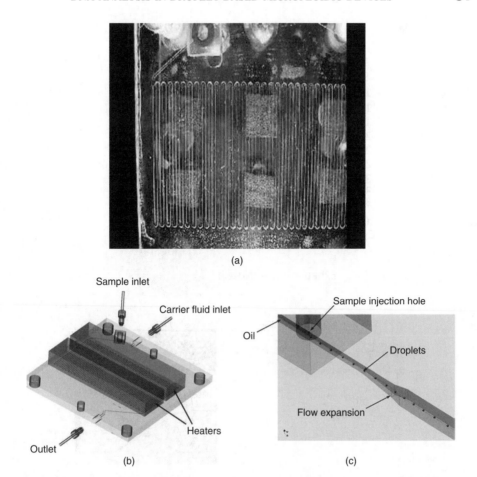

Figure 4.4 (a) The continuous-flow microPCR chip with PCR mixture contained in droplets; (b) layout of the PCR chip; (c) the aqueous droplets are generated in the carrier oil phase. (Reproduced, with permission, from Mohr, S., Zhang, Y.-H., Macaskill, A. *et al.* (2007) Numerical and experimental study of a droplet-based PCR chip. *Microfluid Nanofluidic* **3**: 611–621. © 2007 Springer Verlag)

high-throughput clonal DNA amplification (Figure 4.7). F. Wang and Burns (2009) fabricated a glass–silicon chip, generating droplets at a 45° junction neck which connects to an expansion of 400 µm wide to 3 mm long. By focusing on the critical reagent concentration, they managed to achieve no volume loss of the PCR solution carrying droplets even after 40 thermal cycles. A chip with a rapid detection capacity using plug-based microfluidics was demonstrated by Boedicker *et al.* (2008), which is a very sophisticated example of integration of the droplet reactor device that was earlier demonstrated by Song, Tice and Ismagilov (2003).

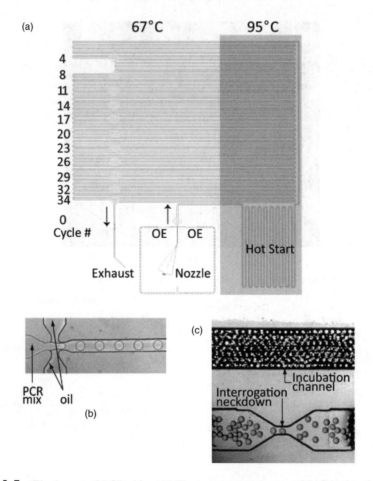

Figure 4.5 The layout of PCR chip. (A) The temperature is set at 95 °C in the pink-shaded regions, and 67 °C in the other nonshaded regions. The yellow regions are the detection zones with channel neckdowns, and the corresponding cycle numbers are noted on the left. The PCR mixture is injected in the nozzle which is highlighted in red, and the carry fluid oil is injected through the side nozzles acting as oil extractor (OE) which is in blue. (B) Droplet generation at the nozzle. (C) Uniform picoliter droplets in the downstream and flowing through one of the channel neckdowns. (Reprinted with permission from Kiss, F.M., Ortoleva-Donnelly, L., Beer, N.R. *et al.* (2008) High-throughput quantitative polymerase chain reaction in picoliter droplets. *Anal Chem* **80**: 8975–8981. Copyright 2008 American Chemical Society)

Beer *et al.* (2007, 2008) reduced the droplet size to pictolitre with real-time measurement of fluorescence detection for PCR amplification (Figure 4.8). Similar to the previous device, this device benefits from the advantages of using two phase flows; aqueous droplets and an immiscible oil-phase as the carrier fluid. This silicon-based device has an off-chip valving system, which traps, stops and shunts the PCR mixture to perform thermal cycles within the microchannels.

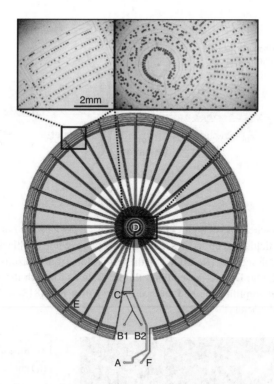

Figure 4.6 The schematic diagram of the radial PCR device, where the carrier fluid oil is injected at the inlet A and the aqueous phase at two inlet channels (B1 and B2), thus droplets are generated at a T-junction (C). The channels are 75 μm deep, and 500 μm wide in the hot zone (D) to ensure initial denaturation of the template. The channels are 200 μm wide in the periphery (E) where primer annealing and template extension occur. After 34 thermal cycles, the droplets are collected at the exit F. The underlying copper rod (Ø: 1.2 cm), highlighted in orange, provides heat and the Peltier module (inner Ø: 2.7 cm, outer Ø: 5.5 cm) are used to adjust thermal gradient (blue area). (Reprinted with permission from Schaerli, Y., Wootton, R.C., Robinson, T. *et al.* (2009) Continuous-flow polymerase chain reaction of single-copy DNA in microfluidic microdroplets. *Anal Chem* **81**: 302–306. Copyright 2009 American Chemical Society)

This system required only 18 cycles for single-copy real-time detection, using TaqMan-based FRET probes, and was used for reverse transcription PCR (RT-PCR) to amplify cDNA from a complementary RNA template, which demonstrates a successful example of single-copy target nucleic acids from a complex environment. However, this is a well-based continuous-flow PCR chip where the whole chip and fluids need to be heated up and cooled down for each thermal cycle, thus leading to a lengthy PCR process. A single DNA amplifying PCR device of Musyanovych, Mailänder and Landfester (2005) generated water nanodroplets with a diameter of 100–300 nm. Schaerli and Hollfelder (2009) discussed about how droplets provide monoclonality. Monoclonality is cloning the single template of DNA in a droplet

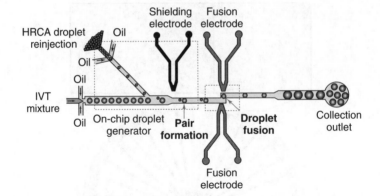

Figure 4.7 The schematic diagram of a droplets fusion device. The blue and red electrodes generate 600 V AC field to electrocoalesce the droplet pairs, while the black electrode prevents unwanted electrocoalescence. (Reprinted with permission from Mazutis, L., Araghi, A.F., Miller, O.J. *et al.* (2009) Droplet-based microfluidic systems for high-throughput single DNA molecule isothermal amplification and analysis. *Anal Chem* **81**: 4813–4821. Copyright 2009 American Chemical Society)

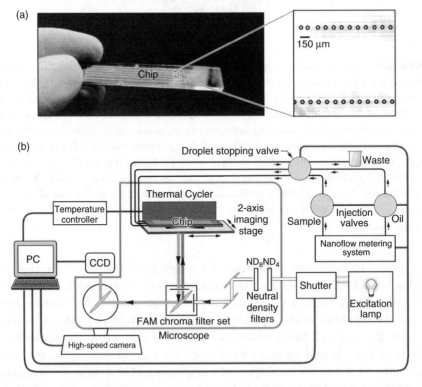

Figure 4.8 Schematic diagram of the on-chip RT-PCR device: (a) fused-silica device with an inset of monodisperse ~70-pl droplets trapped and ready for subsequent PCR; (b) system setup. (Reprinted with permission from Beer, N.R., Wheeler, E.K., Lee-Houghton, L. *et al.* (2008) On-chip single-copy real-time reverse-transcription PCR in isolated picoliter droplets. *Anal Chem* **80**: 1854–1858. Copyright 2008 American Chemical Society)

which leads to single-molecule PCR. Trapping the specific sample to a particular droplet can be achieved by a creative droplet generation system. One of the early solutions to this problem was reported by Song, Tice and Ismagilov (2003) where a reaction time controlled droplet generation system was demonstrated. Another example of single molecule template PCR was achieved by M. Nakano *et al.* (2005).

The droplet-based continuous-flow microPCR devices demonstrated the feasibility and great potential, particularly for high throughput single-molecule and single-cell PCR (Mohr *et al.* 2006, 2007; Tsuchiya *et al.* 2008; Kiss *et al.* 2008; Schaerli *et al.* 2009; Beer *et al.* 2007, 2008). There are still a number of challenges, for example how to produce reliable monodispersed droplets, how to reduce the droplet interactions with each other and with the device walls and how to control droplet trajectories. Significant effort on novel design of droplet-based continuous-flow microPCR devices is required before their full potential can be realized. A summary of the current work on chip-based PCR in droplets is shown in Table 4.2.

Conclusions

Microfluidic technology has been applied to DNA amplification and microfluidic PCR has demonstrated its advantages. Due to its small thermal mass, microfluidic PCR can significantly reduce PCR processing time so that instant medical diagnosis and in-field detection become possible. It has the potential for massive parallel operation to increase throughput dramatically. Experimental costs can be significantly reduced because of the reduced consumption of expensive reagents. PCR inhibition and carryover contamination, which often occur in continuous-flow microPCRs, can be overcome by using droplet technology. PCR in droplets can provide a suitable platform for single DNA molecule and single-cell amplification, which can also effectively eliminate the preference for amplifying short fragments and the production of short chimeric molecules associated with single-phase PCR. However, droplet-based continuous-flow microPCR devices are just emerging, despite their great potential for becoming the next generation PCR for DNA amplification.

Until now, microfluidic technology has not yet revolutionized current practice in biological and chemical analyses. One major hurdle is that the devices are not easy for users to operate. To enable users to embrace this technology, future development will be essential in system integration, device design optimization, device manufacture and system automation. For the PCR process, an integrated functional component for a μTAS device where all the required processing functions such as droplet generation can be achieved on chip is required. In addition, to improve device performance and reduce running costs, we need to optimize device design and operation. Although bespoke design may still be necessary at this stage, many components could be standardized to reduce manufacture and maintenance costs.

Table 4.2   On-chip PCR in droplets

Device	Droplet Generation	Detection	Efficiency/ Achievements	Other Features
Mohr et al. (2006, 2007)	Droplets (100–155 μm diameter) generated at an integrated T-junction	Online monitoring fluorescence level within each droplet	60-base pair fragment from the RNase P gene amplified in about 8 min for 32 cycles	Numerical simulations carried out to optimize flow rates; Continuous-flow PCR chip
Beer et al. (2007, 2008)	Droplets (averaged diameter of 51, 29, 31, 27 and 24 μm) generated at an integrated T-junction	Online fluorescence detection within each droplet	The device used for RT-PCR	Well-based PCR chip
Kiss et al. (2008)	Monodisperse droplets with 50-μm diameter generated by a focused flow at a cross-junction	Online fluorescence detection within each droplet	A 245-base pair adenovirus product amplified and quantified in 35 min at initial template concentrations as low as 1 template molecule/167 droplets	The device can produce millions droplets per hour and is able to perform single-molecule PCR; Continuous-flow PCR chip
Schaerli et al. (2009)	Droplets generated on-chip at a T-junction	34 thermal cycles for only 17 min to amplify a DNA fragment with 85-base pairs.	The temperatures of droplets were measured by fluorescence lifetime imaging inside the droplets; Continuous-flow PCR chip with radial layout.	
Hartung et al. (2009)	Teflon FEP tubing, segment formation	Custom made Flow-through fluorimeter (UV-LED and spectrometer)	392 base-pair amplicon in 24 min with 32 thermal cycles.	Continuous-flow PCR device made of Teflon FEP tubing coiled around three asymmetrical electrical heater

Table 4.2 (*continued*)

Device	Droplet Generation	Detection	Efficiency/ Achievements	Other Features
Mazutis *et al.* (2009)	PDMS microfluidic chip, 20-μm depth	Fluoresence detection	High-throughput chip up to 8 kHZ	A microfluidic chip with a capability of manipulating droplets
Wang and Burns (2009)	Glass microfluidic chip	Fluorescence detection	No volume loss up to 40 cycles,	PCR in nanoliter droplets, outcome product quality improved
Okochi *et al.* (2010)	Droplets generated using magnetic beads	Fluorescence detection	Detects 100 to 1000 copies of mRNA templates	Reverse transcript-PCR achieved
Markey, Mohr and Day (2010)	Droplet generation at a T-junction chip	High-throughput device	Continuous-flow PCR device made of PTFE tubing coiled around heater	

References

Auroux, P.A., Day, P.J.R. and Manz, A. (2005) *Quantitative study of the adsorption of PCR reagents during on-chip bi-directional shunting PCR.* In Proceedings of the 9th International Conference on Miniaturized Systems for Chemistry and Life Sciences, 9–13 October, Boston, MA, USA.

Auroux, P.-A., Koc, Y., deMello, A. *et al.* (2004) Miniaturised nucleic acid analysis. *Lab Chip* **4**: 534–546.

Auroux, P.-A., Reyes, D.R., Iossifidis, D. and Manz, A. (2002) Micro total analysis systems. 2. Analytical standard operations and applications. *Anal Chem* **74**: 2637–2652.

Bartlett, J.M. and Stirling, D. (2003) A short history of the polymerase chain reaction. *Methods Mol Biol* **226**: 3–6.

Beer, N.R., Hindson, B.J., Wheeler, E.K. *et al.* (2007) On-chip, real-time, single-copy polymerase chain reaction in picoliter droplets. *Anal Chem* **79**: 8471–8475.

Beer, N.R., Wheeler, E.K., Lee-Houghton, L. *et al.* (2008) On-chip single-copy real-time reverse-transcription PCR in isolated picoliter droplets. *Anal Chem* **80**: 1854–1858.

Belgrader, P., Benett, W., Hadley, D. *et al.* (1999) PCR detection of bacteria in seven min. *Science* **284**: 449–450.

Belgrader, P., Elkin, C.J., Brown, S.B. *et al.* (2003) A reusable flow-through polymerase chain reaction instrument for the continuous monitoring of infectious biological agents. *Anal Chem* **75**: 3446–3450.

Boedicker, J.Q., Li, L., Kline, T.R. and Ismagilov, R.F. (2008) Detecting bacteria and determining their susceptibility to antibiotics by stochastic confinement in nanoliter droplets using plug-based microfluidics. *Lab Chip* **8**: 1265–1272.

Braun, D., Goddard, N.L. and Libchaber, A. (2003) Exponential DNA replication by laminar convection. *Phys Rev Lett* **91**: 1581031–1581034.

Bruckner-Lea, C.J., Tsukuda, T., Dockendorff, B. *et al.* (2002) Renewable microcolumns for automated DNA purification and flow-through amplification: from sediment samples through polymerase chain reaction. *Anal Chim Acta* **469**: 129–140.

Bu, M.Q., Tracy, M., Ensell, G. *et al.* (2003) Design and theoretical evaluation of a novel microfluidic device to be used for PCR. *J Micromech Microeng* **13**: S125.

Chabert, M., Dorfman, K.D., de Cremoux, P. *et al.* (2006) Automated microdroplet platform for sample manipulation and polymerase chain reaction. *Anal Chem* **78**: 7722–7728.

Chen, L., Manz, A. and Day, P.J.R. (2007) Total nucleic acid analysis integrated on microfluidic devices. *Lab Chip* **7**: 1413–1423.

Chen, L., West, J., Auroux, P.-A. *et al.* (2007) Ultrasensitive PCR and real-time detection from human genomic samples using a bidirectional flow microreactor. *Anal Chem* **79**: 9185–9190.

Chen, Z.Y., Qian, S.Z., Abrams, W.R. *et al.* (2004) Thermosiphon-based PCR reactor: experiment and modelling. *Anal Chem* **76**: 3707–3015.

Chiou, J., Matsudaira, P., Sonin, A. and Ehrlich, D. (2001) A closed-cycle capillary polymerase chain reaction machine. *Anal Chem* **73**: 2018–2021.

Chung, K.H., Park, S.H. and Choi, Y.H. (2010) A palmtop PCR system with a disposable polymer chip operated by the thermosiphon effect. *Lab Chip* **10**: 202–210.

Crews, N., Wittwer, C. and Gale, B. (2008) Continuous-flow thermal gradient PCR. *Biomed Microdevices* **10**: 187–195.

Curcio, M. and Roeraade, J. (2003) Continuous segmented-flow polymerase chain reaction for high-throughput miniaturized DNA amplification. *Anal Chem* **75**: 1–7.

Dorfman, K.D., Chabert, M., Codarbox, J.-H. *et al.* (2005) Contamination-free continuous flow microfluidic polymerase chain reaction for quantitative and clinical applications. *Anal Chem* **77**: 3700–3704.

Dressman, D., Yan, H., Traverso, G. *et al.* (2003) Transforming single DNA molecules into fluorescent magnetic particles for detection and enumeration of genetic variations. *Proc Natl Acad Sci USA* **100**: 8817.

Erill, I., Campoy, S., Erill, N. *et al.* (2003) Biochemical analysis and optimization of inhibition and adsorption phenomena in glass–silicon PCR-chips. *Sens Actuators B Chem* **96**: 685–692.

Felbel, J., Bieber, I. and Köhler, J. M. (2002) Chemical surface management for micro PCR in silicon chip thermocyclers. *Proc. SPIE* **4937**: 34.

Felbel, J., Bieber, I., Pipper, J. and Köhler, J. M. (2004) Investigations on the compatibility of chemically oxidized silicon (SiO_x)-surfaces for applications towards chip-based polymerase chain reaction. *Chem Eng J* **101**: 333–338.

Frey, O., Bonneick, S., Hierlemann, A. and Lichtenberg, J. (2007) Autonomous microfluidic multi-channel chip for real-time PCR with integrated liquid handling. *Biomed Microdevices* **9**: 711–718.

Friedman, N.A. and Meldrum, D.R. (1998) Capillary tube resistive thermal cycling. *Anal Chem* **70**: 2997–3002.

Fukuba, T., Yamamoto, T., Naganuma, T. and Fujii, T. (2004) Microfabricated flow-through device for DNA amplification –towards in situ gene analysis. *Chem Eng J* **101**: 151–156.

Gonzalez, A., Grimes, R., Walsh, E.J. *et al.* (2007) Interaction of quantitative PCR components with polymeric surfaces. *Biomed Microdevices* **9**: 261–266.

Gulliksen, A., Solli, L., Karlsen, F. *et al.* (2004) Real-time nucleic acid sequence-based amplification in nanoliter volumes. *Anal Chem* **76**: 9–14.

Günther, A. and Jensen, K.F. (2006) Multiphase microfluidics: from flow characteristics to chemical and materials synthesis. *Lab Chip* **6**: 1487–1503.

Hardt, S., Dadic, D., Doffing, F. *et al.* (2004) Development of a slug-flow PCR chip with minimum heating cycle times. *Nanotech.* **1**: 55–58.

Hartung, R., Brosing, A., Sczcepankiewicz, G. *et al.* (2009) Application of an asymmetric helical tube reactor for fast identification of gene transcripts of pathogenic viruses by micro flow-through PCR. *Biomed Microdevices* **11**: 685–692.

Hashimoto, M., Barany, F., Xu, F. and Soper, S. A. (2007) Serial processing of biological reactions using flow-through microfluidic devices: coupled PCR/LDR for the detection of low-abundant DNA point mutations. *Analyst* **132**: 913–921.

Hashimoto, M., Chen, P.-C., Mitchell, M. W. *et al.* (2004) Rapid PCR in a continuous flow device. *Lab Chip* **4**: 638–645.

Huebner, A., Sharma, S., Srisa-Art, M. *et al.* (2008) Microdroplets: a sea of applications? *Lab Chip* **8**: 1244–1254.

Kelly, B.T., Baret, J.-C., Taly, V., Griffiths, A.D. (2007) Miniaturizing chemistry and biology in microdroplets, *Chem. Commun.* (18): 1773–1788.

Kim, J.A., Lee, J.Y., Seong, S. *et al.* (2006) Fabrication and characterization of a PDMS–glass hybrid continuous-flow PCR chip. *BioChem Eng J* **29**: 91–97.

Kiss, M.M., Ortoleva-Donnelly, L., Beer, N.R. *et al.* (2008) High-throughput quantitative polymerase chain reaction in picoliter droplets. *Anal Chem* **80**: 8975–8981.

Kojima, T., Takei, Y., Ohtsuka, M. *et al.* (2005) PCR amplification from single DNA molecules on magnetic beads in emulsion: application for high-throughput screening of transcription factor targets. *Nucleic Acids Res* **33**: e150.

Kolari, K., Satokari, R., Kataja, K. *et al.* (2008) Real-time analysis of PCR inhibition on microfluidic materials. *Sens Actuators B* **128**: 442–449.

Kopp, M.U., de Mello, A.J. and Manz, A. (1998) Chemical amplification: continuous-flow PCR on a chip. *Science* **280**: 1046–1048.

Kricka, L.J., Wilding, P. (2003) Microchip PCR, *Anal. BioAnal Chem* **377**: 820–825.

Krishnan, M., Burke, D.T. and Burns, M.A. (2004) Polymerase chain reaction in high surface-to-volume ratio SiO_2 microstructures. *Anal Chem* **76**: 6588–6593.

Krishnan, M., Ugaz, V.M. and Burns, M.A. (2002) PCR in a Rayleigh-Bénard convection cell. *Science* **298**: 793.

Li, M., Diehl, F., Dressman, D. *et al.* (2006) BEAMing up for detection and quantification of rare sequence variants. *Nat Methods* **3**: 95–97.

Li, S., Fozdar, D.Y., Ali, M.F. *et al.* (2006) A continuous-flow polymerase chain reaction microchip with regional velocity control. *J MEMS* **15**: 223–236.

Liu, J., Enzelberger, M. and Quake, S. (2002) A nanoliter rotary device for polymerase chain reaction. *Electrophoresis* **23**: 1531–1536.

Marcus, J.S., Anderson, W.F. and Quake, S.R. (2006) Parallel picoliter RT-PCR assays using microfluidics. *Anal Chem* **78**: 956–958.

Margulies, M., Egholm, M., Altman, W.E. *et al.* (2005) Genome sequencing in microfabricated high-density picolitre reactors. *Nature* **437**: 376–380.

Markey, A.L., Mohr, S. and Day, P.J.R. (2010) High-throughput droplet PCR. *Methods* **50**: 277–281.

Mazutis, L., Araghi, A.F., Miller, O.J. *et al.* (2009) Droplet-based microfluidic systems for high-throughput single DNA molecule isothermal amplification and analysis. *Anal Chem* **81**: 4813–4821.

Mohr, S., Zhang, Y.-H., Macaskill, A. *et al.* (2006) *Optimal design and operation for a droplet-based PCR chip.* Proceedings of the 4th International Conference on Nanochannels, Microchannels and Minichannels, 19–21 June, Limerick Ireland. Paper ICNMM2006-96131.

Mohr, S., Zhang, Y.-H., Macaskill, A. *et al.* (2007) Numerical and experimental study of a droplet-based PCR chip. *Microfluid Nanofluidic* **3**: 611–621.

Musyanovych, A., Mailänder, V. and Landfester, K. (2005) Miniemulsion droplets as single molecule nanoreactors for polymerase chain reaction. *Biomacromolecules* **6**: 1824–1828.

Nakano, H., Matsuda, K., Yohda, M. *et al.* (1994) High speed polymerase chain reaction in constant flow. *Biosci Biotechnol Biochem* **58**: 349–352.

Nakano, M., Nakai, N., Kurita, H. *et al.* (2005) Single-molecule reverse transcription polymerase chain reaction using water-in-oil emulsion. *J Biosci Bioeng* **99**: 293–295.

Nakayama, T., Kurosawa, Y., Furui, S. *et al.* (2006) Circumventing air bubbles in microfluidic systems and quantitative continuous-flow PCR applications. *Anal BioAnal Chem* **386**: 1327–1333.

Nisisako, T., Torii, T. and Higuchi, T. (2002) Droplet formation in a microchannel network. *Lab Chip* **2**: 24–26.

Northrup, M.A., Ching, M.T., White, R.M. and Watson, R. T. (1993) DNA amplification with a microfabricated reaction chamber. *Transducers* **93**: 924–926.

Obeid, P.J. and Christopoulos, T.K. (2003) Continuous-flow DNA and RNA amplification chip combined with laser-induced fluorescence detection. *Anal Chim Acta* **494**: 1–9.

Obeid, P.J., Christopoulos, T.K., Crabtree, H.J. and Backhouse, C. J. (2003) Microfabricated device for DNA and RNA amplification by continuous-flow polymerase chain reaction and reverse transcription-polymerase chain reaction with cycle number selection. *Anal Chem* **75**: 288–295.

Ohashi, T., Kuyama, H., Hanafusa, N. and Togawa, Y. (2007) A simple device using magnetic transportation for droplet-based PCR. *Biomed Microdevices* **9**: 695–702.

Okochi, M., Tsuchiya, H., Kumazawa, F. *et al.* (2010) Droplet-based gene expression analysis using a device with magnetic force-based-droplet-handling system. *J Biosci Bioeng* **109** (2): 193–197.

Ong, S.-E., Zhang, S., Du, H. and Fu, Y. (2008) Fundamental principles and applications of microfluidic systems. *Frontiers Biosci* **13**: 2757–2773.

Panaro, N.J., Lou, X.J., Fortina, P. *et al.* (2004) Surface effects on PCR reactions in multichip microfluidic platforms. *Biomed Microdevices* **6**: 75–80.

Park, N., Kim, S. and Hahn, J. H. (2003) Cylindrical compact thermal-cycling device for continuous-flow polymerase chain reaction. *Anal Chem* **75**: 6029–6033.

Prakash, A.R., Amrein, M. and Kaler, K.V.I. S. (2008) Characteristics and impact of Taq enzyme adsorption on surfaces in microfluidic devices. *Microfluid Nanofluid* **4**: 295–305.

Sadler, D.J., Changrani, R., Roberts, P. *et al.* (2003) Thermal management of BioMEMS: temperature control for ceramic-based PCR and DNA detection devices. *IEEE Trans Compon Packag Technol* **26**: 309–316.

Schaerli, Y. and, Hollfelder, F. (2009) The potential of microfluidic water-in-oil droplets in experimental biology, *Mol Biosyst* **5**: 1392–1401.

Schaerli, Y., Wootton, R.C., Robinson, T. *et al.* (2009) Continuous-flow polymerase chain reaction of single-copy DNA in microfluidic microdroplets. *Anal Chem* **81**: 302–306.

Schneegaß, I. and Köhler, J.M. (2001) Flow-through polymerase chain reactions in chip thermocyclers. *Rev Mol Biotechnol* **82**: 101–121.

Schneegaß, I., Bräutigam, R. and Köhler, J. M. (2001) Miniaturized flow-through PCR with different template types in a silicon chip thermocycler. *Lab Chip* **1**: 42–49.

Shendure, J., Porreca, G.J., Reppas, N.B. *et al.* (2005) Accurate multiplex polony sequencing of an evolved bacterial genome. *Science* **309**: 1728–1732.

Shoffner, M.A., Cheng, J., Hvichia, G.E. *et al.* (1996) Chip PCR. I. Surface passivation of microfabricated silicon-glass chips for PCR. *Nucleic Acids Res* **24**: 375–379.

Song, H., Tice, J.D. and Ismagilov, R.F. (2003) A microfluidic system for controlling reaction networks in time. *Angew Chem Int Ed* **42**: 767–772.

Sun, K., Yamaguchi, A., Ishida, Y. *et al.* (2002) A heater-integrated transparent microchannel chip for continuous-flow PCR. *Sens Actuators B Chem* **84**: 283–289.

Sun, Y., Kwoka, Y.C. and Nguyen, N.T. (2007) A circular ferrofluid driven microchip for rapid polymerase chain reaction. *Lab Chip* **7**: 1012–1017.

Sun, Y., Yien C. and Kwok, C. (2006) Polymeric microfluidic system for DNA analysis. *Anal Chim Acta* **556**: 80–96.

Taly, V., Kelly, B.T. and Griffiths, A.D. (2007) Droplets as microreactors for high-throughput biology. *Chembiochem* **8**: 263–272.

Taylor, T.B., Winn-Deen, E.S., Picozza, E. *et al.* (1997) Optimization of the performance of the polymerase chain reaction in silicon-based microstructures. *Nucleic Acids Res* **25**: 3164–3168.

Teh, S.-Y., Lin, R., Hung, L.-H., Lee, A. P. (2008) Droplet microfluidics, *Lab. Chip*, **8**: 198–220.

Terry, S. C. (1975) A gas chromatographic air analyser fabricated on silicon wafer using integrated circuit technology. PhD thesis, Stanford University, USA.

Tsuchiya, H., Okochi, M., Nagao, N. *et al.* (2008) On-chip polymerase chain reaction microdevice employing a magnetic droplet-manipulation system. *Sens Actuators B* **130**: 583–588.

Vilkner, T., Janasek, D. and Manz, A. (2004) Micro total analysis systems. Recent developments. *Anal Chem* **76**: 3373–3386.

Wang, F. and Burns, M.A. (2009) Performance of nanoliter-sized droplet-based microfluidic PCR. *Biomed Microdevices* **11**: 1071–1080.

Wang, H., Chen, J., Zhu, L., Shadpour, H., Hupert, M. L., Soper, S. A. (2006) Continuous Flow Thermal Cycler Microchip for DNA Cycle Sequencing, *Anal Chem* **78**: 6223–6231.

Wang, W., Li, Z.-X., Luo, R., Lü, S.-H., Xu, A.-D., Yang, Y.-J. (2005) Direct, Highly Enantioselective Pyrrolidine Sulfonamide Catalyzed Michael Addition of Aldehydes to Nitrostyrenes, *J Micromech Microeng* **15**: 1369–1371.

Wang, W., Wang, H. B., Li, Z. X., Guo, Z. Y. (2006) Silicon inhibition effects on the polymerase chain reaction: A real-time detection approach, *J. Biomed. Mater. Res. A*, **77**: 28–34.

West, J. Karamata, B. Lillis, B., Gleeson, J.P., Alderman, J., Collins, J.K., Lane, W., Mathewson, A., Berney, H. (2002) Application of magnetohydrodynamic actuation to continuous flow chemistry, *Lab Chip* **2**: 224–230.

Wetmur, J. G., Kumar, M., Zhang, L., Palomeque, C., Wallenstein, S., Chen, J. (2005) Molecular haplotyping by linking emulsion PCR: Analysis of paraoxonase 1 haplotypes and phenotypes, *Nucleic Acids Res* **33**: 2615–2619.

Wilding, P., Shoffner, M. A., Cheng, J., Hvichia, G. E., Kricka, L. J. (1995) Thermal cycling and surface passivation of micromachined devices for PCR, *Clin. Chem.*, **41**: 1367–1368.

Wilding, P., Shoffner, M. A., Kricka, L. J. (1994) PCR in a silicon microstructure, *Clin. Chem.*, **40**: 1815–1818.

Williams, R., Peisajovich, S. G., Miller, O. J., Magdassi, S., Tawfik, D. S., Griffiths, A. D. (2006) Amplification of complex gene libraries by emulsion PCR, *Nat. Methods*, **3**: 545–550.

Wittwer, C.T., Hermann, M.G. (1999) Rapid Thermal Cycling and PCR kinetics, *In PCR Applications: Protocols for Functional Genomics* (eds. M.A. Innis, D.H. Gelfand, J.J. Sninsky), Academic, San Diego, pp 211–229.

Xia, Y.-M., Hua, Z.-S., Srivannavit, O., Ozel, A. B., Gulari, E. (2007) Minimizing the surface effect of PDMS-glass microchip on polymerase chain reaction by dynamic polymer Passivation, *J. Chem. Tech. Biotech.*, **82**: 33–38.

Yang, M., Pal, R., Burns, M.A. (2005) Cost-effective thermal isolation techniques for use on microfabricated DNA amplification and analysis devices, *J. Micromech. Microeng.* **15**: 221.

Zhang, C. and Xing, D. (2009) Parallel DNA amplification by convective polymerase chain reaction with various annealing temperatures on a thermal gradient device, *Anal. Biochem.*, **387**: 102–112.

Zhang, C., Xing D., Li Y. (2007) Micropumps, microvalves, and micromixers within PCR microfluidic chips: Advances and trends, *Biotech. Adv.*, **25**: 483–514.

Zhang, C.S., Xu, J.L., Ma, W.L., Zheng, W.L. (2006) PCR microfluidic devices for DNA amplification, *Biotech. Adv.* : **24**: 243–284.

Zhang, C. S. and Xing D (2007) Miniaturized PCR chips for nucleic acid amplification and analysis: latest advances and future trends, *Nucleic Acids Res* **35**: 4223–4237.

Zhang, K., Zhu, J., Shendure, J., Porreca, G.J., Aach, J.D., Mitra, R.D., Church, G.M. (2006) Long-range polony haplotyping of individual human chromosome molecules, *Nat. Genet.*, **38**: 382–387.

Zhang, Y.H., and Ozdemir, P. (2009) Microfluidic DNA amplification –a review, *Anal. Chimica Acta*, **638**: 115–125.

5

High-Resolution Melt Profiling

Steven F. Dobrowolski and Carl T. Wittwer

Introduction

Thermal denaturation (melting) is a fundamental property of nucleic acids that has been exploited by many molecular methods. Sequential melting, primer annealing and extension are the basis of polymerase chain reaction (PCR), which revolutionized our ability to assess specific regions of the genome. Early applications of melting, as a means to assess specific nucleotides within a DNA sequence, used PCR followed by hybridization of fluorescently labelled oligonucleotide probe(s) to one strand of the PCR product. The thermal stability of the probe:amplicon hybrid is decreased by any sequence variant under with probe. The characteristic melting temperature of a fully hybridized probe is higher than the probe melting temperature when a base-pair mismatch is present. Among the earliest applications of genotyping via probe-based thermal denaturation was the c.1691G>A Factor V Leiden mutation (Lay and Wittwer 1997) which is still widely used in clinical analysis (Lyon and Wittwer 2009).

When real-time PCR is performed with a dsDNA binding dye (e.g. SYBR® Green), it is common to perform a post-PCR product melt. The product created during a real-time PCR produces a characteristic melting curve. Should an undesired product (e.g. primer dimer, non-specific product) be present, these are likely to melt at a temperature different from the specific product (Ririe, Rasmussen and Wittwer 1997). In this way, the results of a real-time assay are validated when the only melting signature is that of the desired product. Until very recently, real-time PCR instruments

Molecular Analysis and Genome Discovery, Second Edition. Edited by Ralph Rapley and Stuart Harbron.
© 2012 John Wiley & Sons, Ltd. Published 2012 by John Wiley & Sons, Ltd.

had inadequate melting resolution to assess sequence variation from the amplification product melting curve.

High-resolution melt profiling (HRMP) is a relatively new technique that was first described in 2003 (Wittwer *et al.* 2003), whereby fine details concerning sequence characteristics of a dye-stained PCR product are revealed in the shape of the melting profile. HRMP was made available to the greater scientific community in 2003 with the introduction of the first commercial instrument and melting dye. There are now several options for dyes and instrument platforms so both clinical laboratories and basic scientists have adequate choices to match their needs with an appropriate platform (Reed, Kent and Wittwer 2007; Farrar, Reed and Wittwer 2010). The major contribution of HRMP is to enable the identification of sequence variation at any location throughout the length of a PCR product, a technique which has been termed 'gene-scanning'. As analysing coding regions and the sequences flanking them is a common practice in both clinical and research laboratories, HRMP offers a means to greatly increase throughput while reducing turnaround time and cost (Erali, Voelkerding and Wittwer 2008). By prospectively identifying regions where sequence variation exists, DNA sequencing may be targeted to only appropriate regions. Triaging regions where DNA sequencing is both unnecessary and uninformative can reduce the overall sequencing burden by 90% or more. A less well adopted, yet major contribution of HRMP, is the enabling of specific genotyping tests without the necessity of an oligonucleotide probe to discriminate alternative genotypes (Liew *et al.* 2004). So-called amplicon-based genotyping is immensely simple, highly robust and, given that probes are not required, inexpensive to perform. The transition from low-resolution melting (e.g. probe:amplicon or confirmatory product melt) to high-resolution melting required advances in three areas: (1) instrument platforms; (2) dsDNA binding dyes; and (3) software. This chapter will explore factors that lead to the creation of effective HRMP assays, applications of high-resolution melting and the adaptations necessary to successfully use melt profiling, as well as the strengths and limitations of melt profiling.

Basic concepts of melt profiling

The workflow and basis of HRMP are exceedingly simple. PCR is performed in the presence of an appropriate dsDNA binding dye that stains the newly synthesized amplification product. The dye has greater fluorescence when bound to DNA and upon thermal denaturation the dye is released leading to a loss of the fluorescent signal. Figure 5.1 provides a simple melting profile of a PCR product that flanks the 11th exon of the human ACADM gene. At low temperature, the fluorescence is high and as the temperature increases the fluorescence decreases until it reaches a baseline as the curve approaches the origin of the y-axis. Some profiles in Figure 5.1 were created with samples having a wild-type sequence while others were created by samples with

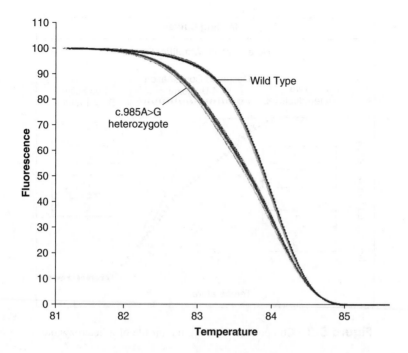

Figure 5.1 Basic melting profile

a heterozygous copy of the c.985A>G mutation (McKinney *et al.* 2004). Note how
the profiles generated by samples heterozygous for the mutation are shaped differently than those with the wild-type sequence. In an HRMP protocol, following PCR
a denaturation and re-annealing step causes a random redistribution of DNA strands.
When a heterozygous mutation is present, it leads to a population of PCR products
where DNA strands, derived from the wild-type allele and mutant allele, combine
to create double strand molecules having a base pair mismatch – these are so-called
heteroduplex molecules. Figure 5.2 illustrates the overall complement of molecular
species generated when a heterozygous sequence variation is present and the melting
profile created by each of these species. Two populations of heteroduplex molecules
are present as are two populations of fully base-paired molecules. Thus, the combined melting characteristics of the four molecular species contribute to the overall
shape of the melting profile. In Figure 5.1, the profiles generated by the wild-type
samples are created by a homogenous population of PCR products represented by the
A:T base-paired product identified on the upper right of Figure 5.2. While the basic
premise underlying HRMP is straightforward, its applications in gene-scanning and
site-specific genotyping are more complex. The following sections will cover ways
to establish functional HRMP assays, scenarios that will be encountered, applications
and interpretation of melting data.

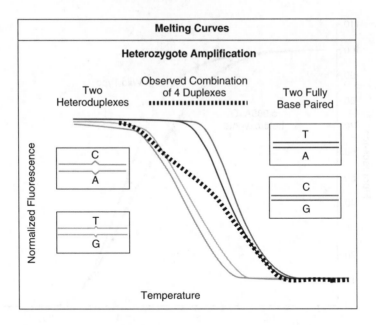

Figure 5.2   Components creating the profile of a heterozygote

HRMP and polymerase chain reaction

Low-resolution melting assays that interrogate a PCR product with a fluorescent probe are tolerant to the presence of extraneous amplification products other than the intended product. Extraneous PCR products (primer-dimers, etc.) are invisible in low-resolution probe-based assays as the probe generates a signal only when hybridized to the complementary sequence on a strand of the target amplicon. As the probe does not hybridize to extraneous products, their presence does not contribute to the fluorescent signal. Figure 5.2 shows how the presence of four distinct molecular species combine to create the melting profile observed for heterozygous specimens. The dsDNA binding dyes used in HRMP are ultimately promiscuous and provide equivalent staining to every dsDNA molecule that is present. The presence of extraneous products combines their melting characteristics with that of the specific product, impacting the overall shape of the melting profile. To obtain consistent melt profiling results, PCR must be robust and highly optimized. Furthermore, the PCR optimization process must be performed with melting dye included in the reaction. When HRMP was emerging, many investigators assumed that their existing PCR reactions could be instantaneously converted to HRMP assays with the simple addition of dye. When dye was added to established PCR reactions, it was not an uncommon event to observe reduced amplification efficiency or complete amplification failure. This was initially explained by stating that 'the dye inhibits PCR'. In at least 99% of all such instances, this interpretation was erroneous. The correct interpretation was that the PCR conditions used without dye (cycling, primer concentration, etc.) were not compatible with amplifying the product in the presence of dye. A few simple procedures are typically all that is necessary to translate an established PCR assay to accept dye in the reaction. As specific PCR is the single most important factor to effective melt profiling, this section will address aspects of PCR that lead to effective application for HRMP.

Reagents used in HRMP assays are largely identical to those used in other PCR-based applications. The preferred amplification buffer for HRMP is Tris base (Wittwer *et al.* 1997). Deoxynucleotide triphosphates from essentially all manufacturers provide equivalent results and should be used at standard working concentrations (200 μM each dNTP). Deoxyuracil triphosphate is compatible with HRMP when uracil n-glycosylase pre-digestion is indicated. While dUTP does not alter the shape of the melting profile, the overall melting temperature will decrease by ∼1.5–2.0 °C compared to that when dTTP is used. Magnesium chloride is compatible at standard concentrations (1.5–4.0 mM) and there is no evidence to suggest that melting dye alters the requirement for $MgCl_2$. Our experiences lead us to recommend the global application of hot-start PCR protocols (Dobrowolski *et al.* 2005). Hot-start protocols promote specificity of PCR that, in turn, will generate the highest quality melt profiles. We also observe that antibody mediated hot-start generally provides superior results to thermally activated polymerases. This is particularly true when amplifying template DNA derived from suboptimal sources such as dried blood on filter paper or formalin-fixed paraffin-embedded tissue. Primers are obviously an important component of any PCR reaction and a debatable issue involves the appropriate level of primer purification required for HRMP assays. In our applications, we have almost exclusively used unpurified, salted-out primer preparations for both gene-scanning and genotyping applications. While we have not realized a benefit from more highly purified primers (e.g. HPLC, gel), each investigator should determine what provides optimal results in their systems.

There are no special design requirements for primers being applied to HRMP. To facilitate the design of amplification products, manufacturers of HRMP platforms typically provide a primer design software tool. Our experience indicates that these seemingly specialized design tools offer no advantage over other primer design software. The following tenants dictating effective primer design for all PCR applications apply equally to designing products used in HRMP: (1) avoid stable hairpins; (2) avoid extendable homomeric/heteromeric 3′ cross hybrids; (3) avoid primers that have extraneous 3′ stable binding sites that may potentially create an extraneous product; (4) match Tm and ΔG^0 values of reverse and forward primers. We do not discourage using primer design tools provided by the manufacturers of HRMP platforms, but suggest that individual investigators will realize the greatest level of success using a primer design tool with which they are comfortable and have a history of success. To streamline post-melting DNA sequence analysis of samples that generate deviant melting profiles, the use of primers having a 5′ universal DNA sequencing tail is recommended. These tailed primers are well tolerated by HRMP.

A major benefit of HRMP, when applied to gene-scanning, is the ability to avoid DNA sequencing of regions where no variation exists. The use of tailed primers further streamlines the overall process by allowing fragments that generate deviant profiles to be recovered from the melting plate and used as templates to support DNA sequencing with a common set of sequencing primers. Co-utilization of an amplification product for melting analysis and follow-on DNA sequencing is discussed in the section on gene-scanning.

A hotly debated issue involves the relationship between the length of a PCR product and the sensitivity of HRMP to detect sequence variation. An early systematic study investigated this question using a plasmid-based system to alter G:C content, the location and class of the single base substitution within the fragment and the length of a fragment (Reed and Wittwer 2004). Up to 300 bp, the sensitivity and specificity were 100%. From 400–1000 bp, the sensitivity and specificity were 96.1% and 99.4%, respectively, generally decreasing at longer lengths. Another plasmid study later revealed similar results on a different platform (McKinney *et al.* 2010). However, small insertions, small deletions and the effect of repeated sequences were not investigated. Furthermore, only three plasmid backgrounds were studied, limiting the range of G:C content and melting domain complexity studied. Nevertheless, a comprehensive compilation of published studies that includes various instruments and dyes suggest similar high sensitivity (99.3%) and specificity (98.8%) (Farrar, Reed and Wittwer 2010). From our experience, fragments up to 550 bp are a very safe bet for identification of heterozygous sequence variants. A study of sequence variation in the mitochondrial genome demonstrated high detection sensitivity with fragments over 800 bp. However, the mitochondrial genome is low in G:C content and the effect of G:C content on scanning accuracy is not clear.

To optimize the amplification strategy for a newly designed product or to re-optimize amplification conditions for an existing product to accommodate the melting dye, the best first step is temperature gradient PCR. Figure 5.3 shows agarose gel electrophoresis of PCR products that amplify the first exon of the human N-acetyglutamate synthase gene. A two-temperature cycling parameter (30 s, 95 °C; 30 s anneal/extension) is used with a gradient of annealing/extension temperature from 62 to 74 °C. The theoretical Tm for these primers is ~62 °C and each primer was used at 0.2 µM. Figure 5.3 clearly shows these primers are not specific at the theoretical Tm but generate a robust highly specific product above 69 °C and are effective up to 74 °C. Figure 5.3 highlights two points: (1) the presence of melting dye in a PCR reaction alters the apparent Tm of oligonucleotide primers; and (2) Tm prediction based upon nearest neighbour pairs (independent of the software package used) must be confirmed by functional assessment, especially so when dye is included in the PCR reaction. Temperature gradient PCR greatly facilitates effective amplification and is the single most useful tool to optimize cycling conditions when performing PCR with melting dyes. Secondarily, primer concentration affects PCR specificity and should be modified to suite the characteristics of each individual primer set. Our experience indicates concentrations from 0.075 to 0.5 µM

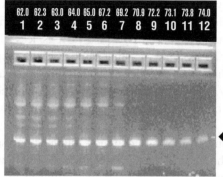

Arrow identifies specific product

Figure 5.3 Temperature gradient PCR

are generally effective. Going higher than 0.5 μM increases the probability of generating undesired products (so-called non-specific products). As every product in a PCR reaction contributes to the melting profile, extraneous products are always deleterious to effective analysis, and limiting primer concentration is a useful means to avoid extraneous products. We typically prefer a two-temperature PCR cycling protocol with a combined primer annealing/enzyme extension stage. In our experience, when dye is part of the PCR reaction, the optimal annealing temperature for primers is typically higher than predicted by primer design programs (see Figure 5.3). When primers anneal at temperature >60 °C, using a combined annealing/extension step in PCR is effective as Taq polymerase provides a high level of activity over a broad range of temperatures. Optimized three-temperature protocols are also effective.

Glycerol, betain, dimethylsulfoxide (DMSO) or other adjuvants are commonly used to weaken the dsDNA hybrid to improve amplification efficiency in regions of high G:C content. All PCR adjuvants are compatible with melting dyes. Indeed, the fragment of the N-acetyglutamate synthase gene in Figure 5.3 has a 67% G:C content and the PCR reaction contained both 10% DMSO and 1X LCGreen Plus dye (Mitchell *et al.* 2009). We have found DMSO to be the most effective agent to weaken strand affinity and concentrations from 5% to 15% v/v are effective. The dyes used in HRMP increase the stability of the DNA:DNA hybrid and exacerbate issues of hybrid stability and G:C content. As a consequence, adjuvants may be required to improve PCR efficiency at G:C content lower than typically associated with a need to weaken hybrid stability. In general, fragments with G:C content below 57% do not require adjuvant. Fragments with G:C content ranging from 58% to 64% should be assessed both with and without 10% DMSO to determine which condition provides a more robust and specific amplification. Except in some rare instances, fragments with G:C content ≥65% will require adjuvant for optimally robust PCR.

Some investigators are so attached to PCR protocols established in their labs that they are unwilling to re-optimize a protocol to accommodate amplification in the presence of melting dye. As an alternative, dye has been added post-PCR along with one or more cycles of melting and re-annealing to facilitate staining of the dsDNA molecule. Generally, to add dye post-PCR is less effective for both systematic and practical reasons. First, opening PCR reactions, especially so within the context of clinical applications, provides an opportunity to create amplicon contamination in the laboratory, a situation that must always be avoided. Second, essentially all amplifications may be re-optimized with minimal effort to accommodate melting dyes, so the potentially dangerous extra step of adding dye post-PCR is unnecessary. Figure 5.4 shows melting profiles of a common fragment and sample set where PCR products are stained throughout the cycling protocol (Figure 5.4a) as opposed to fragments stained by post-PCR addition of dye followed by a round of denaturation and re-annealing (Figure 5.4b). The quality of the melt profiles between the continuously stained group and the post-PCR stained group is obvious. The data in

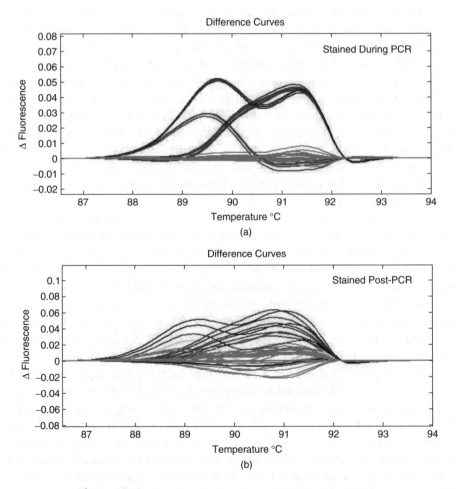

Figure 5.4 Products stained during PCR and post-PCR

Figure 5.4 is not unique to this fragment as similar results have been observed with a variety of PCR products. Although we cannot offer a mechanism to explain superior staining and subsequently superior melt profiles when dye is present throughout a cycling protocol as opposed to post-PCR staining, we do not recommend adding the melting dye after PCR.

DNA specimens and HRMP

For analysis of human genomic sequences, 10–50 ng of genomic DNA provides an adequate amplification template. HRMP requires consistency and this extends not only to the amount of template used in a reaction, but also to the chemistry used to prepare the DNA (Seipp *et al.* 2007). Figure 5.5 shows an assessment of exon 1 of

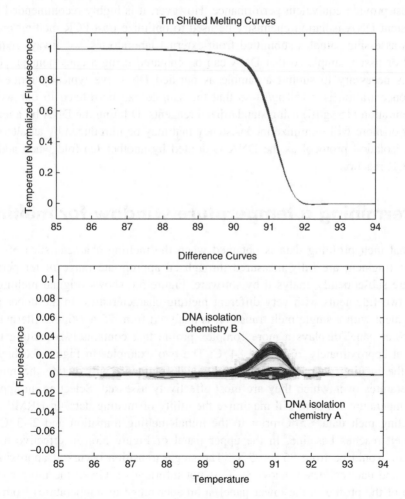

Figure 5.5 Influence of DNA isolation chemistry on the melting profile

the human phenylalanine hydroxylase gene using 94 unique DNA specimens. In the melt profile (upper panel), but far more pronounced in the subtractive difference plot (lower panel), there are two clusters of melting profiles. All the DNA samples in the alternative clusters have an identical wild-type sequence; however, those showing a deviation were prepared using an alternative DNA isolation chemistry than those clustered about the origin of the y-axis. Deviation between the two groups is a consequence of salts used in the isolation procedure. The DNA sample represented 10% of the reaction volume (1 µl DNA in a 10 µl PCR cocktail) and this demonstrates that small changes in ionic strength have the potential to alter the melt profile. The deviation between the two groups is not large, but sufficient that a sequence variation could be suspected. While a heterozygous variant would rarely create so modest a deviation, a homozygous variant may very well create a deviation of the magnitude observed. There is no recommended DNA isolation chemistry for use with HRMP as most provide equivalent performance. However, it is highly recommended that a consistent DNA isolation chemistry be used to optimize post-PCR melting profiles. When assessing samples submitted from external laboratories, we always request a blood or tissue sample so that DNA can be prepared using a consistent procedure. If it is necessary to submit a sample as purified DNA, we typically request that the concentration is $\geq$150 ng/µl so that the sample can be diluted to our working concentration (15 ng/µl) using standardized reagents. Diluting the DNA by a ten-fold factor or more will minimize inconsistency that may be introduced by an alternative DNA isolation protocol as the DNA is diluted by another ten-fold when added to the PCR reaction.

Determining a temperature window for melting

Optimal melt profiling data is obtained when the melting characteristics of a particular fragment are fully measured through an appropriate range of temperatures that are subsequently analysed by software. Figure 5.6 shows original melting data from two fragments with very different melting characteristics. In the upper panel, a fragment with a single melt transition is analysed from 75 to 96 °C. Alternatively, the lower panel displays a more complex profile that contains two melting transitions at approximately 76 °C and 84 °C. The two examples in Figure 5.6 highlight how the melting characteristics of individual fragments will dictate the range of temperatures over which they are most effectively assessed. Selecting an appropriate temperature window will maximize the utility of melting data. In HRMP, begin collecting melt data $\sim$5 °C prior to the initial melting transition and 2–3 °C after the melt reaches baseline. In the upper panel of Figure 5.6, an effective melting window would be from 85 to 95 °C. The single transition occurs at approximately 90 °C and data collected below 85 °C is not informative. Given the more complex shape of the profile in the lower panel, a broader range of temperatures from 69 to 88 °C is used. In this case, data collected above 88 °C is uninformative. Some data

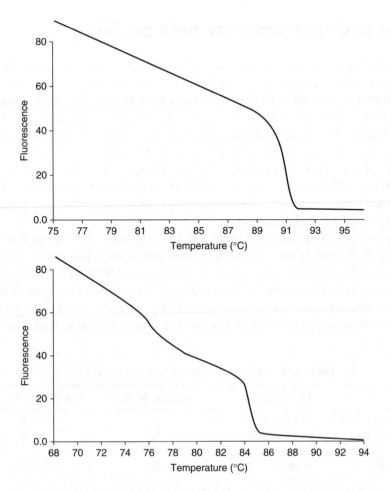

Figure 5.6 Profiles with differing melt characteristics

collected prior to the first melting transition and after the melt is complete provides the analysis software with a region to establish lower and upper baselines. Inadequate data to establish these baselines will compromise the quality of the melting profiles. Many HRMP assays concurrently assess multiple fragments (e.g. coding regions of a gene) and therefore, by necessity, a compromise melting window is required. If the PCRs of those fragments being co-assessed are robust and well optimized, the use of a compromise melting window will be adequate to generate effective data. Our assay of the human phenylalanine hydroxylase gene co-assesses a panel of 15 fragments (13 gene-scanning, two multiplex genotyping), using a wide compromise melting window of 70–95 °C, with no loss of sensitivity (Dobrowolski *et al.* 2007a). When assessing a single fragment, a window fitted to the fragment is optimal; however, when co-assessment of multiple fragments is performed, a compromise window is acceptable.

Dyes and platforms for melt profiling

Dyes used in melt profiling possess 'saturating characteristics' that enable these dyes to support HRMP (Dujols *et al.* 2006). Table 5.1 lists the seven dyes currently used in melt profiling. While all seven dyes support melt profiling, there is significant debate as to whether any particular dye provides superior results under all situations where HRMP is applied. Manufacturers tout the benefits of individual products; however, very few comparisons have been published (Farrar *et al.* 2010) and some dyes are only available in a complete PCR master mix. The authors have experience with LCGreen Plus dye and are satisfied with its performance; however individual investigators should consider the available options to determine which product is most effective in their laboratories.

There are now instruments with diverse characteristics that support HRMP and therefore most investigators should be able to match the requirements of their laboratories with an instrument platform. Table 5.2 lists HRMP instrument platforms and some of their characteristics. Most platforms are multifunctional providing PCR, real-time detection and probe-based post-PCR genotyping in addition to HRMP. For the high-throughput laboratory interested in adding HRMP to their repertoire of

Table 5.1   Dyes promoted for high resolution melting

Dye	Manufacturer	Available as Dye	Available in Master Mix
SYBR® GreenER	Invitrogen		X
LCGreen® Plus	Idaho Technology	X	X
STYO® 9	Invitrogen	X	
EvaGreen®	Biotium	X	X
ResoLight	Roche		X

Table 5.2   Platforms promoting high resolution melting

Instrument	Manufacturer	Real Time PCR	Format
Lightcycler® 480	Roche	Yes	96-, 384-well plates
LightCycler Nano	Roche	Yes	4 × 8-tube strips
Rotor-Gene Q HRM	Qiagen	Yes	36-, 72-, 100-tube rotors
LightScanner®	Idaho Technology	No	96-, 384-well plates
LS32™	Idaho Technology	Yes	32 capillaries
7500 Fast	Life Technology	Yes	96-well plate
7900HT Fast	Life Technology	Yes	96-, 384-well plates
StepOne™	Life Technology	Yes	48-, 96-well plate
ViiA™ 7	Life Technology	Yes	96-, 384-well plate
CFX	BioRad	Yes	96-, 384-well plate
Eco™	Illumina	Yes	48-well custom plate

techniques, one dedicated melting instrument (LightScanner) will easily manage the output of several thermacyclers. The LightCycler 480 and the 7500/7900 Fast Real-Time PCR Systems are plate-based and perform real-time PCR in addition to HRMP. For laboratories requiring multiple functions in a single platform, the LightCycler 480 and the 7500/7900 Fast Real-Time PCR Systems are good choices. The LightScanner 32 has characteristics very similar to the Roche carousel LightCycler and uses a 32-sample rotor to perform rapid PCR in capillaries. To accommodate HRMP, the LightScanner 32 has an ingot that individually interrogates each capillary, a system which is unique among HRMP platforms. Its resolution is high, evidenced by the fact that it is the only system to report genotyping of up to four SNPs in one reaction (Seipp *et al.* 2009). However, the throughput is less than similar systems such as the Rotorgene that supports up to 100 samples. The LightScanner 32 and Rotorgene may appeal to similar users where a multifunctional rapid instrument is desired.

Scanning PCR products for sequence variation

Antiquated techniques such as sequence specific conformational polymorphism, denaturing gradient gel electrophoresis, and denaturing HPLC are traditional means by which to screen for the presence of sequence variation. HRMP in essence does the same thing, but the workflow is simplified, turnaround time is reduced and detection sensitivity is more robust. Using HRMP to screen for sequence variation may be divided into two broad categories: (1) analysis of constitutional genetic changes; and (2) analysis of sequence variation within a variable background of normal DNA. The former involves analysis of autosomal or sex-linked genes to identify disease associated mutations, disease predisposition, carrier frequencies, etc. DNA obtained from tumour tissue contains a mixture of DNA derived from both cancerous cells and stromal cells. Studies of tumour DNA have been the most common application of HRMP to heterogeneous DNA specimens; however analysis of germline/somatic mosaicism and heteroplasmic variants in mitochondrial DNA are also instances where a heterogeneous DNA sample is assessed. While the means to identify sequence variants in a homogeneous vs. heterogeneous population of DNA molecules is similar, the results obtained and sensitivity to identify variation will differ, thus these scenarios are addressed individually. However, before delving into these topics, there are generic issues that one should be familiar with to better understand situations encountered in gene-scanning.

When scanning for sequence variation, each PCR product will have a unique sequence as well as a unique distribution of G:C content through the product. These characteristics combine with other parameters, such as product length, causing each amplification product to generate a unique melting profile. Characteristics unique to a melting profile include the rate of the melting transition, the number of melting transitions and the temperature(s) where melting transition(s) are observed. Furthermore, given that amplification products generate characteristic profiles, the impact of sequence variation and its impact on the shape of the profile are most often unique as well. Figure 5.7 displays four of the most commonly encountered types of melting profiles. Figure 5.7a displays a single transition profile, the type often perceived by new users to be the most common and generally desired. Figures 5.7b and 5.7c are variations on the two transition profile. It is notable that an exceedingly large range of shapes are observed in two transition profiles. The profile in the upper right is a classic example of two converging transitions; however such profiles may be even more subtle. A profile where the transitions are exceedingly close can appear as a single transition profile where the melting transition proceeds in a gradual manner. The profile labelled as 'Two transition distinctive melts' is another classic form, but once again there is immense diversity in this style of profile. In some two transition profiles with distinctive melts, following the low temperature transition, the profile will achieve a reduced baseline before the final transition. Overall, two transition profiles are quite common and will account for ~45–50% of all profiles observed in gene-scanning. Figure 5.7d is a complex melting profile having three distinct transitions. Complex profiles are not always associated with long amplification fragments, but arise from multiple segments of differing G:C content. For example, a three domain melting curve includes regions of relatively low, intermediate and high GC content. Complex profiles are more likely to be observed in regions of low G:C content. Complex profiles having more than three apparent transitions are quite rare but have been observed. Given the overall rarity of complex profiles, when one is observed, the presence of extraneous PCR products that may be contributing to the complexity of the profile should be suspected.

Scanning for DNA sequence variants in autosomal and sex-linked genes is the most common application of HRMP (Ebberink *et al.* 2010; Fadhil *et al.* 2010; Shih *et al.* 2010; Yan *et al.* 2010; Lo *et al.* 2009). Analysis of autosomal genes is straightforward; however, analysis of sex-linked genes, especially in hemizygous males, involves certain complicating issues that are addressed later (Dobrowolski *et al.* 2007b). For both research and clinical applications, assessment of all coding regions within a gene is a common practice. Some genes are compact (e.g. globin, connexin) and therefore easily assayed while others, such as the collagen genes, have large numbers of coding regions making their assessment a significant undertaking. Complete DNA sequence analysis of large complex genes is immensely labour intensive at the levels of PCR, product clean-up, sequencing and data analysis. Post-PCR melt profiling, to prospectively identify regions where sequence variation exists, streamlines this analysis and significantly reduces the materials and labour associated with PCR clean-up and DNA sequencing. Probably the most significant savings involve the time and effort spent in analysis of uninformative DNA sequence data. The 27 coding regions of the large and complex CFTR gene can be assayed with 37 PCR products (Montgomery *et al.* 2007). Selected data from analysis of the CFTR gene in 95 patient samples are shown in Figure 5.8 displaying subtractive difference plots for exons 2, 5, 7, 8, 10 and 24. In each set of samples a control specimen, whose gene is known to have a consensus wild-type sequence, is assessed to define the shape of

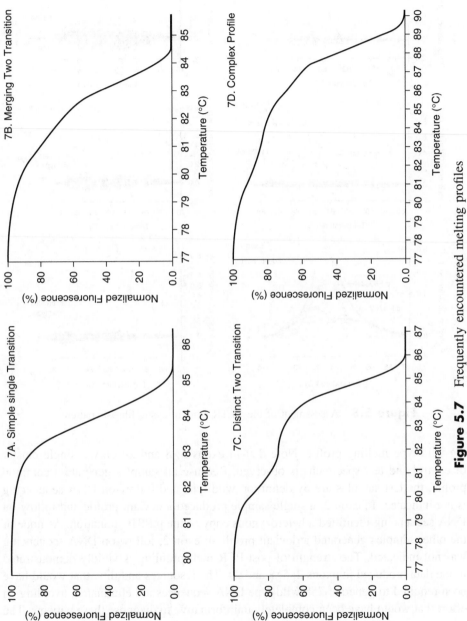

Figure 5.7 Frequently encountered melting profiles

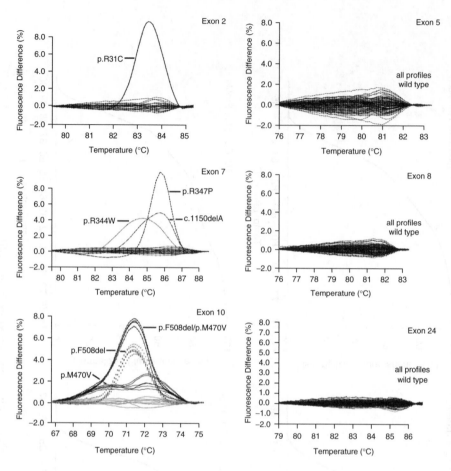

Figure 5.8   Assessment of the CFTR gene in cystic fibrosis patients

the wild-type melting profile. Note that in exons 5, 8 and 24 only a single cluster of control and test specimens is observed. Because all samples generate a common profile, the test samples are by definition wild type and follow-on DNA sequencing is not indicated. In exon 2, a single sample produced a deviant profile and follow-on DNA sequencing identified a heterozygous copy of the p.R31C mutation. As none of the other samples generated a deviant profile in exon 2, follow-on DNA sequencing was not indicated. The strength of post-PCR melt profiling is vividly demonstrated in the data produced in exons 2, 5, 8 and 24. The resources and time that would have been required to generate 758 wild-type DNA sequences are eliminated, avoiding an effort that would have been completely uninformative to diagnose these patients. The data in exon 7 also provides a classic example of what investigators will encounter during melt profiling. There are three deviant melting profiles, each displaying a distinctive shape, and follow-on DNA sequencing identified heterozygous copies of c.1150delA, p.R344W, and p.R347P. In most situations, unique mutations will create distinctive melting profiles. A further lesson is demonstrated in exon 10 that houses the common p.M470V polymorphism and the common p.F508del mutation. The profiles created by heterozygocity for p.M470V are clustered together. Also present in exon 10 are samples with the p.F508del mutation and those that are compound heterozygous for both p.M470V and p.F508del. The profiles generated by each genotype are distinctive, readily distinguished from each other and distinguishable from the wild-type cluster. However, although scanning identifies heterozygous variants, it does not distinguish between benign polymorphisms and disease-causing mutations. Another caveat is that although unique variants typically produce distinctive melting profiles, while rare, two unique variants can produce deviating melt profiles that share a common shape. Investigators and, more importantly, clinical geneticists must remain vigilant to situations where unlike variants create deviant melting profiles with an identical shape. In our own investigations of phenylalanine hydroxylase deficient phenylketonuria, the pathological mutation c.441+5G>T and the common polymorphism c.353−22C>T are situated on opposite ends of the fragment that assays PAH exon 4 (Dobrowolski *et al.* 2007a, 2009). In heterozygous form, the profiles from these two very different sequence variants create deviant melting profiles that are virtually indistinguishable. In a clinical setting, follow-on DNA sequencing is indicated for every deviant melting profile. Although experienced investigators may recognize the profile generated by a particular variant, follow-on confirmation is still necessary to characterize the variant with 100% certainty. It is unwise to become complacent concerning profile shape to identify a mutation without follow-on confirmation even though experienced investigators will be able to do so correctly >99% of the time.

Polymorphic, non-pathological variation is present in essentially every gene. Common polymorphisms can lead to frequent indications for follow-on DNA sequencing, which defeats the purpose of post-PCR melt profiling. To definitively identify a polymorphic variant such that follow-on DNA sequencing is not indicated requires the concordance of a control melting profile and a site-specific genotyping test (Dobrowolski *et al.* 2007a, 2009). The control melting profile is generated with a sample containing a heterozygous copy of the polymorphism. Should a sample produce a melting profile that is identical to the control, this suggests that the variants are alike, but it is not definitive proof and must be confirmed. The confirmatory assay is a specific genotyping test for the polymorphism. If these independent assays (scanning control and genotyping) are concordant, the polymorphism is identified and follow-on DNA sequencing is not indicated. If the melting profile does not match that of the control but the genotyping test indicates the presence of the polymorphism, a second variant is likely in addition to the polymorphism. In some but not all cases, the *cis* or *trans* nature of two variants within the same amplicon can be distinguished by melting (Tindall *et al.* 2009; Wittwer 2009). If the melting profile matches the control profile for the polymorphism and the genotyping test does not indicate the presence of the polymorphism, the deviating profile is an example of different sequence changes creating a common profile. Although this is uncommon for SNPs, it is more common for small insertions or deletions, for example p.F508del and p.I507del. As large genes and even some small genes have many common polymorphisms, the combined use of control melt profiles and genotyping tests has great utility to limit the amount of DNA sequencing required to complete the evaluation of a gene.

Thus far, all of the deviant melting profiles displayed (Figures 5.1 and 5.8) have contained heterozygous sequence variants. Heterozygous variants are by far the most frequently encountered and are easily identified by scanning. Homozygotes are not typically distinguished by traditional scanning techniques (dHPLC, DDGE, etc.) because heteroduplexes are not formed. However, HRMP is an exception in that many homozygous sequence variants are resolved. In various studies across different platforms and genes, the sensitivity to detect homozygous variants has ranged from 30% to 91% (McKinney *et al.* 2004; Dobrowolski *et al.* 2005). In small amplicons, (<200 bp), most homozygous variants can be recognized by Tm differences. However, as the amplicons get larger, the Tm differences between homozygotes get smaller and shape comparisons (after temperature overlay) become more reliable than Tm for homozygote detection, especially when multiple domains are present. Figure 5.9 shows melting profiles derived from the 12th exon of the phenylalanine hydroxylase gene where the c.1222C>T (p.R408W) mutation is readily detected by melting profiles in homozygous, heterozygous and compound heterozygous form. The large shape deviation observed in Figure 5.9 for homozygous c.1222C>T is not uncommon. However, such deviations depend on the individual homozygous variant and range from pronounced to subtle to no deviation at all. Approximately 80% of homozygous variants in autosomal genes can be identified by shape deviation of the melt profile in fragments ≤550 bp.

To identify that minority of homozygous variants that do not alter the melting profile, combining the test DNA specimen with wild-type DNA and using the combined sample in co-PCR is required. For example, when a sample is submitted with an indication of hyperphenylalaninemia and analysis of the PAH gene identifies no deviant melting profile(s), the specimen is reassessed with a co-PCR protocol (Dobrowolski *et al.* 2009). The test DNA specimen is combined 1:1 with a DNA specimen sequence characterized to possess a consensus wild-type sequence and the mixed DNA sample is amplified. Co-PCR with wild-type DNA artificially creates heterozygosity such that heteroduplex molecules are formed facilitating identification of the variant. Co-PCR facilitates the identification of all homozygous variants. Analysis of X-linked genes in hemizygous males is similar to identification of homozygous mutations in autosomal genes. When assessing X-linked genes, such as ornithine transcarbamylase in hemizygous males, co-PCR should prospectively be performed (Dobrowolski *et al.* 2007a). Figure 5.10 provides an assessment of the tenth exon of ornithine transcarbamylase where three samples from affected hemizygous males are assessed with and without co-PCR and displayed as difference plots (Dobrowolski *et al.* 2007a). In Figure 5.10a the samples are assessed without co-PCR. The specimen containing p.P336S generates a readily identified deviant profile while two other samples from hemizygous males generate mild, albeit recognizable, deviations. A sample from an

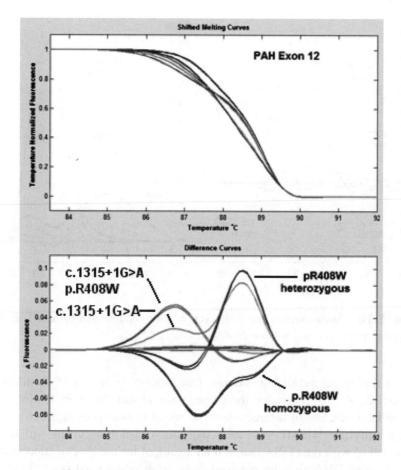

Figure 5.9 Identification of homozygous p.R408W

affected heterozygous female is shown for comparison. Figure 5.10b shows profiles from the same samples after co-PCR with wild-type DNA. Note that all of the samples from hemizygous males generate more pronounced deviant profiles. A potential pitfall in co-PCR is that the sources and isolation methods of the test DNA and the control DNA used to create heteroduplexes must be considered. Usually, standardized wild-type DNA samples are high-quality reagents prepared from liquid blood or cultured cells. However, DNA prepared from suboptimal sources (e.g. dried blood on filter paper), while fully adequate to support robust PCR, is difficult to accurately quantify by spectrophotometry as it is typically fragmented, partially single stranded and contaminated with RNA. When co-PCR is performed by mixing suboptimal DNA with high-quality reagent DNA, their amplification efficiencies may differ to a point that the high quality reagent DNA overwhelms the reaction. This is a problem as the test DNA and standard DNA would ideally be derived from a common source, but suboptimal DNA sources typically provide small volume, dilute samples that are

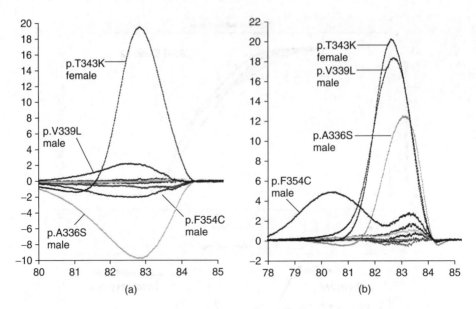

Figure 5.10 Assessment of the 10th exon of the ornithine transcarbamylase gene in hemizygous males with and without co-amplification

prone to further degradation in storage. One strategy is to test DNA dilutions of the standard DNA to determine the proper ratio so that the co-PCR protocol will facilitate the formation of heteroduplexes to reveal homozygous mutations.

The sensitivity of HRMP to identify sequence variation occurring within a large background of wild-type DNA makes it an ideal tool for the investigation of tumour tissue. Tumour DNA may be obtained from fresh or frozen tissue. However, in clinical settings, tumour DNA is most often prepared from formalin-fixed paraffin-embedded tissue recovered from blocks or slides. When preparing DNA from such tissue, a pathologist identifies an area within a tissue slice containing tumour cells, tissue from this region is scraped from the slide and DNA is prepared. This crude means of obtaining tumour-derived DNA creates an advantage for HRMP. Tumour tissue is heterogeneous, containing malignant cells, infiltrating lymphocytes, vasculature and other stromal cells. When DNA is prepared, both tumour-derived DNA and normal DNA is co-isolated. The presence of DNA from both sources assures heteroduplex molecules are created post-PCR even if there is loss of heterozygosity in the tumour. The lower limit for detection of variants in heterogeneous DNA depends on the variant and the tissue studied (Montgomery, Sanford and Wittwer 2010). Although reports have documented the detection of sequence variants in the range of a 1–5% allele fraction, the sensitivity of detection at these fractions is seldom clear (Nomoto *et al.* 2006; Bastien *et al.* 2008; Li *et al.* 2009; Milbury, Li and Makrigiorgis 2009). Factors such as fragment length, nature of the sequence variant, shape of the profile, G:C content and type of fixation are likely to influence sensitivity. In most circumstances, allele fractions ≥10% are easily detected (Krypuy *et al.* 2006; Takano *et al.* 2007; Polakova *et al.* 2008). Below an allele fraction of 10%, sensitivity decreases. An unanticipated consequence of identifying low allele fraction variants by melt profiling is that the variant eludes identification by follow-on DNA sequencing. Software-based analysis of DNA sequence generally ignores heterozygous peaks with an area <20% of the major peak. As melt profiles have identified variants at single digit allele frequencies, HRMP is generally more sensitive than DNA sequencing. Although PCR products can be cloned and individually sequenced to identify a low-level variant allele, manual inspection of the sequencing trace may provide evidence of low-level variants. However, the quality of sequencing traces varies and there are clearly difficulties associated with identification of low-level sequence variants. Assessment of the platelet derived growth factor receptor A (PDGFA) gene is used to categorize gastrointestinal stromal tumours. Figure 5.11 assays the 18th exon of the PDGFA gene using DNA obtained from gastrointestinal stromal tumour tissue that was scraped off slides. Note that two deviant melting profiles are observed. These deviant profiles result from a common sequence variant c.2525A>T (p.D842V), but the ratio of DNA derived from tumour vs. stroma is different in the two samples and thus their profiles are unalike. As tumour-derived DNA contains the somatic activating mutation and the stroma-derived DNA does not, the ratios between the two populations of DNA dictate the magnitude of the profile deviations. When assessing autosomal or X-linked genes, overlapping profiles suggest a common variant is involved while divergent profiles indicate different variants. When assessing heterogeneous DNA specimens, the magnitude of the deviation varies with the allele fraction. The shape of deviation on the subtractive difference plots may

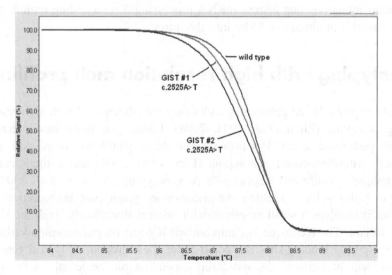

Figure 5.11 Assessing a heterogeneous DNA sample: gastrointestinal stromal tumour and PDGFRA exon 18

correlate with the specific variant; however, this does not necessarily provide surety that deviations are derived from a common variant.

The necessity to follow-up a deviant melting profile with DNA sequencing to characterize the variant has been discussed. Among the most convenient aspects of melt profiling is that the PCR product used in melt profiling may serve 'double duty' as a template for DNA sequencing. There is no indication that the dyes listed in Table 5.1 interfere with downstream DNA sequencing. Prior to sequence analysis, PCR nucleotides, primers and dye are removed before cycle sequencing is initiated. Alternative approaches include filtration and enzymatic digestion. Although filtration may be most commonly used, digestion using shrimp alkaline phosphatase/exonuclease (ExoSAP) is effective in the presence of melting dye. This ability to further use the PCR product employed for melting in follow-on DNA sequencing streamlines the process.

The examples provided in the previous section on gene-scanning involved analysis of coding regions and their flanking sequence. Primer design should assure that not only coding sequence is assessed but also critical regions of non-coding sequence to identify mutations that disrupt mRNA processing (Erali and Wittwer 2010). Upstream of all coding regions (except for the first exon), there is typically a canonical AG splice acceptor motif and a poly CT tract that is variable in length. The branch site, with a consensus motif of CTSAY, is located 20–50 bases upstream of the exon. Although not always considered in primer design, upstream primers flanking 50 bases of intron will avoid most pathological mutations that disrupt mRNA processing in this region. The intronic sequence 3′ to a coding region is less complex with the consensus splice donor site of CAG/ GTAAGT, proceeding only six bases into the intron. Primers should be placed outside of these critical intronic sequences and, to avoid most mutations that disrupt mRNA processing 3′ to a coding region, primers can be placed a minimum of 15 bp into the intron.

Genotyping with high-resolution melt profiling

A variety of probe-based genotyping techniques have been used with high-resolution melting platforms (Zhou *et al.* 2004, 2008). Given that probe-based techniques may be performed using low-resolution melting platforms, these will not be discussed. Amplicon-based genotyping (Liew *et al.* 2004) and a modification of the technique – 'calibrated amplicon-based genotyping' (Gundry *et al.* 2008) – are unique to high-resolution melting. Amplicon-based genotyping is theoretically very simple as it involves a small amplicon with primers that closely flank the site of a known variant. The technique is compromised if common polymorphic variants are within the primer binding sites and could cause allele dropout. A typical genotyping amplicon will be between 38 and 50 bp depending on the length of the primers and any short intervening sequences. PCR is performed in the presence of melting dye. Post-PCR melting generates profiles indicative of the alternative genotypes,

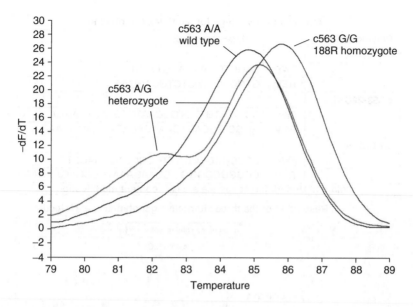

Figure 5.12 Amplicon-bases Analysis of GALT c.563A>G

usually shown as derivative plots. Figure 5.12 shows an amplicon-based genotyping assay for the common galactose-1-phosphate uridyltransferase mutation c.563A>G (p.Q188R) that is observed in patients affected with classical galactosemia. Note how the alternative homozygotes create characteristic peaks while the heterozygote creates a composite peak from the alternative fully base paired fragments and a second, lower temperature peak from heteroduplex molecules (see Figure 5.2).

The separation between alternative homozygotes is typically about 1.0 °C. Those more accustomed to probe-based assays (that typically provide 4–10 °C of separation) may view this as an uncomfortably small difference between the peaks of the alternative homozygotes; however, these assays are very robust, reliable and highly reproducible. Amplicon genotyping assays are sensitive to the same issues as gene-scanning assays. That is, PCR must be robust and specific and template DNA should be used at a standard concentration and obtained with a common isolation chemistry.

Multiplexing amplicon-based genotyping assays is generally feasible to three products and occasionally to four products (Dobrowolski et al. 2007b; Seipp, Pattison et al. 2008). The factor that limits multiplexing is the range of available temperatures, usually 70–90 °C. The small amplicons used in genotyping melt at relatively low temperature but seldom below 70 °C. To put this into perspective, 20-mer oligonucleotide PCR primers often have Tm values over 60 °C, therefore, it is not unexpected for short amplicons of 38–50 bp to melt at temperatures ≥70 °C. In a multiplex assay, individual fragments must melt at unique temperatures such that their profiles are not overlapping. To achieve this, fragments are modified with 5′ G:C tails to increase

Amplification Strategy for Multiplex Genotyping		
Polymorphism		Primer Sequence
c.696G>A		
	FWD	CAGCTGGAAGACGTTTCTCA
	REV	GTGGACTTACTCTGCAGGAA
c.353-22C>T		
	FWD	**CC** CCATGTTCTGCCAATCTGTACTCAGGA
	REV	**GGGC** GGCACTGAAACACAGAGAAGGCAAC
c.735G>A		
	FWD	**CCCCCCGCCC** TTTCCGCCTCCGACCTG
	REV	**CCGGGCCGGG** CCCAAGAAATCCCGAGAGGAAA
Bold and Underlined sequence are G:C content modification		

Resolution of the three alternative genotypes at each locus

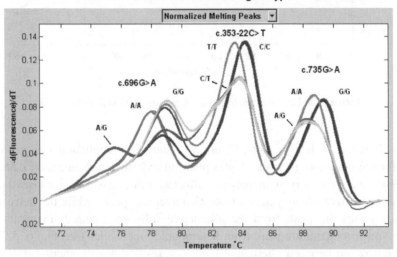

Figure 5.13 Multiplex amplicon-based genotyping

melting temperature so that individual melts are spaced over the range of available temperatures. Figure 5.13 displays a multiplex genotyping assay that concurrently assesses three common polymorphic sites in the human phenylalanine hydroxylase gene. The upper panel displays the primer sets and identifies the G:C content modification used to alter melting characteristics of the intermediate and high melting temperature fragments. Note that the G:C content modification to the intermediate melting fragment is asymmetric because the lengths of the G:C tail added to the forward and reverse primers are not equal. For the high temperature melting fragment, each primer is modified with a 10 base pair G:C tail. The lower panel shows an assessment of 16 unique DNA specimens and at each assay there are profiles with the three alternative genotypes (A/A, A/B, B/B). Compound heterozygotes are also displayed. Note that the G:C content modification to the high temperature fragment (20 G:C base pairs) represents 33% of its total length (60 bp), yet does not impede resolution of the alternative homozygotes. The G:C content modification reduces the separation of the peaks representing the alternative homozygotes in comparison to the same fragment without content modification. We have never observed a scenario where added G:C content disallowed the resolution of alternative homozygotes or rendered them 'too close to call'. While neither our experiences nor that in the published literature have required the use of G:C tails longer than 10 base pairs, it is likely that even longer modifications would preserve the resolution between alternative homozygotes.

The examples shown in Figures 5.12 and 5.13 have assayed variants where the nucleotide change modifies the G:C content of the fragment (e.g. C>T or G>A) allowing resolution of the alternative homozygotes. While the majority of SNPs do change the G:C content, what about SNPs where the G:C content of the alternative homozygotes remains unchanged such as A>T, T>A, G>C, and C>G? A variety of Tm prediction programs based upon nearest neighbour pairs conclude that alternative homozygotes that change the orientation of the base pair on the DNA strands are not resolvable by melting if flanked by symmetrical sequence. A modification of amplicon-based genotyping that employs temperature calibration enables these so-called base-pair neutral changes to be resolved. Calibration uses synthetic, self-complementary, non-extendable dsDNA molecules that are included in the amplification reaction. During the post-PCR melt, the synthetic molecule provides an invariant melting signal(s) that is used to 'calibrate' the temperatures of the overall profile. A software application aligns the signal of the calibrator(s). In doing so, the overall melting profile shifts to remove variance owing to position in the plate and even variations in chemistry such as different salt concentrations from evaporation. Figure 5.14 displays the resolution of base pair neutral alternative homozygotes for the c.1155C>G in the phenylalanine hydroxylase gene. While the separation of the alternative homozygotes is not large, it is reliable and fully reproducible. There is debate about whether one (as shown in Figure 5.14) or two calibration signals are required.

Some base pair neutral variants are reliably resolved using a single calibrator but in many cases two calibrators will be necessary. Given that the body of published data on the resolution of base pair neutral variants is small, determining the need for one or two calibrators must be dealt with on a case by case basis. Calibration may be used with any amplicon-based genotyping assay such as those shown in Figures 5.12 and 5.13 as improved clustering of specimens with common genotypes will be observed.

The actual calibrator molecule is simply a pair of self-complimentary oligonucleotides that may be readily designed and developed by individual investigators. A critical aspect of the calibrator molecule is that it must be of high purity as the presence of failure sequences and synthesis errors will cause mis-pairing, the result being a broadening of the calibration peak, which deteriorates its utility. When building a calibration molecule, gel purified oligonucleotides should be used. For the laboratory that will be using calibrated amplicon-based genotyping extensively,

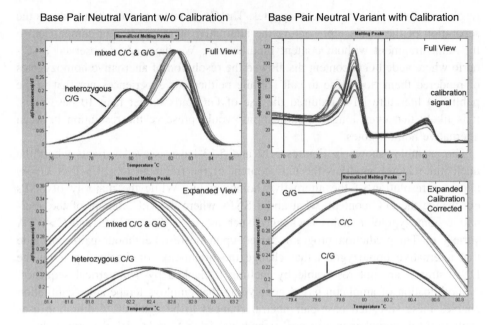

Figure 5.14   Calibrated amplicon-based genotyping

creating your own reagents is the best option. There is a commercially available mastermix that contains calibrators, which is an acceptable option. The only HRMP instrument platforms whose software will support calibrated melting are those from Idaho Technology (see Table 5.2). For those having the necessary skill set, it would be relatively easy to write their own application to use calibration on platforms that do not currently support the methodology.

Just as scanning for sequence variation may be performed with heterogeneous DNA specimens, genotyping may be performed with heterogeneous samples as well. Heteroplasmic mutations in mitochondrial DNA are associated with several disorders. In Figure 5.15, the m.8344A>G and m.3243A>G mutations that are causative in MERRF syndrome and MELAS syndrome respectively are assayed. In Figures 5.15, individual patient specimens with unique levels of heteroplasmy are resolved. The MELAS m.8342G assay resolved heteroplasmy down to 5%. The MERRF m.8344G assay resolved heteroplasmy down to10%. It is likely that both assays could resolve lower levels of heteroplasmy but appropriate patient samples were not available. These assays use calibration as described earlier for tests that resolve the base pair neutral variants. To resolve heteroplasmic profiles that differ by a few percent in the ratio of mutant to wild-type sequence requires the added resolving power provided by calibration. In situations such as genotyping specific mutations in tumour tissue, calibration is also suggested. Genotyping provides some advantage when dealing with low allele fraction variants compared to similar assessments made with gene-scanning assays. Genotyping assays query a very small region of sequence, often only

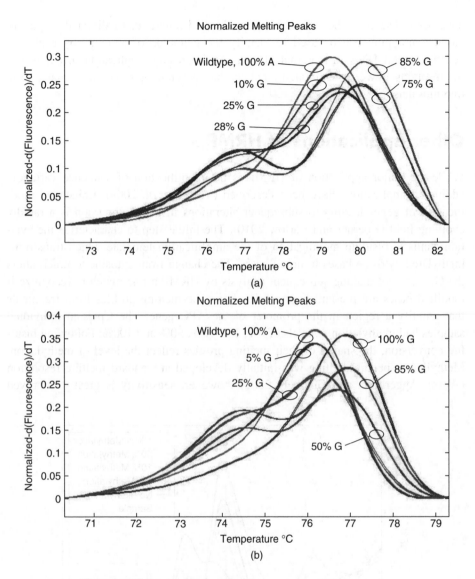

Figure 5.15 Genotyping in a heterogeneous DNA sample

a single nucleotide. If a specific genotyping assay generates a deviant profile, then the mutation of interest is present. While the shape of the profile gives some indication of the allele fraction, the current technology is at very best semi-quantitative. It is feasible to include one or more controls with known allele fractions to which the profile from the test specimen is compared to estimate an allele fraction. Similar to the observation that HRMP resolves many homozygous variations, it is quite surprising that HRMP is able to resolve sequence variation at the low allele fractions that have been observed. There have been unpublished reports that calibrated amplicon-based genotyping can be used to identify the genotype of a foetus by analysing cell-free foetal DNA in maternal plasma. An interesting application that may test the resolving power of calibrated amplicon-based genotyping would be male, germ line mosaicism.

Other applications of HRMP

While the major application of HRMP is the identification of sequence variation, additional applications have been described (Vossen *et al.* 2009). Epigenetic modification of genes leading to subsequent alterations in gene expression is a rapidly evolving field (Koerner and Barlow 2010). The initial step to characterize methylation status is bisulfite modification of the amplification template which transforms methylated cytosine bases to uracil bases. The change from cytosine to uracil alters the G:C content making subsequent analysis by HRMP a fine match to recognize if modified bases are present. Figure 5.16 shows the melting profiles from fragments that amplify a region in the promoter of the APC gene. The CpGs in individual samples had methylation at levels of 0%, 1%, 10%, 50% and 100%. Following bisulfite conversion, the shapes of their melting profiles reflect the level of methylation. Methylation-specific melting was initially developed at standard melting resolution (Worm, Aggerholm and Guldberg 2001); however, sensitivity is greatly improved

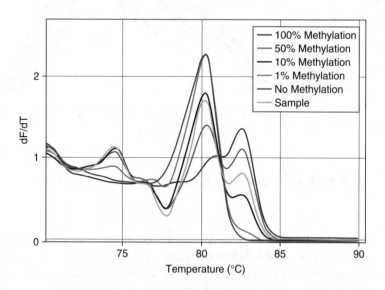

Figure 5.16 Methylation-sensitive HRMP

by HRMP and it is now a popular option for analysis of epigenetic changes in tumour tissue (Yu *et al.* 2010; Zhang *et al.* 2010; Wojdacz and Dobrovic 2007; Kristensen and Hansen 2009). Given the realization of the critical nature played by epigenetic modification of gene expression in embryopathies, type II diabetes and endocrine disorders, it is likely that methylation-specific HRMP will be applied to these disorders.

A truly unique application of HRMP was reported by Pepers *et al.* (2009) where melting was used to characterize clones isolated from a phage display library. Clones were selected against an epitope of the Huntington protein. To examine the diversity of the clones, the inserted sequence was amplified and melt profiling used to discriminate clones that contained a common insert from those containing unique inserts. This 'out of the box' application of HRMP credits the creativity of the authors and suggests future applications of HRMP.

Final notes

HRMP has established itself as a powerful way to genotype and screen for sequence variation. Given that HRMP has been widely available for less than 5 years suggests that its true potential has not yet been realized. Many clinical laboratories have adopted a 'just sequence it all' attitude toward full gene analysis which has led to exorbitant pricing to assess genes of even modest complexity. HRMP has immense potential for clinical applications in full gene analysis that will significantly reduce material costs, data analysis and turnaround time. Full gene analysis, which was once the purview of research laboratories, is now a common part of the diagnostic regimen and in these days of scarce health care resources, the time and cost savings that HRMP offers is timely indeed. With the enthusiastic adoption of gene-scanning, it is puzzling that investigators have been far slower to employ amplicon-based genotyping. Given the simplicity of amplicon-based genotyping, its utility for both clinical and basic science applications are apparent. Screening for sequence variation was clearly the 'sexy' application of HRMP and genotyping followed with significantly less fanfare. Broader application of amplicon-based genotyping is probably yet to come.

The most intriguing aspects of HRMP will be realized in specialized applications of the technology. Gene imprinting is an emerging field and the explosion of new data would suggest this mechanism of gene modification will be involved in many biological systems. The relative ease by which HRMP assesses methylation make it an ideal method to identify imprinted regions that are heritable (e.g. Prader-Willi/Angelman Syndromes) or induced by exposure to environmental factors (nutrition, illicit drugs, etc.). The work of Pepers *et al.* (2009) begs one to question if HRMP is destined for routine laboratory procedures. PCR using primers located in vector sequence that cross a cloning site could be used to determine if an insert has been successfully cloned. The melt profile of a product from an empty vector compared to that containing an insert would be trivial to discern. It is likely that HRMP will find applications in routine procedures that will save investigators an immense amount of effort.

Some have predicted it is imminent that so-called next generation sequencing technologies will replace PCR and traditional DNA sequencing. The authors do not see this happening in the short or mid-term. Analysis of small regions of the genome (gene-centric, exon-centric) will remain critical to both research and clinical diagnostics for years to come and HRMP has only begun its contribution to making these assessments more effective while at the same time providing savings in both resources and time.

References

Bastien, R., Lewis, T.B., Hawkes, J.E. *et al.* (2008) High-throughput amplicon scanning of the TP53 gene in breast cancer using high-resolution fluorescent melting curve analyses and automatic mutation calling. *Hum Mutat* **29**: 757–764.

Dobrowolski, S.F., Borski, K., Ellingson, C.C. *et al.* (2009) A limited spectrum of phenylalanine hydroxylase mutations is observed in phenylketonuria patients in western Poland and implications for treatment with 6R tetrahydrobiopterin. *J Hum Genet* **54**: 335–339.

Dobrowolski, S. F., Ellingson, C., Coyne, T. *et al.* (2007a) Mutations in the phenylalanine hydroxylase gene identified in 95 patients with phenylketonuria using novel systems of mutation scanning and specific genotyping based upon thermal melt profiles. *Mol Genet Metab* **91**: 218–227.

Dobrowolski, S.F., Ellingson, C.E., Caldovic, L. *et al.* (2007b) Streamlined assessment of gene variants by high-resolution melt profiling utilizing the ornithine transcarbamylase gene as a model system. *Hum Mutat* **28**: 1133–1140.

Dobrowolski, S.F., McKinney, J.T., Amat di San Filippo, C. *et al.* (2005) Validation of dye-binding/high-resolution thermal denaturation for the identification of mutations in the SLC22A5 gene. *Hum Mutat* **25**: 306–313.

Dujols, V. E., Kusukawa, N., Dobrowolski, S.F. *et al.* (2006) High-resolution melting analysis for scanning and genotyping. In *Real-Time PCR* (ed. M.T. Dorak), Garland Science, New York, pp. 157–171.

Ebberink, M.S., Kofster, J., Wanders, R.J. *et al.* (2010) Spectrum of PEX6 mutations in Zellweger syndrome spectrum patients. *Hum Mutat* **31**: E1058–1070.

Erali, M., Voelkerding, K.V. and Wittwer, C. (2008) High-resolution melting applications for clinical laboratory medicine. *Exp Mol Pathol* **85**: 50–58.

Erali, M. and Wittwer, C.T. (2010) High-resolution melting analysis for gene-scanning. *Methods* **50**: 250–261.

Fadhil, W., Ibrahem, S., Seth, R. and Ilyas, M. (2010) Quick-multiplex-consensus (QMC)-PCR followed by high-resolution melting: a simple and robust method for mutation detection in formalin-fixed paraffin-embedded tissue. *J Clin Pathol* **63**: 134–140.

Farrar, J.S., Reed, G.H. and Wittwer, C.T. (2010) High-resolution melting curve analysis for molecular diagnostics. In *Molecular Diagnostics*, 2nd edn (eds G.P. Patrinos and W. Ansorge), Elsevier, London, pp. 229–245.

Gundry, C.N., Dobrowolski, S.F., Martin, Y.R. *et al.* (2008) Base-pair neutral homozygotes can be discriminated by calibrated high-resolution melting of small amplicons. *Nucleic Acids Res* **36**: 3401–3408.

Koerner, M.V. and Barlow, D.P. (2010) Genomic imprinting-an epigenetic gene-regulatory model. *Curr Opin Genet Dev* **20**: 164–170.

Kristensen, L.S. and Hansen, L.L. (2009) PCR-based methods for detecting single-locus DNA methylation biomarkers in cancer diagnostics, prognostics, and response to treatment. *Clin Chem* **55**: 1471–1483.

Krypuy, M., Newnham, G.M., Thomas, D.M. *et al.* (2006) High-resolution melting analysis for the rapid and sensitive detection of mutations in clinical samples: KRAS codon 12 and 13 mutations in non-small cell lung cancer. *BMC Cancer* **6**: 295.

Lay, M.J. and Wittwer, C.T. (1997) Real-time fluorescence genotyping of factor V Leiden during rapid-cycle PCR. *Clin Chem* **43**: 2262–2267.

Li, J., Milbury, C.A., Li, C. and Makrigiorgis, G.M. (2009) Two-round coamplification at lower denaturation temperature-PCR (COLD-PCR)-based Sanger sequencing identifies a novel spectrum of low-level mutations in lung adenocarcinoma. *Hum Mutat* **30**: 1583–1590.

Liew, M., Pryor, R., Palais, R. *et al.* (2004) Genotyping of single-nucleotide polymorphisms by high-resolution melting of small amplicons. *Clin Chem* **50**: 1156–1164.

Lo, F.S., Luo, J.D., Lee, Y.J. *et al.* (2009) High-resolution melting analysis for mutation detection for PTPN11 gene: applications of this method for diagnosis of Noonan syndrome. *Clin Chim Acta* **409**: 75–77.

Lyon, E. and Wittwer, C.T. (2009) LightCycler technology in molecular diagnostics. *J Mol Diagn* **11**: 93–101.

McKinney, J.T., Longo, N., Hahn, S.H. *et al.* (2004) Rapid, comprehensive screening of the human medium chain acyl-CoA dehydrogenase gene. *Mol Genet Metab* **82**: 112–120.

McKinney, J.T., Nay, L.M., De Koeyer *et al.* (2010) Mutation scanning and genotyping in plants by high resolution DNA melting. In *The Handbook of Plant Mutation Screening: Mining of Natural and Induced Alleles* (eds K. Meksem and G. Kahl. Weinheim), Wiley-VCH, pp. 149–165.

Milbury, C.A., Li, J. and Makrigiorgis, G.M. (2009) COLD-PCR-enhanced high-resolution melting enables rapid and selective identification of low-level unknown mutations. *Clin Chem* **55**: 2130–43.

Mitchell, S., Ellingson, C., Coyne T. *et al.* (2009) Genetic variation in the urea cycle: a model resource for investigating key candidate genes for common diseases. *Hum Mutat* **30**: 56–60.

Montgomery, J., Wittwer, C.T., Kent, J.O. and Zhou, L. (2007) Scanning the cystic fibrosis transmembrane conductance regulator gene using high-resolution DNA melting analysis. *Clin Chem* **53**: 1891–1898.

Montgomery, J.L., Sanford, L.N. and Wittwer, C.T. (2010) High-resolution DNA melting analysis in clinical research and diagnostics. *Expert Rev Mol Diagn* **10**: 219–240.

Nomoto, K., Tsuta, K., Takano, T. *et al.* (2006) Detection of EGFR mutations in archived cytologic specimens of non-small cell lung cancer using high-resolution melting analysis. *Am J Clin Pathol* **126**: 608–615.

Pepers, B.A., Schut, M.H., Vossen, R.H. *et al.* (2009) Cost-effective HRMA pre-sequence typing of clone libraries; application to phage display selection. *BMC Biotechnol* **9**: 50.

Polakova, K.M., Lopotova, T., Klamova, H. and Moravcova, J. (2008) High-resolution melt curve analysis: initial screening for mutations in BCR-ABL kinase domain. *Leuk Res* **32**: 1236–1243.

Reed, G.H., Kent, J.O. and Wittwer, C.T. (2007) High-resolution DNA melting analysis for simple and efficient molecular diagnostics. *Pharmacogenomics* **8**: 597–608.

Reed, G.H. and Wittwer, C.T. (2004) Sensitivity and specificity of single-nucleotide polymorphism scanning by high-resolution melting analysis. *Clin Chem* **50**: 1748–1754.

Ririe, K.M., Rasmussen, R.P. and Wittwer, C.T. (1997) Product differentiation by analysis of DNA melting curves during the polymerase chain reaction. *Anal Biochem* **245**: 154–160.

Seipp, M T., Durtschi, J.D., Liew, M.A. *et al.* (2007) Unlabeled oligonucleotides as internal temperature controls for genotyping by amplicon melting. *J Mol Diagn* **9**: 284–249.

Seipp, M.T., Durtschi, J.D., Voelkerding, K.V. and Wittwer, C.T. (2009) Multiplex amplicon genotyping by high-resolution melting. *J Biomol Tech* **20**: 160–164.

Seipp, M.T., Pattison, D., Durtschi, J.D. *et al.* (2008) Quadruplex genotyping of F5, F2, and MTHFR variants in a single closed tube by high-resolution amplicon melting. *Clin Chem* **54**: 108–115.

Shih, H.C., Er, T.K., Chang, Y.S. *et al.* (2010) Development of a high-resolution melting method for the detection of hemoglobin alpha variants. *Clin Biochem* **43**: 671–676.

Takano, T., Ohe, Y., Tsuta, K. *et al.* (2007) Epidermal growth factor receptor mutation detection using high-resolution melting analysis predicts outcomes in patients with advanced non small cell lung cancer treated with gefitinib. *Clin Cancer Res* **13**: 5385–5390.

Tindall, E.A., Petersen, D.C., Woodbridge, P. *et al.* (2009) Assessing high-resolution melt curve analysis for accurate detection of gene variants in complex DNA fragments. *Hum Mutat* **30**: 876–883.

Vossen, R.H., Aten, E., Roos, A. and Dunnen, J.T. (2009) High-resolution melting analysis (HRMA): more than just sequence variant screening. *Hum Mutat* **30**: 860–866.

Wittwer, C.T. (2009) High-resolution DNA melting analysis: advancements and limitations. *Hum Mutat* **30**: 857–9.

Wittwer, C.T., Herrmann, M.G., Moss, A.A. and Rasmussen, R.P. (1997) Continuous fluorescence monitoring of rapid cycle DNA amplification. *Biotechniques* **22**: 130–131, 134–138.

Wittwer, C.T., Reed, G.H., Gundry, C.M. *et al.* (2003) High-resolution genotyping by amplicon melting analysis using LCGreen. *Clin Chem* **49** (6 Pt 1): 853–860.

Wojdacz, T.K. and Dobrovic, A. (2007) Methylation-sensitive high-resolution melting (MS-HRM): a new approach for sensitive and high-throughput assessment of methylation. *Nucleic Acids Res* **35**: e41.

Worm, J., Aggerholm, A. and Guldberg, P. (2001) In-tube DNA methylation profiling by fluorescence melting curve analysis. *Clin Chem* **47**: 1183–1189.

Yan, J.B., Xu, H.P., Xiong, C. *et al.* (2010) Rapid and reliable detection of glucose-6-phosphate dehydrogenase gene mutations in Han Chinese using high-resolution melting analysis. *J Mol Diagn* **12**: 305–311.

Yu, B., Yang, H., Zhang, C. *et al.* (2010) High-resolution melting analysis of PCDH10 methylation levels in gastric, colorectal and pancreatic cancers. *Neoplasma* **57**: 247–252.

Zhang, W., Li, T., Shao, Y. *et al.* (2010) Semi-quantitative detection of GADD45-gamma methylation levels in gastric, colorectal and pancreatic cancers using methylation-sensitive high-resolution melting analysis. *J Cancer Res Clin Oncol* **136**: 1267–1273.

Zhou, L., Errigo, R.J., Lu, H. *et al.* (2008) Snapback primer genotyping with saturating DNA dye and melting analysis. *Clin Chem* **54**: 1648–1656.

Zhou, L., Myers, A.N., Vandersteen, J.G. *et al.* (2004) Closed-tube genotyping with unlabeled oligonucleotide probes and a saturating DNA dye. *Clin Chem* **50**: 1328–1335.

6
Massively Parallel Sequencing

Tracy Tucker, Marco Marra and Jan M. Friedman

Sanger sequencing

DNA sequencing was first described by two separate groups in 1977, Maxam and Gilbert (1977) and Sanger, Nicklen and Coulson (1977), each using different methods. Sanger sequencing became the more commonly used technique and is discussed below. Two different methods of DNA preparation can be performed for Sanger sequencing, depending on the application. If there is a known region of interest in the genome, it can be polymerase chain reaction (PCR) amplified with flanking primers. For *de novo* sequencing, the DNA is fragmented and cloned into a high-copy number plasmid that is transformed into *Escherichia coli,* and the DNA fragments are purified from bacterial colonies.

In either case, the DNA is then sequenced using a 'cycle sequencing' method that involves several rounds of template denaturation, primer annealing and extension. Each round of extension is randomly terminated by the incorporation of fluorescently labelled dideoxynucleotides (ddNTPs). With different mixtures of DNA fragments terminated with different ddNTPs, the sequence is determined by size separation using high-resolution capillary electrophoresis. Laser excitation of fluorescent labels provides the readout of each base. Software then translates the readout into a linear sequence.

Massively parallel sequencing

Evolving massively parallel sequencing (MPS) approaches have addressed many of the throughput-limiting obstacles inherent in the Sanger sequencing paradigm.

Molecular Analysis and Genome Discovery, Second Edition. Edited by Ralph Rapley and Stuart Harbron.
© 2012 John Wiley & Sons, Ltd. Published 2012 by John Wiley & Sons, Ltd.

MPS has eliminated the hands-on DNA preparation required for Sanger sequencing by removing the reliance on cloning and amplification in bacteria or sequence-specific PCR amplification (see below). MPS protocols often use tiny reaction volumes, permitting reactions to occur in shallow wells or flow cells, facilitating automated reagent exchange. MPS is also able to sequence millions of templates simultaneously in parallel, compared to Sanger sequencers that are limited to analysis of 96 samples at a time.

Although improvements have been consistently made, read lengths are generally shorter for MPS systems than for Sanger sequencing. The development of paired-end reads (reads from either end of both strands of a double-stranded DNA molecule) and mate-paired reads (tandem reads of one of the strands of a DNA molecule) (see below) have greatly increased the utility of MPS, particularly in the context of complex genomes.

There are several different types of commercially available massively parallel sequencers, each using different approaches, which are discussed below. In addition, there are several other technologies that are in the proof-of-principle stage but are not yet available commercially.

Commercially available massively parallel sequencers

454 genome sequencer

For sequencing using the 454 instrument (www.454.com), the DNA is sheared to produce fragments of 1–1.2 Kb, which are each ligated to two adapters, one of which is biotinylated. The fragments are attached to agarose beads via oligomers complementary to the adaptor sequences, with each agarose bead binding a single DNA fragment. PCR amplification of the fragment is performed in emulsion droplets (Dressman *et al.* 2003). After amplification, each emulsion droplet contains millions of copies of the original DNA fragment immobilized to the bead surface by primers complementary to the biotinylated adapter.

Each capture bead is put into an individual well on a fibre optic slide, which is placed in the sequencer. Pyrosequencing is then performed by attaching the necessary enzymes to the beads in each well. Reagents for sequencing, each containing a single nucleotide, are washed over the plates in a sequential manner. The sequencing process uses an enzymatic cascade to generate light from inorganic phosphate molecules released by the incorporation of nucleotides as the polymerase replicates the template DNA (Figure 6.1) (Ronaghi *et al.* 1996). The light emitted is directly proportional to the amount of a particular nucleotide incorporated. After each nucleotide is added, all unincorporated nucleotides are washed away and fresh reagents are added with a different nucleotide. The pattern and magnitude of light emission is interpreted to provide the sequence of each DNA fragment.

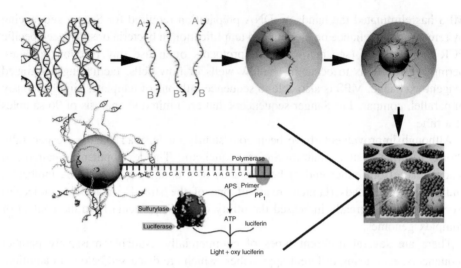

Figure 6.1 GS-FLX 454 sequencer workflow. DNA is fragmented and adaptors, one of which is biotinylated, are ligated to each end. Fragments are coupled to agarose beads by oligonucleotides complementary to the adaptor sequences and contained within an emulsion droplet for amplification. The beads are put into an individual well on a fibre optic slide and placed in the sequencer. Nucleotides are sequentially added, and, when incorporation into the growing complementary strand occurs, light is generated and recorded. (Reproduced by kind permission of 454 Sequencing © 2010 Roche Diagnostics)

The GS-FLX 454 sequencer produces an average read length of 700 bp and generates ~700 Mb of sequence data per run with a raw base accuracy of >99% (www.454.com).

Applied Biosystems SOLiD sequencer

The SOLiD sequencer technology involves fragmentation of the DNA and ligation of adaptor oligonucleotides to the ends. The DNA fragments are coupled to magnetic beads with bound oligonucleotides that are complementary to the adaptors, and clonal amplification is performed on the bound DNA in an emulsion. After amplification, the oil emulsion is broken and beads containing amplification products are attached to a glass slide that is placed into a fluid cassette within the sequencer.

Sequencing begins with the addition of a universal primer that is complementary to the adaptor oligonucleotides. Each cycle of sequencing involves the hybridization of fluorescently labelled degenerate octomers to the DNA fragment sequence adjacent to the 3′ end of the universal primer, followed by the addition of DNA ligase to seal the phosphate backbone (Shendure *et al.* 2005). The octomer mixture is structured so that each position corresponds to one of four fluorescent labels. Incorporation of each octomer is followed by imaging, collecting information on the same base position across all DNA fragments. Then the octomer is cleaved between bases 5 and 6, removing the fluorescent tag, and a new octomer is added. After several rounds of ligation, every fifth base is sequenced. After several cycles, the extended primer is removed and the process is repeated with a universal primer that is offset by one base from the adaptor-fragment position (Figure 6.2). Offsetting the universal primer in five sets of cycles permits the entire fragment to be sequenced and provides an error-correction scheme because each base position is queried twice (once as a first base and again as the second base in the next (or preceding) set of cycles (Shendure and Ji 2008).

The SOLiD sequencer has read lengths of 50 bp and produces >60 Gb of sequence per run. The raw base accuracy of the SOLiD System is 99.94% (www. solid.appliedbiosystems.com).

Complete Genomics

Complete Genomic sequencers (www.completegenomics.com) are not commercially available, but the company is currently offering sequencing as a service. These sequencers are able to generate 70 bp reads with an accuracy of 99.9% (Drmanac *et al.* 2010). In this approach, the DNA is fragmented and adaptors are ligated to the ends of the fragment. Then a Type II restriction enzyme is used to digest the fragment and another adaptor is ligated to the fragment. The process of adaptor ligation and restriction enzyme digestion is repeated until there are 4 adaptors per DNA fragment. DNA amplification is then performed to generate head-to-tail con-catemers consisting of more than 200 copies of a circular template formed into a 'DNA nanoball'. The nanoball is added to an array that can hold 1 billion individual DNA nanoballs. Sequencing is performed by a combinatorial probe-anchor ligation method in which the probes consist of an anchor sequence complementary to the adaptors and nine bases. Sequencing is done with a pool of probes, each labelled with four dyes, one for each base, to read the base adjacent to each adaptor. This method is able to read 10 bases from each end of the adaptors by washing away each bound probe and then adding another pool of probes to read the next base as the cycle continues (Figure 6.3) (Drmanac *et al.* 2010).

Heliscope Helicos sequencer

Unlike the other sequencers described, above the Helicos sequencer uses a highly sensitive fluorescence detection system to interrogate each nucleotide directly as it is synthesized. The target does not require prior amplification because sequencing is performed on single DNA molecules (www.helicosbio.com).

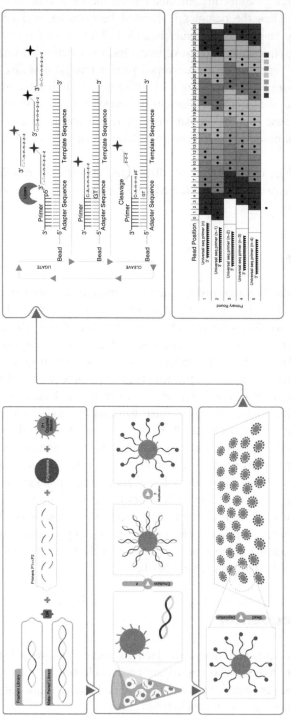

Figure 6.2 Applied Biosystems SOLiD sequencer workflow. DNA is fragmented and oligonucleotide adaptors are ligated to each end. These adaptors are used to hybridize the DNA fragments to complementary oligonucleotides attached to magnetic beads. The beads are contained within an oil emulsion, where amplification is performed. The emulsion is then broken, and the beads are attached to a glass surface and placed within the sequencer. A universal sequencing primer, complementary to the adaptor sequence, is added, followed by subsequent ligation cycles with fluorescently labelled degenerate octomers. After each cycle, the glass surface is imaged. The octomer is then cleaved between bases 5 and 6, removing the fluorescent tag, and a new octomer is added. After several rounds of sequencing, the extended universal primer is removed, and a new universal primer is added that is offset by one base. (Reproduced by kind permissions of Applied Biosystems Inc.)

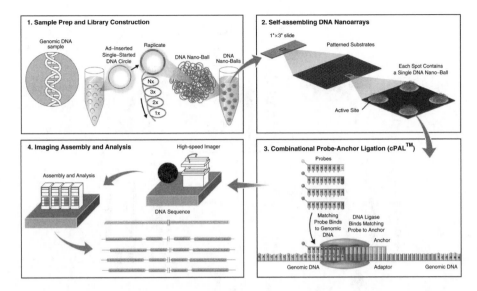

Figure 6.3   Complete Genomics workflow. DNA is fragmented, and adaptors are ligated to each end. A restriction enzyme digests the DNA, another adaptor is added, and the process is repeated until there are four adaptors. The DNA is then amplified repeatedly to form a DNA nanoball. The nanoball is added to an array, with one ball per spot on the array, and sequencing is performed with a pool of probes, each labelled with four dyes, one for each base. A separate pool of probes is used for each read. (Reproduced by kind permission of Complete Genomics Inc.)

The DNA is prepared for analysis by fragmentation and attachment of poly-A tails. The fragments are captured by poly-T oligomers that are tethered to an array. At each sequencing cycle, polymerase and a single fluorescently labelled nucleotide are added. The array is imaged and fragments that incorporated a nucleotide are recorded. The fluorescent tag is then removed from the nucleotide and the cycle is repeated with each of the other three fluorescently labelled nucleotides, one at a time (Figure 6.4).

The Helicos sequencer produces read length of 25–55 bp and generates 21–35 Gb of sequence per run with a raw base accuracy greater than 95% (www.helicosbio.com).

Illumina Genome Analyzer

For sequencing on the Illumina Genome Analyzer (www.illumina.com), DNA is fragmented, denatured and ligated to sequencing adaptors. The fragments are introduced into a glass flow cell in a microfluidics chamber. The interior surfaces of the flow cell have covalently attached oligonucleotides complementary to the specific adaptors that have been ligated onto both ends of the DNA fragments. The adaptors

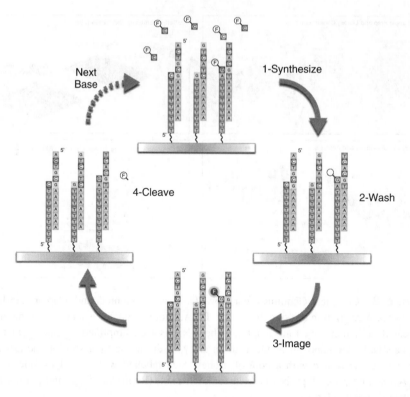

Figure 6.4 Heliscope Helicos sequencer workflow. Fragments are captured by poly-T oligomers tethered to an array. At each sequencing cycle, polymerase and a single fluorescently labelled nucleotide are added, and the array is imaged. The fluorescent tag is then removed, and the cycle is repeated. (Reproduced by kind permission of Helicos BioSciences © 2010. Helicos BioSciences Corporation. All Rights Reserved)

hybridize to the oligonucleotides on the flow cell surface, forming a 'bridge'. A new strand of DNA is primed from the 3′ end of the surface oligonucleotides and copied until it reaches the other end of the 'bridge' at the 5′surface oligonucleotide. The original strand is then removed by denaturation. The process is repeated as the adaptor sequence at the 3′ end of each copied strand is annealed to new surface-bound oligonucleotides, forming a bridge and generating a new site for synthesis of a second strand. After multiple rounds of amplification, millions of discrete clusters of identical strands are formed on the flow cell surface.

The DNA in each cluster is linearized by cleaving one of the adaptor sequences, and sequencing primers are attached. The flow cell is placed into a fluidics cassette in the sequencer and supplied with polymerase and four differentially labelled fluorescent nucleotides that have their 3′OH chemically inactivated to ensure that only a single base is incorporated per cycle. Each base incorporation is followed by an imaging step to identify the incorporated nucleotide at each cluster (Figure 6.5).

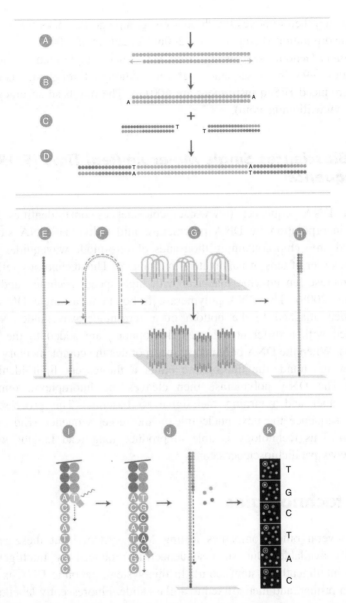

Figure 6.5 Illumina Genome Analyzer workflow. Sequencing libraries are generated by fragmenting genomic DNA, denaturation and adaptor ligation. Fragments are added to the flow cell chamber, which is coated with oligonucleotides complementary to the adaptors. Hybridization forms a 'bridge', and amplification is primed from the 3′ end and continues until it reaches the 5′ surface. After several rounds of amplification, discrete clusters of a single sequence are formed. The clusters are denatured, and sequencing primers, polymerase and fluorescently labelled nucleotides with their 3′OH chemically inactivated are added one at time. After each base is incorporated, the surface is imaged, the 3′OH is removed, and the process is repeated with a different labelled nucleotide. (Reproduced by kind permission of Illumina Inc.)

A chemical step then removes the fluorescent group and deblocks the 3′ end for the next base incorporation (Fedurco *et al.* 2006; Turcatti *et al.* 2008).

The Illumina Genome sequencer currently produces single reads of up to 150 bp and generates ∼95 Gb of sequence per run, although throughput claims for the recently introduced HiSeq instrument are 600 Gb. The raw base accuracy is greater than 98% (www.illumina.com).

Pacific Biosciences Single Molecule Real Time (SMRT) DNA sequencer

The SMRT DNA sequencer (www.pacificbiosciences.com) identifies fluorescent nucleotide incorporation by DNA polymerase into individual DNA strands. This is performed on a chip containing thousands of zero-mode waveguides, which are pores on the order of tens of nanometers in diameter. This technology relies on Φ29 DNA polymerase, an enzyme that acts with high speed, accuracy and efficiency (Harris *et al.* 2008). The DNA polymerase is bound to a single DNA molecule that has been attached to the bottom of a zero-mode waveguide. Nucleotides, each labelled with a different coloured fluorophore, are added to the zero-mode wavelengths. When the DNA polymerase incorporates the complementary nucleotide in the growing strand, the fluorophore emits a fluorescent light identifying the nucleotide. The DNA polymerase then cleaves the fluorophore, removing the fluorescent label and returning the signal to baseline. The processes is then repeated to sequence the next nucleotide in the target sequence (Eid *et al.* 2009) (Figure 6.6). This technology is able to produce long read lengths with 99.5% accuracy (www.pacificbiosciences.com).

Future technologies

There are several other companies testing MPS systems, but these are not yet commercially available. One such sequencer is produced by Intelligent Biosystems (www.intelligentbiosystems.com). In this system, genomic DNA is amplified, attached to a primer, and then affixed to a glass slide. Fluorescently labelled dyes are added and incorporated. After each base is added, the array is imaged at each spot. The dye is then cleaved and the process is repeated. A similar system by Visigen Biotechnology uses DNA polymerase immobilized on to a glass slide and modified with a fluorescent donor molecule. The DNA polymerase incorporates colour-coded nucleotides with an acceptor fluorescent moiety that allows real-time detection of a growing DNA strand.

Another method involves the use of nanopores, a technique used by Oxford Nanopore. When a voltage is applied across a nanopore, it creates an electrical current. When a DNA fragment is pulled electrophoretically through the nanopore,

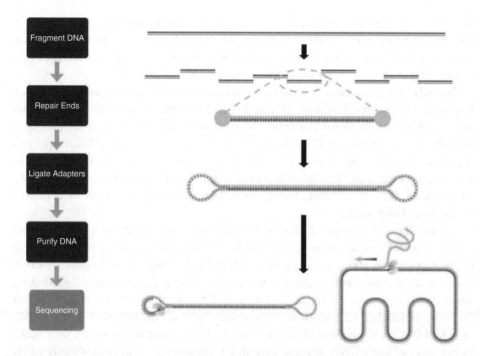

Figure 6.6 Pacific Biosciences Single Molecule Real Time (SMRT) DNA sequencer workflow. Sequencing by synthesis is performed on a chip with thousands of zero-mode waveguides (ZMW), each of which contains DNA polymerase bound to a single DNA molecule. Differentially labelled nucleotides are added to the ZMW, and when the DNA polymerase incorporates the complementary nucleotide, a fluorophore emits coloured light identifying the nucleotide. Incorporation of the nucleotide into the growing DNA strand results in cleavage of the fluorescent label and consequent loss of the signal. The process is then repeated with the next nucleotide. (Reproduced by kind permission of Pacific Biosciences Inc.)

each base creates a unique change in the magnitude of the electrical current (Branton *et al*. 2008). It has been suggested that this technology could sequence as many as 25 bases per second without interruption (Clarke *et al*. 2009).

Paired-end or mate-paired reads

Each technology is slightly different and uses different terminology. For the 454 Genome Sequencer, the terms mate-paired and paired-end are synonymous. To generate the paired-end library, the DNA is fragmented to 2.5 kb and blunt ends are generated. Biotinylated adaptors are then added that contain both an *Eco*RI and two *Mme*I restriction enzyme sites. The fragment is digested with *Eco*RI to generate sticky ends to allow the fragment to be circularized. The circularized DNA is then cleaved with a *Mme*I restriction enzyme which cleaves the DNA 20 nucleotides away from the restriction site. This results in a fragment with the adaptor in the middle of the strand and the target regions to be sequenced on the ends. Sequencing adaptors are then added to the ends of each fragment which generates sequence reads of both strands of a double-stranded DNA molecule (paired-end reads).

The ABI SOLiD sequencer only uses mate-paired reads. In this protocol, DNA is fragmented to 500 bp. Adaptors with an *Eco*R151 restriction enzyme site are added to the ends of the fragment to permit circularization. The fragment is then digested with *Eco*R151 which cuts 27 bp from the restriction enzyme site. This results in a fragment with the adaptor in the middle of the strand and the target regions to be sequenced on the ends. The DNA is then denatured and sequencing adaptor primers are added to the ends of each strand, which generates sequence reads of both ends of the same DNA strand.

For the Illumina Genome Analyser, "paired-end reads" refers to sequencing both strands of a DNA molecule from opposite directions. It is accomplished by following the protocol given above and generates 200 bp reads. "Mate-paired reads" refers to longer reads but does require circularization. The DNA is fragmented to 2–5 kb long. Blunt ends are generated using biotinylated dNTPs. The fragment is circularized then digested, generating fragments with the biotinylated dNTP fragment in the middle and the DNA to be sequenced on the outside. Sequencing adaptors are then added to the ends of both DNA strands, permitting sequencing from both strands of the double-stranded DNA molecule (paired-end reads).

Target-enrichment strategies for MPS

There are times when sequencing entire genomes is not the goal or is not cost effective for a project, and it may be of more benefit to sequence only exons or other regions of interest. There are a number of methods used to enrich for desired target DNA sequences, but these capture systems can be divided into two types of enrichment: microarray based and solution based. Microarray-based methods can enrich for large genomic fragments but require a large amount of input DNA compared to solution-based methods. Solution-based methods work better than microarray capture if the purpose of the experiment is to enrich for smaller fragments (Mamanova *et al.* 2010). Sequence capture systems are commercially available from Agilent Technologies, Inc. (www.chem.agilent.com), Roche NimbleGen, Inc. (www.nimblegen.com) and Raindance Technologies, Inc. (www.raindancetechnologies.com).

Microarray-based enrichment methods

Currently, Agilent Technologies produces an array containing 244 000 or 1 million 60-mer oligo probes, and Roche NimbleGen produces arrays containing 385 000 or 2.1 million 60-mer oligo probes to be used for target selection. Both approaches use similar methods; the DNA is fragmented and overhangs removed, and sequencing adaptors are ligated to the ends of each fragment. The DNA is then hybridized to the array and any unbound DNA is washed away. The hybridized DNA can now be released, and the adaptors used to PCR amplify the fragments for sequencing.

Solution-based enrichment methods

Kits for solution-based enrichment of most human exons are commercially available from Agilent Technologies ('SureSelect') and Roche NimbleGen ('SeqCap EZ'). Both systems use a similar method, although the genomic sequence selection probes are different. The Agilent Technologies system has 55 000 unique 120-mer RNA probes, while the Roche NimbleGen system contains 2.1 million 60-mer DNA probes. Both systems involve DNA fragmentation and blunt end generation, followed by ligation of sequencing adaptors to the ends of each fragment. The fragmented DNA is hybridized to biotinylated probes (either RNA probes for Agilent or DNA probes for NimbleGen). The DNA–probe complex is selected using streptavidin magnetic beads that bind the biotinylated probes. The DNA is removed from the probes and the adaptors are used to PCR amplify the DNA for MPS. The Agilent system is designed to work with the Illumina and SOLiD sequencers, while the Roche NimbleGen system is designed for the 454 sequencer.

Raindance Technologies offer RainStorm, a droplet-based technology in which each droplet contains all of the components needed for PCR, with each droplet containing a different pair of PCR primers. Currently, this method permits 20,000 different polymerase chain reactions to be performed simultaneously. After the PCR amplification, the droplets are broken, releasing the PCR products, which can then be used for library construction and sequencing. An advantage of this technology is that no competition among sequences or cross-primer hybridization occurs during the polymerase chain reactions because each primer pair is physically separated from all of the others (Tewhey et al. 2009).

Applications of MPS

Mutation and CNV detection

Genetic gains and losses smaller than the resolution of conventional cytogenetics are called copy number variants (CNVs). Use of array genomic hybridization (also called array-comparative genomic hybridization, cytogenomic array or chromosomal microarray) has shown that CNVs are a frequent cause of disease, including developmental delay and autism (de Smith et al. 2007, 2008; Friedman et al. 2006; Henrichsen, Chaignat and Reymond 2009; Marshall et al. 2008; Yang et al. 2008). However, CNVs also occur within normal individuals, affecting thousands of bases without any obvious phenotypic effect. Although many polymorphic or pathogenic CNVs have been identified by array genomic hybridization, microarrays are limited in their resolution and cannot identify most Mendelian mutations or balanced structural rearrangements (e.g., inversions or translocations). In addition, CNV boundaries are estimated based on statistical comparisons of probe intensity, and the actual location of these changes in the genomic sequence can be tens of kilobases away from the actual boundary (Kidd *et al.* 2008). Clearly defining the boundaries of the CNVs is essential for genotype–phenotype correlations and elucidating the function of genes.

MPS offers the opportunity to identify most, if not all CNVs, Mendelian sequence mutations and structural variants that cause disease. The power of this technology has greatly reduced the need for linkage studies involving very large families or large numbers of smaller families (Lupski *et al.* 2010). Once a mutation is identified, MPS allows the identification of genomic structures or sequences in the region of the disease gene that may provide insight into the mechanism of mutation.

Disease risk and rare variant studies

MPS has distinct advantages as a means of recognizing variants that may predispose to, or protect against, the development of common complex diseases (Gratacos *et al.* 2008; Need *et al.* 2009). Genome-wide association studies (GWAS) are one mechanism used to identify disease mutations or sequence variants that contribute to a complex phenotype. The process involves typing thousands of single nucleotide polymorphisms (SNPs) in patients and controls. Computer programs are then used to compare the frequencies of single SNPs, haplotypes or genotypes between patient and control samples. The location of each tested SNP is known, so a SNP or region found to occur more often in patients than controls suggests involvement of this region or a closely liked genomic region in the pathogenesis of the disease. However, many of the significant SNPs identified are not the causal variant and provide little information about the impact on the disease.

MPS offers the opportunity to identify sequence variants directly in candidate loci, all genomic coding regions or anywhere in the genome. In the latter case, a large number of sequence variants are likely to be identified, and to reduce such lists to manageable numbers, several studies have focused on sequence variants that are predicted to affect protein expression or function (Lupski *et al.* 2010). Resequencing candidate loci by MPS is likely to identify mutations that cause or contribute to many well-studied disorders. However, looking at the genome at such a high resolution, there is a high probability that variants will be identified that are unique to particular ethnic backgrounds. It is becoming increasingly important to get accurate background information on both patients and controls to determine if they are well matched. It is likely that MPS will provide more insight into the increased risk of particular disorders in particular ethnic populations.

Cancer

Classical cellular characteristics of cancers include evasion of apoptosis, uncontrolled proliferation, invasion of adjacent tissues, and metastasis to distant sites (Hanahan and Weinberg 2000). Depending on the cancer type, genome instability, which results in genome rearrangements, and accelerated mutation rates may also frequently be observed. Increasingly, our view of cancer is that it is heterogeneous, in that there are many different cancers, and that even within one type of cancer there may be multiple different subtypes that differ with respect to at-risk populations, predisposing factors, histopathology markers, molecular pathogenesis and response to therapy. Further, even within a single individual, heterogeneity can be observed. Human tumours are rarely uniform masses consisting of cells of a single type but instead can be thought of as communities of cells in which families differ at the levels of mutational load and the programs of gene expression they execute.

Over the past few decades, gain-of-function and loss-of-function mutations contributing to cancer phenotypes and affecting hundreds of genes have been identified, and it is clear in at least several cancer types that mutation is a key event in oncogenesis (Kinzler and Vogelstein 1996). This fact and the speculation that not all mutations that drive cancers have been identified, have led recently to international large-scale efforts to identify all genes that, when mutated, can drive cancer progression (www.cancergenome.nih.gov, www.icgc.org). These efforts have focused on the use of MPS because such technologies provide the cost-efficient throughput required to survey thousands of complete cancer genomes and transcriptomes for candidate cancer-driving mutations. Equally important, such technologies for the first time facilitate the detection of rare mutations within a population of cells. Using these technologies, it is now possible to measure the abundance of a mutation in a tumour sample and measure how this prevalence changes over time, for example in response to therapy (Shah *et al.* 2009). With such information, it is likely that new insights into the mechanisms of therapy resistance will arise, and hopefully this insight will be translated into new approaches for treating resistant disease.

Population genetics

Whole genome sequencing using MPS methods offers substantial advantages for population studies of genetic relationships, migration or selection. MPS can identify rare as well as common variants and patterns of linkage disequilibrium that differ from those seen in the major populations used for HapMap studies. The ability to identify rare variants is especially important for studies in African populations, among whom variability is particularly great (McKernan *et al.* 2009; Need and Goldstein 2009; Teo, Small and Kwiatkowski 2010). Identifying rare or novel variants also provides a useful genetic tool for tracing recent patterns of migration or selection.

Whole genome sequencing can be used to identify the full range of informative variants, including structural and copy number variants, insertions/deletions (InDels) and single nucleotide variants (Snyder, Du and Gerstein 2010). Recognizing some of these variants requires *de novo* assembly of sequenced individual human genomes rather than resequencing based on matching short MPS reads to a reference genome, as has been done for most individual human genome sequences reported to date (Li *et al.* 2010; Snyder, Du and Gerstein 2010). The challenges of *de novo* assembly and other bioinformatic analyses of the large numbers of individual genome sequences required for population genetic studies are major bottlenecks to the efficient use of MPS data for such investigations (McPherson 2009; Mir 2009; Snyder, Du and Gerstein 2010).

Pharmacogenomics

Substantial progress has been made over the past decade in the use of genetic variants to recognize individuals in whom a particular treatment is likely to be efficacious (or not) or who are likely to suffer serious adverse effects if given certain treatments (Brockmoller and Tzvetkov 2008; Huang and Ratain 2009; Peters and McLeod 2008; Shin, Kayser and Langaee 2009). Both candidate gene studies and GWAS have made important contributions to this progress, and whole genome or exome MPS offers the promise of making both of these approaches more cost effective.

MPS will also enable multiple pharamacogenomic loci to be screened simultaneously for rare or novel variants (Mardis 2009). The ability of MPS to detect such variants will be particularly valuable in African populations, where genetic variability is especially great and the complex population structure does not favour approaches based on standard HapMap polymorphisms and estimates of linkage disequilibrium (Marsh, 2008). The identification of pharmacogenomic variants in genetic isolates or groups with frequent consanguinity will also benefit from MPS because polymorphisms that are most informative for GWAS in these populations may not be well represented on standard commercial genotyping arrays.

Unfortunately, however, interpreting genome-wide pharmacogenomic data for any particular individual is likely to be compromised for some time by our limited knowledge of the clinical utility of the variants that may be observed (Via, Gignoux and Burchard 2010). Nevertheless, one can envision whole genome MPS being done at the time of diagnosis of a chronic disease or as a condition for a health or life insurance policy, or even routinely on every newborn infant, with subsequent storage of the data and pharmacogenomic re-analysis whenever a new treatment is being considered. This would enable each new treatment decision to be made on the basis of all currently available therapeutic options as well as all of the relevant pharmacogenomic knowledge that has been accumulated to that point, without the need for any additional testing.

Metagenomics

The term 'metagenomics' is frequently used to describe the application of genomic approaches to characterize the composition of complex microbial and viral communities. Such communities can reside in soil samples, seawater, polluted environments or even within the human gut. A main advantage of metagenomics is that it does not rely on isolation and culturing of individual microbes prior to characterizing them, but instead uses sequencing directly as a way to catalogue the biological diversity within the sample of interest. Because the majority of microbes within such samples are thought to be either unculturable or lack established culture conditions (Handelsman 2004), metagenomics offers one of the best opportunities to survey the genetic diversity of the complex community of organisms within a sample. Metagenomic approaches have been used in searches for enzymes that degrade xenobiotics (Eyers *et al.* 2004), in searches for new taxa living in marine environments (Venter *et al.* 2004), in efforts to define T-cell receptor beta chain diversity (Freeman *et al.* 2009) and very recently to produce a human gut microbial gene catalogue (Qin *et al.* 2010).

Throughput is the major advantage conferred by MPS to metagenomic studies. Because 'metagenomes' may represent hundreds to millions of individual organisms that may reside within the sample in variable proportion, many sequences are required to comprehensively catalogue them, a task that was at best daunting and at worst not feasible until the advent of the new sequencing approaches.

A significant disadvantage of metagenomic approaches, at least early on, was the short read lengths produced by several of the sequencing platforms. Ideally, sequencing readouts should be of sufficient length to allow even a single sequence from a metagenomics study to be assigned to a particular gene in a particular species (or at least family or genus). Of the instruments first to the market place, the 454 was the only one to offer reads of sufficient length for this to be realized. However, increasing read lengths on other platforms are beginning to address this early disadvantage (Qin *et al.* 2010).

Transcriptional analysis

The repertoire of transcripts that resides within cells is often referred to as the 'transcriptome'. The term 'transcriptome' can be used to denote transcripts from all classes of RNA, including small RNAs (e.g., miRNAs or snoRNAs), ribosomal RNAs, and messenger RNAs. Although there are several tools for measuring the abundance of one or a few transcripts (e.g., Northern blots and RT-PCR), there are two main classes of technology that have emerged as suitable for profiling large numbers of transcripts. These are, broadly, sequencing-based approaches and microarray-based approaches.

The Expressed Sequence Tag, or EST, approach is a sequencing method originally used to sample the repertoire of mRNAs expressed in tissues (Adams *et al.*
1991). In this approach, a single Sanger read hundreds of base-pairs in length,
generated from the ends of cDNAs, was used to sequence a portion of the transcript. EST data were used for differential gene expression analysis, gene discovery
and, as genome sequence data became available, for genome annotation and gene
prediction. Despite the utility of EST data, various limitations prevented the approach
from becoming as popular as microarrays. These limitations included the expense
associated with large-scale cDNA cloning, cDNA clone arraying and sequencing and
the requirement for large facilities specialized in such activities. The cost efficiency
of EST generation was in part addressed by the advent of Serial Analysis of Gene
Expression (SAGE) (Velculescu *et al.* 1995) in which ESTs were made shorter (e.g.
17–21 bp) and concatenated such that ~30 or more transcripts could be tagged in
a single Sanger sequencing read. Although this increased the number of genes that
could be surveyed without increasing costs, SAGE tags could not be used to survey
the entire length of transcripts and SAGE was therefore not robust for assessing transcript structure, exon–exon connectivity, quantifying RNA abundance at the level of
individual exons or detection of mutations in expressed genes.

Many of the limitations of ESTs and SAGE tags have been addressed with the
introduction of MPS technology. Two similar approaches, RNA-Seq (Wang, Gerstein
and Snyder 2009) and whole transcript shotgun sequencing (WTSS) (Morin *et al.*
2008) have been used to quantify mRNA abundance, measure exon–exon connectivity and detect mutations in expressed transcripts (Morin *et al.* 2008; Shah *et al.*
2009; Wang, Gerstein and Snyder 2009). In general, protocols involve purification
of total RNA from samples, construction of cDNA and then random fragmentation
of the cDNA library. These random fragments serve as the template for sequencing
reactions on MPS instruments. Tens of millions of sequences can easily be generated
at reasonable cost, yielding deeply redundant sequencing data for thousands of genes.
Small RNAs can also be detected (miRNA-Seq) (Tang *et al.* 2009) using massively
parallel sequencers, as can other classes of RNA.

Although still in their infancy, bioinformatics approaches for *de novo* assembly of
transcript sequences are becoming available (e.g., AbySS) (Birol *et al.* 2009). Using
such approaches, it is possible to assemble tens of millions of sequence reads to
reconstruct sequence contigs representing multiple exons and their junctions. This
exercise can reveal both the expression level of individual exons and the structure
and, in some cases, the likely function of transcripts.

Epigenetics

Epigenetics is the study of somatically heritable gene regulation that does not
involve the modification of DNA. There are two main types of epigenetic
modifications – DNA methylation and post-translational modifications of histone

tails. Recognizing the importance of epigenetics in cancer genetics and normal development, it is the aim of the Human Epigenome Consortium to identify, catalogue and interpret genome-wide methylation patterns of all human genes in all major tissues (Esteller 2006).

There are three main techniques used to detect DNA methylation: restriction digestion, bisulfite sequencing and immunoprecipitation. Digesting the DNA with a methylation-sensitive restriction enzyme (i.e., an enzyme that is unable to digest methylated DNA) will determine if the DNA is methylated at a particular site but requires prior knowledge of the site under investigation. Using PCR primers that flank the restriction enzyme site, if the DNA is methylated, the restriction enzyme will not be able to cut the DNA and a PCR product will be produced. If there is no methylation, the restriction enzyme can cut the DNA and there will not be a PCR product produced. This type of analysis is not likely to be used for MPS.

Bisulfite sequencing, the method used by the Human Epigenome Consortium, converts all non-methylated cytosines to uracils, while all methylated cytosines remain intact. Sequencing identifies the proportion and position of all methylated cytosines. Bisulfite sequencing can be combined with MPS to examine methylation patterns in tumours or constitutional genetic conditions caused by improper methylation (e.g., mutations of the gene for DNA methyltransferase 3B in patients with ICF syndrome).

Histone modifications include acetylation, methylation, phospohorylation and ADP ribosylation, and certain marks are characteristic of active or inactive expression. Histone modifications are usually identified using chromatin immunoprecipitation (CHiP). CHiP involves cross-linking protein to DNA, followed by immunoprecipitation of a protein of interest with a specific antibody. The DNA is dissociated from the protein, and the immunoprecipitated DNA is used to construct sequencing libraries. Combining MPS with CHiP (CHiP-Seq) provides a high-resolution look at these modifications. In CHiP-Seq the number of sequence reads in a particular area is proportional to the number of histone modifications of the type being assayed (Barski *et al.* 2007; Mikkelsen *et al.* 2007).

Summary

MPS has greatly increased our capacity to identify sequence variants faster than previously possible and at a fraction of the cost. The cost is likely to continue to drop as there are commercial incentives to move MPS forward. In addition, the US National Human Genome Research Institute offers Advanced Sequencing Technology Development Awards for novel ideas and approaches to develop low-cost sequencing with the goal of reaching the $1000 genome (www.genome.gov). The Archon X Prize is also offering $10 million to the first team to sequence 100 human genomes in 10 days (http://genomics.xprize.org).

However, the increase in the volume of sequence data brings the need for analysis software and more functional analysis tools to help interpret the identified sequence variants. These programs will need to be user friendly because, as the cost of MPS decreases, the number of small research, clinical and other service laboratories using them will increase, and most of these smaller labs will not have bioinformaticians to interpret the results.

Overall, as the price decreases, the number of whole genomes sequenced will increase, as will the number of recognized disease-causing or disease-predisposing variants, our knowledge about the organisms that inhabit our bodies and our understanding of the patterns of migration and selection that have shaped our species.

References

Adams, M.D., Kelley, J.M., Gocayne, J.D. *et al.* (1991) Complementary DNA sequencing: expressed sequence tags and human genome project. *Science* **252**: 1651–1656.

Barski, A., Cuddapah, S., Cui, K. *et al.* (2007) High-resolution profiling of histone methylations in the human genome. *Cell* **129**: 823–837.

Birol, I., Jackman, S.D., Nielsen, C.B. *et al.* (2009) De novo transcriptome assembly with ABySS. *Bioinformatics* **25**: 2872–2877.

Branton, D., Deamer, D.W., Marziali, A. *et al.* (2008) The potential and challenges of nanopore sequencing. *Nat Biotechnol* **26**: 1146–1153.

Brockmoller, J. and Tzvetkov, M.V. (2008) Pharmacogenetics: data, concepts and tools to improve drug discovery and drug treatment. *Eur J Clin Pharmacol* **64**: 133–157.

Clarke, J., Wu, H.C., Jayasinghe, L. *et al.* (2009) Continuous base identification for single-molecule nanopore DNA sequencing. *Nat Nanotechnol* **4**: 265–270.

de Smith, A.J., Tsalenko, A., Sampas, N. *et al.* (2007) Array CGH analysis of copy number variation identifies 1284 new genes variant in healthy white males: implications for association studies of complex diseases. *Hum Mol Genet* **16**: 2783–2794.

de Smith, A.J., Walters, R.G., Froguel, P. and Blakemore, A.I. (2008) Human genes involved in copy number variation: mechanisms of origin, functional effects and implications for disease. *Cytogenet Genome Res* **123**: 17–26.

Dressman, D., Yan, H., Traverso, G. *et al.* (2003) Transforming single DNA molecules into fluorescent magnetic particles for detection and enumeration of genetic variations. *Proc Natl Acad Sci USA* **100**: 8817–8822.

Drmanac, R., Sparks, A.B., Callow, M.J. *et al.* (2010) Human genome sequencing using unchained base reads on self-assembling DNA nanoarrays. *Science* **327**: 78–81.

Eid, J., Fehr, A., Gray, J. *et al.* (2009) Real-time DNA sequencing from single polymerase molecules. *Science* **323**: 133–138.

Esteller, M. (2006) The necessity of a human epigenome project. *Carcinogenesis* **27**: 1121–1125.

Eyers, L., George, I., Schuler, L. *et al.* (2004) Environmental genomics: exploring the unmined richness of microbes to degrade xenobiotics. *Appl Microbiol Biotechnol* **66**: 123–130.

Fedurco, M., Romieu, A., Williams, S. *et al.* (2006) BTA, a novel reagent for DNA attachment on glass and efficient generation of solid-phase amplified DNA colonies. *Nucleic Acids Res* **34**: e22.

Freeman, J.D., Warren, R.L., Webb, J.R. *et al.* (2009) Profiling the T-cell receptor beta-chain repertoire by massively parallel sequencing. *Genome Res* **19**: 1817–1824.

Friedman, J.M., Baross, A., Delaney, A.D. *et al.* (2006) Oligonucleotide microarray analysis of genomic imbalance in children with mental retardation. *Am J Hum Genet* **79**: 500–513.

Gratacos, M., Costas, J., de Cid, R. *et al.* (2008) Identification of new putative susceptibility genes for several psychiatric disorders by association analysis of regulatory and non-synonymous SNPs of 306 genes involved in neurotransmission and neurodevelopment. *Am J Med Genet B Neuropsychiatr Genet* **150B**: 808–816.

Hanahan, D. and Weinberg, R.A. (2000) The hallmarks of cancer. *Cell* **100**: 57–70.

Handelsman, J. (2004) Metagenomics: application of genomics to uncultured microorganisms. *Microbiol Mol Biol Rev* **68**: 669–685.

Harris, T.D., Buzby, P.R., Babcock, H. *et al.* (2008) Single-molecule DNA sequencing of a viral genome. *Science* **320**: 106–109.

Henrichsen, C.N., Chaignat, E. and Reymond, A. (2009) Copy number variants, diseases and gene expression. *Hum Mol Genet* **18**: R1–8.

Huang, R.S. and Ratain, M.J. (2009) Pharmacogenetics and pharmacogenomics of anticancer agents. *CA Cancer J Clin* **59**: 42–55.

Kidd, J.M., Cooper, G.M., Donahue, W.F. *et al.* (2008) Mapping and sequencing of structural variation from eight human genomes. *Nature* **453**: 56–64.

Kinzler, K.W. and Vogelstein, B. (1996) Lessons from hereditary colorectal cancer. *Cell* **87**: 159–170.

Li, R., Li, Y., Zheng, H. *et al.* (2010) Building the sequence map of the human pan-genome. *Nat Biotechnol* **28**: 57–63.

Lupski, J.R., Reid, J.G., Gonzaga-Jauregui, C. *et al.* (2010) Whole-Genome Sequencing in a Patient with Charcot-Marie-Tooth Neuropathy. *N Engl J Med* **362**: 1181–1191.

Mamanova, L., Coffey, A.J., Scott, C.E. *et al.* (2010) Target-enrichment strategies for next-generation sequencing. *Nat Methods* **7**: 111–118.

Mardis, E.R. (2009) New strategies and emerging technologies for massively parallel sequencing: applications in medical research. *Genome Med*, **1**, 40.

Marsh, S. (2008) Pharmacogenetics: global clinical markers. *Pharmacogenomics* **9**: 371–373.

Marshall, C.R., Noor, A., Vincent, J.B. *et al.* (2008) Structural variation of chromosomes in autism spectrum disorder. *Am J Hum Genet* **82**: 477–488.

Maxam, A.M. and Gilbert, W. (1977) A new method for sequencing DNA. *Proc Natl Acad Sci USA* **74**: 560–564.

McKernan, K.J., Peckham, H.E., Costa, G.L. *et al.* (2009) Sequence and structural variation in a human genome uncovered by short-read, massively parallel ligation sequencing using two-base encoding. *Genome Res* **19**: 1527–1541.

McPherson, J.D. (2009) Next-generation gap. *Nat Methods* **6**: S2–5.

Mikkelsen, T.S., Ku, M., Jaffe, D.B. *et al.* (2007) Genome-wide maps of chromatin state in pluripotent and lineage-committed cells. *Nature* **448**: 553–560.

Mir, K.U. (2009) Sequencing genomes: from individuals to populations. *Brief Funct Genomic Proteomic* **8**: 367–378.

Morin, R., Bainbridge, M., Fejes, A. *et al.* (2008) Profiling the HeLa S3 transcriptome using randomly primed cDNA and massively parallel short-read sequencing. *Biotechniques* **45**: 81–94.

Need, A.C., Ge, D., Weale, M.E. *et al.* (2009) A genome-wide investigation of SNPs and CNVs in schizophrenia. *PLoS Genet* **5**: e1000373.

Need, A.C. and Goldstein, D.B. (2009) Next generation disparities in human genomics: concerns and remedies. *Trends Genet* **25**: 489–494.

Peters, E.J. and McLeod, H.L. (2008) Ability of whole-genome SNP arrays to capture 'must have' pharmacogenomic variants. *Pharmacogenomics* **9**: 1573–1577.

Qin, J., Li, R., Raes, J. *et al.* (2010) A human gut microbial gene catalogue established by metagenomic sequencing. *Nature* **464**: 59–65.

Ronaghi, M., Karamohamed, S., Pettersson, B. *et al.* (1996) Real-time DNA sequencing using detection of pyrophosphate release. *Anal Biochem* **242**: 84–89.

Sanger, F., Nicklen, S. and Coulson, A.R. (1977) DNA sequencing with chain-terminating inhibitors. *Proc Natl Acad Sci USA* **74**: 5463–5467.

Shah, S.P., Morin, R.D., Khattra, J. *et al.* (2009) Mutational evolution in a lobular breast tumour profiled at single nucleotide resolution. *Nature* **461**: 809–813.

Shendure, J. and Ji, H. (2008) Next-generation DNA sequencing. *Nat Biotechnol* **26**: 1135–1145.

Shendure, J., Porreca, G.J., Reppas, N.B. *et al.* (2005) Accurate multiplex polony sequencing of an evolved bacterial genome. *Science* **309**: 1728–1732.

Shin, J., Kayser, S.R. and Langaee, T.Y. (2009) Pharmacogenetics: from discovery to patient care. *Am J Health Syst Pharm* **66**: 625–637.

Snyder, M., Du, J. and Gerstein, M. (2010) Personal genome sequencing: current approaches and challenges. *Genes Dev* **24**: 423–431.

Tang, F., Barbacioru, C., Wang, Y. *et al.* (2009) mRNA-Seq whole-transcriptome analysis of a single cell. *Nat Methods* **6**: 377–382.

Teo, Y.Y., Small, K.S. and Kwiatkowski, D.P. (2010) Methodological challenges of genome-wide association analysis in Africa. *Nat Rev Genet* **11**: 149–160.

Tewhey, R., Warner, J.B., Nakano, M. *et al.* (2009) Microdroplet-based PCR enrichment for large-scale targeted sequencing. *Nat Biotechnol* **27**: 1025–1031.

Turcatti, G., Romieu, A., Fedurco, M. and Tairi, A.P. (2008) A new class of cleavable fluorescent nucleotides: synthesis and optimization as reversible terminators for DNA sequencing by synthesis. *Nucleic Acids Res* **36**: e25.

Velculescu, V.E., Zhang, L., Vogelstein, B. and Kinzler, K.W. (1995) Serial analysis of gene expression. *Science* **270**: 484–487.

Venter, J.C., Remington, K., Heidelberg, J.F. *et al.* (2004) Environmental genome shotgun sequencing of the Sargasso Sea. *Science* **304**: 66–74.

Via, M., Gignoux, C. and Burchard, E.G. (2010) The 1000 Genomes Project: new opportunities for research and social challenges. *Genome Med* **2**: 3.

Wang, Z., Gerstein, M. and Snyder, M. (2009) RNA-Seq: a revolutionary tool for transcriptomics. *Nat Rev Genet* **10**: 57–63.

Yang, T.L., Chen, X.D., Guo, Y. *et al.* (2008) Genome-wide copy-number-variation study identified a susceptibility gene, UGT2B17, for osteoporosis. *Am J Hum Genet* **83**: 663–674.

7

Aptamers for Analysis: Nucleic Acids Ligands in the Post-Genomic Era

Pedro Nadal, Alessandro Pinto, Marketa Svobodova and Ciara K. O'Sullivan

Introduction

At the beginning of the 1990s, three different laboratories (Tuerk and Gold 1990; Robertson and Joyce 1990; Ellington and Szostak 1990) reported their results on the development of an *in vitro* selection and amplification technique for the isolation of specific nucleic acids able to bind to target molecules with high affinity and specificity. The technique was coined SELEX (Systematic Evolution of Ligands by EXponential enrichment) and the resulting oligonucleotides were named aptamers, taken from the latin 'aptus', to fit. A considerable number of comprehensive reviews regarding the selection of aptamers are available (Baldrich-Rubio, Campas and O'Sullivan 2004; James 2000; Kulbachinskiy 2007; Mairal *et al.* 2008; Stoltenburg, Reinemann and Strehlitz 2007) as well as a practical introduction to the methodology of *in vitro* evolution of RNA ligands (Fitzwater and Polisky 1996), a mathematical description of the process (Levine and Nilsen-Hamilton 2007; Irvine, Tuerk and Gold 1991) and a model for simultaneous selection against multiple targets (Vant-Hull *et al.* 1998).

Molecular Analysis and Genome Discovery, Second Edition. Edited by Ralph Rapley and Stuart Harbron.
© 2012 John Wiley & Sons, Ltd. Published 2012 by John Wiley & Sons, Ltd.

Aptamers are typically composed of RNA (RNA aptamers), single-stranded DNA (DNA aptamers) or a combination of both with unnatural nucleotides, and range in size from approximately 6 to 40 kDa. Due to their three-dimensional structure, characterized by stems, loops, hairpins, bulges, triplexes and quadruplexes, aptamers can bind to a wide variety of targets from single molecules to complex target mixtures or a whole organism. They bind selectively to their target due to structural compatibility, electrostatic and van der Waals interactions, hydrogen bonding, or a combination of these effects (Hermann and Patel 2000), with dissociations constants (Kd) typically in the low nanomolar range, comparable to those observed for monoclonal antibodies. Aptamers, as we outline in this review, are, however, far more flexible to different assay formats than their antibody counterparts, allowing them to address the ever-increasing requirements of analytical applications in the post-human genome era. Several facets of aptamers contribute to this enhanced flexibility; not only can they exploit already developed technologies for genomics due to their nucleic acid nature, but they have also been used in a variety of assays exploiting displacement of labelled targets, or DNA sequences complementary to the aptamer sequence, where the only required end-user intervention is sample addition. Furthermore, the unique nature of aptamers has been taken advantage of to achieve remarkably low detection limits of proteins, where aptamers have been used both as recognition and reporter molecules in an assay coined 'apta-PCR'. These types of assay formats, combined with the stability and ease of immobilizing and labelling aptamers, present as very attractive candidates for analytical applications. However, aptamers do have drawbacks – each aptamer behaves differently and requires a specific set of conditions for optimal operation. Furthermore, there are a limited number of aptamers available, although an increasing number of aptamers are being reported in the literature. In this chapter, we review progress in the selection and isolation of aptamers, advances in approaches for the analytical application of aptamers as well as their application for image analysis.

SELEX

Overview

Aptamers are routinely selected using a process of SELEX, which involves iterative cycles of incubation, recovery of bound nucleic acids (single stranded DNA/RNA), followed by amplification, and generation of an enriched pool of oligonucleotides. A typical SELEX procedure begins with a nucleic acid library, generally consisting of about $10^{13}-10^{15}$ different sequences (James 2000), containing a random region flanked by two primers used for amplification. A randomized single stranded DNA (ssDNA) or RNA pool is incubated directly with target, which is immobilized on a matrix material or is free in solution. Following incubation, the bound complexes are separated from the unbound and weakly bound oligonucleotides, which is one of the most critical steps in the selection process and various techniques have been reported for the isolation of the bound nucleic acids. Target bound oligonucleotides are eluted and amplified by polymerase chain reaction (PCR) in the case of DNA or reverse transcription PCR (RT-PCR) in the case of RNA. From the resulting double stranded DNA (dsDNA), ssDNA or transcribed RNA is generated, and this enriched pool of selected oligonucleotides is used in the next SELEX round. Generally, 8–15 rounds of SELEX are required to obtain a pool of aptamer candidates with the highest binding affinity for the target. The number of rounds depends on different parameters, such as design of the oligonucleotide library, type of target, oligonucleotides to target ratio, selection conditions and the efficiency of the partitioning methodology. The stringency can be increased during the selection process by reducing the target concentration or changing the binding and washing conditions (Marshall and Ellington 2000), and monitoring of the evolution of aptamer selection is performed using a variety of techniques. When affinity saturation of the enriched pool of oligonucleotides is achieved, aptamer candidates are cloned and sequenced, and individual sequences investigated for their ability to bind to the target.

Libraries

A typical SELEX library consists of a multitude of oligonucleotides ($\approx 10^{15}$) comprising a central random region of 20–80 nucleotides (nt) flanked by primer sequences of 18–23 nt. Although any length of randomized region may be used to provide 4^n theoretical random sequences, the maximum number of molecules that can be manipulated in standard molecular biology laboratories is about 10^{15} and errors of synthesis and workup have been estimated to reduce this diversity to $10^{13}–10^{14}$ (James 2000). This number corresponds to the total number of different sequence variants for oligonucleotides that are 25-nt long ($4^{25} \approx 10^{15}$). Libraries with the random region longer than 25 nt are thus underrepresented and contain the same number of different variants as shorter libraries (Kulbachinskiy 2007). An aptamer for the human thrombin was selected by Bock *et al.* (1992) from a DNA library of 96-mer oligonucleotides containing a random sequence with 60 nt. The aptamer was truncated to the minimal size required for binding of the thrombin target and a consensus sequence with 15 nt was obtained, suggesting that libraries with short randomized regions are sufficient for an aptamer selection. On the other hand longer randomized regions of libraries can lead to greater structural complexity and provide more opportunities for the selection of aptamers (Marshall and Ellington 2000). It has been demonstrated with selection of aptamers to isoleucine that both the short (16 nt) and the long (90 nt) randomized regions complicate aptamer selection (Legiewicz *et al.* 2005). In practice, therefore, randomized regions of 30–60 nt are most common. In classical SELEX experiments DNA or RNA libraries are usually used. DNA has no 2'hydroxyl group, which is highly reactive, and accordingly is much more stable to hydrolysis under conditions of higher than neutral pH and elevated temperature (James 2001; Kusser 2000). DNA libraries are thus often used for applications requiring increased aptamer stability while RNA libraries yield aptamers with higher binding affinities due to the ability of RNA to take on a wider variety of conformations than DNA (Hamula *et al.* 2006), but this is not universally true and indeed there is evidence that ssDNA aptamers do not differ in affinity and specificity from RNA aptamers (Breaker 1997).

With the goal of increasing the stability of oligonucleotides, modified libraries are often used in the selection process. Modifications are introduced into the 2'position of the pyrimidines, using 2'-aminopyrimidines (Jellinek *et al.* 1995; Lin *et al.* 1994), 2'fluoropyrimidines (Biesecker *et al.* 1999; Rusconi *et al.* 2002), 2'O-methyl nucleotides (Burmeister *et al.* 2005, 2006), position 5 of pyrimidines (Latham, Johnson and Toole 1994; Masud *et al.* 2004), and position 4 of pyrimidines using thio UTO and CTP (Kato *et al.* 2005), among others. A detailed description of the different modifications can be found in several reviews (Eaton 1997; Eaton, Gold and Zichi 1995; Keefe and Cload 2008; Kusser 2000). In some experiments structurally constrained libraries are used in selection. In this case, the random region is flanked by fixed sequences able to form a particular secondary structure such as a hairpin, G-quartet or pseudoknot (Tuerk and Gold 1990; Hamm, Huber and Lührmann 1997; Hamm, Alessi and Biondi 2002; Biroccio *et al.* 2002).

The libraries in genomic SELEX are derived from the genome of the organism of interest. To create such libraries, genomic DNA is excised into short fragments and appended with the fixed sequences. In all other aspects, *genomic SELEX* is similar to conventional SELEX. This variant of SELEX is used for the identification of important nucleic–acid–protein interactions within any organism (Shimada *et al.* 2005; Shtatland *et al.* 2000).

A variation of the SELEX protocol called *tailored-SELEX* allows isolation of aptamers free of fixed regions with desired properties (Vater *et al.* 2003). In this case, a library with 4 or 6 nt on both sides of the random region is used in the SELEX process. After each selection round, the primer binding sequences are ligated to the library using special adapters and removed before the next round. A very similar technique called *primer-free genomic SELEX* can also avoid the primers' influence on aptamer selection. This method is used to solve a specific problem with genomic SELEX, where fixed regions of the library often form complexes with the central genomic fragments. Using this method, fixed regions are removed from the genomic library before incubation with the target and are subsequently regenerated to allow amplification of the selected genomic fragments. An important point in the regeneration of primer-annealing sequences is to use thermal cycles of hybridization-extension, using sequences from unselected pools as templates (Wen and Gray 2004). More recently, there have been reports of a *minimal primer* and *primer-free* SELEX protocols for screening of aptamers from random DNA libraries (W.H. Pan and Clawson 2009; W.H. Pan, Xin and Clawson 2008).

Target molecules

The evidence that aptamers can be generated against small ions, such as Zn^{2+} (Ciesiolka, Gorski and Yarus 1995), Ni^{2+} (Hofmann *et al.* 1997), to nucleotides such as ATP (Huizenga and Szostak 1995; Sassanfar and Szostak 1993), oligopeptides (Nieuwlandt, Wecker and Gold 1995), and large glycoproteins such CD4 (Kraus, James and Barclay 1998), with the size range from 65 to 150 kDa, demonstrates their tremendous flexibility. The smallest molecular target to date is ethanolamine (Mann *et al.* 2005), and furthermore, the classes of targets are diverse, including organic dyes (Ellington and Szostak 1990; Wilson and Szostak 1998), neutral disaccharides (Q. Yang *et al.* 1998), antibiotics (Famulok and Huttenhofer 1996), neurotransmitters (Mannironi *et al.* 1997), pigments (Y. Li, Geyer and Sen 1996) and vitamins (Wilson, Nix and Szostak 1998).

However, SELEX has been most often exploited for the generation of aptamers against proteins (J.F. Lee *et al.* 2004). When the aptamer-binding sites (aptatopes) on large molecules are mapped, it is usually found that they are coincident, even if the aptamers fall into unrelated sequence families, signifying that only a small proportion of the surface of the macromolecule is accessible to aptamers – most probably regions of high solvent exposure where positively charged residues provide a degree of electrostatic steerage towards the aptatope (James 2000).

SELEX has also been applied towards complex cellular targets – CELL SELEX (Blank *et al.* 2001; Bruno and Kiel 1999; Cerchia *et al.* 2005; Daniels *et al.* 2003; Hicke *et al.* 2001; Vivekananda and Kiel 2006; C.L. Wang *et al.* 2003). For cancer cells, aptamers are able to penetrate tumours rapidly and can be cleared quickly from the blood due to their small size (10–25 kDa), making them an effective tool for molecular medicine, biomarker discovery and cancer biology. Due to the abundance of targets on cell membranes, however, it is often impossible to know the precise target of an aptamer prior to its selection and often aptamers are selected against purified cell surface markers (Cerchia *et al.* 2005; Ohuchi, Ohtsu and Nakamura 2006).

Partitioning methodology

One of the most critical steps in SELEX is the partitioning of the bound oligonucleotides from molecules not bound to the target. A commonly used method is affinity chromatography with immobilization of the target on a sepharose or agarose column (Ciesiolka *et al.* 1995; J.J. Liu and Stormo 2005; Nieuwlandt *et al.* 1995; Tombelli *et al.* 2005a). The use of magnetic beads with magnetic separation is also widely used as a useful tool for the separation of target and nucleic acids (Bruno and Kiel 2002; Lupold *et al.* 2002; Murphy *et al.* 2003), as this method only requires small amounts of target and is simple to handle.

While the immobilization of the target on a specific matrix allows an effective separation, protein immobilization can destroy key features of some aptatopes as well as reducing the mobility of the protein. The use of ultrafiltration without target immobilization is thus an alternative method used for partitioning (Bianchini, Radrizzani and Brocardo 2001; Fitzwater and Polisky 1996; Schneider, Gold and Platt 1993; Tuerk and Gold 1990), and although this nitrocellulose filter technique is very widely reported, there are significant non-specific interactions with the nitrocellulose membrane combined with large losses of target-bound oligonucleotides.

An optional partitioning strategy is based on detecting fluorescence via flow cytometry (Blank *et al.* 2001; K.A. Davis *et al.* 1996; X. Yang *et al.* 2003), or another alternative is based on surface plasmon resonance (SPR), which provides binding efficiency information and online evaluation during the selection process (Misono and Kumar 2005). Atomic force microscopy (AFM) can dynamically detect the adhesion or affinity force between a sample surface and a cantilever, a feature that is useful for the selection of high affinity aptamers (Yusuke *et al.* 2010). Electrophoretic separation (Golden *et al.* 2000; Goodman *et al.* 1999; Jensen *et al.* 1995; W. Yao, Adelman and Bruenn 1997) or centrifugation (Homann and Goringer 1999; Rhie *et al.* 2003) are further examples of tools for partitioning.

Regulation of aptamer specificity

The aptamer specificity to a target can be regulated during selection. Thus, if it is required to obtain an aptamer of a very high specificity, it is necessary to select only sequences that bind the specific target (or its part) but do not interact with the matrix or other structurally closely related molecules. On the other hand, if it is necessary to obtain an aptamer to recognize several related molecules, then sequences are selected with the ability to bind to several protein targets (Kulbachinskiy 2007).

Negative SELEX

This step is used to exclude oligonucleotides absorbed by the support matrix material (e.g. affinity chromatography column, magnetic beads, nitrocellulose filters, etc.) with the aim of selecting aptamers that bind only to the specific target. During negative selection, oligonucleotides are incubated with the matrix to eliminate the absorbed sequences. The unbound part is subsequently incubated with immobilized targets to select desired aptamers, as it has been demonstrated that the use of negative selection during SELEX process reduces the possibility of evolving non-specific binders (Ellington and Szostak 1992).

Counter SELEX

The purpose of counter selection is to gain highly selective aptamers. The aptamer's selectivity is improved by excluding oligonucleotide molecules with affinity to molecules similar in structure to the target (Fitzwater and Polisky 1996; Jenison *et al.* 1994). A prime example of the importance of incorporating a counter SELEX step is the example of the RNA aptamer selected against the small molecule theophylline, using caffeine as the counter target and resulting in a discrimination factor of 10 000 of the aptamer for theophylline over caffeine, even though their structures only differ by a methyl group (Jenison *et al.* 1994). Another excellent example is that of the aptamer that was selected and demonstrated to be highly specific for ADP through the use of a counter selection step against ATP (Srinivasan *et al.* 2004).

Amplification and post-amplification processing

The high degree of diversity of the initial oligonucleotide library results in a very small percentage of the library bound to the target at the end of the first selection step, which consequently requires amplification. RNA oligonucleotides are firstly amplified by (RT)-PCR obtaining the corresponding cDNA, which is amplified in a subsequent PCR using a specific primer with a T7 promoter at the 5'-end. Single stranded DNA aptamers are amplified by straightforward PCR, and it is possible to attach modifications via special primers during the amplification step, which can be used for the monitoring of enrichment of the selection process or for the preparation of ssDNA (Stoltenburg *et al.* 2007).

After PCR, dsDNA is obtained, or in case of RNA, the DNA undergoes transcription with T7 RNA polymerase and the resulting RNA molecules can be used directly in the following SELEX round. For DNA aptamers, ssDNA needs to be prepared. One of the most widely used methods is magnetic separation with streptavidin-coated beads and biotinylated forward or reverse primers (Espelund, Stacy and Jakobsen 1990; Hultman *et al.* 1989; Naimuddin *et al.* 2007). In this technique, immobilization of biotinylated dsDNA on the surface of streptavidin coated beads, followed by denaturation of the dsDNA by alkaline treatment or high temperature is used. An alternative method used for generation of ssDNA exploits a urea-polyacrylamide gel, where PCR is performed with a specific long primer to produce strands of different lengths, followed by strand separation with denaturing gel electrophoresis (Fitzwater and Polisky 1996; Pagratis 1996; Wiliams and Bartel 1995; Mann *et al.* 2005; Stoltenburg, Reinemann and Strehlitz 2005). Another possibility is to perform asymmetric PCR, which is used to amplify one strand of the original DNA more than the other by using an unequal molar ratio of forward and reverse primers. After asymmetric PCR, the PCR products are separated by gel electrophoresis and the ssDNA extracted and purified, representing a lengthy procedure, which is also known to be inefficient (Gyllensten and Erlich 1988; Wu and Curran 1999). Another option for the preparation of ssDNA is the use of enzymes such as T7 Gene 6 exonuclease (Nikiforov *et al.* 1994; Ruan and Fuller 1991) and lambda exonuclease via modified primers (Avci-Adali *et al.* 2009; Higuchi and Ochman 1989; Jones *et al.* 2006;

Kujau and Wölfl 1997). T7 Gene 6 exonuclease hydrolyzes one strand of DNA non-processively in the 5′ to 3′ direction, while the second strand is protected against the hydrolytic activity of the enzyme with several phosphorothioates at its 5′ end, while the lambda exonuclease selectively digests the phosphorylated strand of dsDNA and has greatly reduced activity on ssDNA and non-phosphorylated DNA.

Monitoring of evolution during SELEX

Another important step for the success of SELEX is efficient monitoring of the evolution of the selection process, which allows control and adjustment of the selection pressure and stringency to achieve the desired properties of the selected aptamers. The technique most commonly used for monitoring of evolution uses radioactively labelled nucleic acids (Jeong *et al.* 2001; Sayer *et al.* 2002; Schurer *et al.* 2001), which is a very sensitive method that enables detection of a small amount of nucleic acids. The use of radioactivity is not possible in every laboratory, however, and strict safety precautions have to be taken. Fluorescence, with inserted fluorescent labels via PCR, allows rapid and non-radioactive monitoring of the selection progress (Stoltenburg *et al.* 2005). This method is rather costly and might also interfere with the binding properties of the nucleic acid (Nutiu and Li 2004). The fluorescent dye-linked aptamer assay (OliGreen) is another method used for monitoring of evolution (Wochner and Glokler 2007). The fluorescent dye OliGreen is specific for single-stranded DNA and enables the quantification of as little as 100pg/ml nucleic acid with a standard spectrofluorimeter. An alternative method for monitoring the selection process is denaturing high-performance liquid chromatography (dHPLC) (Muller *et al.* 2008), a technique based on the fact that the diversity of a nucleic acid library is dramatically reduced during the selection process. Other methods used for monitoring aptamer selection include capillary electrophoresis, affinity chromatography and SPR (Mendosa and Bowser 2004a; Misono and Kumar 2005; Mosing and Bowser 2007; Yunusov *et al.* 2009).

Cloning, sequencing and analysis of aptamers

When the affinity saturation of an enriched library is achieved, the final oligonu-cleotide pool is cloned into a bacterial vector and individual colonies are sequenced. Typically 50–100 clones are sequenced and analysed by sequence analysis. Sequence alignments are used to identify aptamers with homologous sequences and compar-ative analysis of the aligned sequences is used to determine the consensus motif, with these regions playing an important role in the specific target binding of the aptamers. Analysis of the conserved regions is used for predicting the secondary structure of aptamer candidates using specialized programs developed for this pur-pose (J.P. Davis *et al.* 1996). Shortened aptamer sequences are determined and synthesized from predicted secondary structure, with the objective that truncation of aptamer size can result in an increase in specificity and affinity. Binding studies determine the specificity and affinity of selected aptamers using techniques such as SPR, enzyme-linked oligonucleotide assays (ELONA) and fluorescent or radioactive binding assays. Finally, aptamer–target interactions, determination of three-dimensional structure and the effect of the aptamer on the properties of the target are studied.

Aptamer stability

One of the main problems with nucleic acids is their high sensitivity to nuclease attack, limiting their lifetime in biological media such as serum to minutes (DNA) or even seconds (RNA). Several approaches are used to stabilize the aptamers, including chemical modifications and the use of *Spiegelmers*.

Chemically modified aptamers

As previously mentioned, modified oligonucleotides can be used directly during the selection process. The main restriction in this case is that polymerases generally used in SELEX are specific for their substrates (wild-type RNA, DNA) and therefore variants of original polymerases have to be used (Huang *et al.* 1997; Padilla and Sousa 1999; Sousa and Padilla 1995). Alternatively, it is possible to use chemical modifications similar to those used in the modified libraries following the selection process. As discussed before, many modifications of aptamers such as 2′-F and 2′-NH$_2$ groups introduced at the 2′-position of the ribose make them nuclease resistant (Eaton 1997; Green *et al.* 1995; Ruckman *et al.* 1998), and other types of modifications added at 3′ and 5′ ends of oligonucleotides, such as amine, phosphate, phosphotiothate, cholesterol and fatty acids, protect oligonucleotides from exonucleases (Klussman 2006; Nikiforov *et al.* 1994; Stoltenburg *et al.* 2007). However, these modifications may result in a change of aptamer affinity and specificity to their targets, but this is not often observed and modified nucleotides have been widely reported.

Spiegelmers

Spiegelmers, whose name was coined from the German word for mirror, *spiegel,* offer an interesting alternative route for producing nuclease-resistant aptamers. Spiegelmers are mirror images of aptamers, composed of L-ribose or L-2′deoxyribose, that bind specifically to a target, but are not recognized by ribonucleases, the chiral inversion leading to high biological stability and long lifetime (Klussman *et al.* 1996; Nolte *et al.* 1996). Spiegelmers are produced using a mirror-image SELEX procedure. Since RNA and DNA polymerases are not able to carry out polymerization of L-nucleotides, selection is perfomed using original D-oligonucleotides that are selected against synthetic enantiomers of a chosen target. It is accepted that spiegelmers do not hybridize with nucleic acids of natural configuration and it has been shown that spiegelmers show the same high affinity for targets as their aptamer analogues (Leva *et al.* 2002). These favourable properties led to the identification of several spiegelmers that bind to various target molecules (Faulhammer *et al.* 2004; Helmling *et al.* 2004; Leva *et al.* 2002; Purschke *et al.* 2003; Vater *et al.* 2003; Williams *et al.* 1997). Moreover, as they are chemically and structurally similar to D-oligonucleotides, they induce minimal immunogenic response (Sooter and Ellington 2002), making spiegelmers ideal candidates for *in vitro* and *in vivo* diagnostics as well as their application as *in vivo* imaging agents (Boisgard *et al.* 2005), as well as for therapeutics.

Alternative SELEX strategies

Photo SELEX

In photochemical SELEX (PhotoSELEX), modified oligonucleotides are capable of photocross-linking the target molecule. The method is based on the incorporation of a modified nucleotide that can be activated by absorption of light, in place of a native base in RNA/DNA libraries. This method is based on the strength of covalent bonding rather than affinity, therefore the specificity of the aptamer can be higher, but the false positives obtained with this method can also be higher, requiring the cross-linking conditions to be optimized to ensure screening validity. An example of the type of modified oligonucleotide used is 5-bromo-2′-deoxyuridine (5-Br-dUTP) or 5-iodouracil (5-IU) used as a substituent for TTP (Meisenheimer and Koch 1997). The modified oligonucleotide absorbs ultraviolet light in the 310 nm range, where nucleic acids and proteins absorb very weakly, resulting in an excited singlet that specifically cross-links with the target's aromatic or sulfur-containing amino acids, when in close proximity (Dietz and Koch 1987, 1989; Ito, Saito and Matsuura 1980). This technique was first used in 1995 by Jensen (Jensen *et al.* 1995), who obtained RNA aptamers for human immunodeficiency virus type 1 Rev protein using 5-IU. Later, with a random 61-mer oligonucleotide library, in which 5-Br-dUTP replaced thymidine, aptamers against the recombinant human basic fibroblast growth factor 155 (bFGF155) were selected with exceptional sensitivity (Golden *et al.* 2000). The kinetic analysis of photoaptamer–protein photocrosslinking reactions was studied by Koch and a mathematical model for this was presented (Koch *et al.* 2004). The model is based on the hypothesis that specific binding of aptamer and target followed by laser excitation can lead to either an aptamer/target complex formed with covalent bonds or irreversible photo damage to the aptamer. This model was used to characterize the photocross-linking between three photoaptamers and their targets and the results obtained from cross-linking data were confirmed by other independent measurements (W. Wang and Jia 2009). PhotoSELEX results in high affinity and highly specific aptamers by improving the separation method; however, a major limitation is that low molecular mass molecules that do not have functional groups to cross-link cannot be used as targets for the photoSELEX procedure.

Toggle SELEX

Toggle SELEX allows isolation of aptamers with a wide range of specificities by selecting against related targets in alternating cycles (Bunka and Stockley 2006). Alternation of the target between homologous proteins of different species ensures that aptamers will bind to both proteins, most likely to domains conserved between the two proteins (Y. Yang *et al.* 2007). Aptamers that bind both human and porcine thrombin were selected with this method (White *et al.* 2001). The same method with another name – target switching – was used to obtain aptamers called 'oligobodies' against ERK2 (Bianchini *et al.* 2001) and protein phosphatase 2A (Radrizzani *et al.* 1999).

Capillary electrophoresis based SELEX

Methods for aptamer selection that exploit capillary electrophoresis report a considerable improvement in the separation of bound and unbound nucleic acids, only requiring a few cycles for the isolation of very high affinity aptamers. Capillary electrophoresis SELEX (CE-SELEX) involves the selection of binding oligonucleotides based on a mobility shift due to complex formation. Sequences bound to the target move through the capillary electrophoresis at a velocity different from those not bound to the target, allowing separation of complex bound sequences from the rest of library. The main advantages of CE-SELEX are the ability to perform the selection in free solution, reduction of non-specific binders and increased capability of separation. Using CE-SELEX, high affinity aptamers against IgE and neuropeptide Y were selected in just a few selection rounds (Mendosa and Bowser 2004a, 2004b, 2005). Two aptamers with the same sequence were selected against cytotoxin ricin using both affinity chromatography SELEX and CE-SELEX, but with markedly shortened selection time with the latter (J.J. Tang *et al.* 2006). The determination of kinetic and thermodynamic constants can be achieved using an efficient selection process based on equilibrium capillary electrophoresis of equilibrium mixtures (ECEEM). Using this method, an aptamer binding MutS protein was selected with a K_d of 15 nM in just three rounds of selection (Drabovich, Berezovski and Krylov 2005). Capillary electrophoresis has also been used in a format called non-equilibrium capillary electrophoresis of equilibrium mixtures (NECEEM). NECEEM enables the 'selection' of aptamers with nanomolar K_d values in a single round of selection (Berezovski, Drabovich and Krylova 2005; Krylov 2006; Krylov and Berezovski 2003), and has the same advantages as ECEEM for the selection of aptamers with pre-defined kinetic parameters (Hamula et al. 2006).

Non-SELEX

Non-SELEX is another method for aptamer screening. This method is based on a process that involves repetitive steps of partitioning with no amplification step. A high affinity aptamer for h-ras was obtained using Non-SELEX with NECEEM as a separation step (Berezovski et al. 2006). The advantage of this method is its potential applicability to non-amplifiable libraries, such as those of DNA tagged small molecules obtained by DNA-template synthesis (Gartner et al. 2004; Sakurai, Snyder and Liu 2005).

Automated selection

The *in vitro* selection process in practice is repetitive, time-consuming and not always applicable for high-throughput selections. To overcome these limitations the SELEX process has been automated, handling multiple targets efficiently since they are processed in parallel (Eulberg et al. 2005). The first robotic workstation was a modified Beckam Biomek 2000 pipetting robot in which a PCR thermal cycler, a magnetic bead separator, reagent trays and a pipette tip station were integrated (Cox, Rudolph and Ellington 1998). The automated selection was later further optimized, substituting the magnetic separator by more efficient vacuum-filtering (Cox and Ellington 2001). The new protocol was applied to the generation of anti-protein aptamers in a matter of days, and demonstrated the automated selection of an anti-lysozyme aptamer that functions as an efficient inhibitor of cell lysis. This automated selection was successful in developing aptamers towards proteins such as CYT-18, MEK1 and Rho with dissociation constants in the picomolar to nanomolar range (Cox et al. 2002a) against proteins that had been transcribed and translated directly on the robotic workstation (Cox et al. 2002b) and against the mirror-image configuration (Spiegelmer) of substance P (Eulberg et al. 2005). The prototype of a microfluidic, microline-based assembly that uses Labview-controlled actuable valves and a PCR machine was used for the selection and synthesis of an anti-lysozyme aptamer (Hybarger et al. 2006).

Overall, there are a variety of techniques available for the isolation of aptamers and the choice of technique is largely dependent on the target molecule and the application and there is no definitive pattern to predict which method is preferred, but it is clear that more efficient methods of partitioning target bound nucleic acids are being reported, leading to higher affinity aptamers.

Aptamers in analysis

Aptamers in affinity-based assays

Since their appearance, aptamers have shown great promise as biocomponents for analysis not only due to their high affinity and specificity but also due to their increased stability, flexibility and versatility as compared to antibodies. Techniques involving antibodies or phage displayed antibody fragments are increasingly being replaced by aptamers in different configurations, taking advantage of the unique properties of aptamers.

Reporter Linked Aptamer Assay (RLAA)

Ellington was the first to exploit aptamers for the quantification of a specific protein in a cell extract using radiolabelled aptamers in a filter binding assay (Conrad and Ellington 1996). By means of previously selected RNA aptamers immobilized on a nitrocellulose filter, βII isozyme of rat protein kinase C was quantified with high reproducibility and specificity in the presence of rat brain extract. While Ellington was demonstrating the concept, at Larry Gold's NeXus Pharmaceutical Inc., Drolet *et al.* reported the first use of aptamers in an enzyme-linked immunosorbent assay (ELISA)-like assay, referred to as ELONA (enzyme-linked oligonucleotide assay), or more correctly RLAA (reporter-linked aptamer assay). In this first reported RLAA (Drolet, Moon-McDermott and Romig 1996), the reporting antibody of a sandwich ELISA was substituted by a fluorescein-tagged RNA aptamer to detect the vesicular endothelial growth factor (VEGF) in serum. A monoclonal antibody specific for VEGF was immobilized and used to capture the target, followed by incubation of a VEGF-binding fluorescein-labelled RNA aptamer with the detection facilitated by an enzyme-labelled anti-fluorescein antibody. This cumbersome assay yielded results very similar to those obtained in typical ELISA, being able to detect concentrations down to 1 pM and without showing any cross-reactivity towards other cytokines. Although this work did not exploit the specific properties of aptamers, it did highlight the possibility of aptamers to compete with, and complete the use of, antibodies in bioanalysis, paving the way for a new approach for detection.

Since its conceptualization in 1971, ELISA (Engvall and Perlmann 1971) has been extensively used in diagnostics as well as in research, due to the robustness, suitability to automation and relative simplicity of the method. Nevertheless, the emerging needs of ever decreasing detection limits, necessitates the search for new assays and methodologies, such as immuno-PCR, that can meet these demands and keep ELISA at the cutting edge. As such, RLAA represents a natural development of ELISA, exploiting the unique flexibility of aptamers to achieve reliable and consistent detection of very low levels of target molecules, but still using a microtitre plate format.

In simple formats, aptamers have been exploited in RLAA in much the same way as antibodies in ELISA (O'Sullivan 2002). Among others, Vivekananda and Keil (2006) reported on a RLAA format that uses aptamers both as capturing and reporting elements. Aiming to detect an antigen associated with *Francisella tularenis japonica* – a bacteria which causes the infectious endemic disease tularaemia – the group selected different DNA aptamers against the target to be used in the RLAA, where the reporting oligonucleotide was labelled with biotin, recognized in turn by streptavidin-HRP. The results obtained demonstrate that for detection, the RLAA sandwich is superior to ELISA, lowering the limit of detection (LoD) by about three times – from 6.9 to 1.7×10^3 bacteria/ml, while also improving the specificity. An exhaustive work on different RLAA sandwich formats was accomplished by Baldrich, Restrepo and O'Sullivan (2004). Different RLAA and mixed antibody/aptamer formats using the Thrombin Binding Aptamer (TBA) were studied, elucidating the critical parameters for optimal aptamer performance and highlighting that the conditions for each aptamer assay must be individually optimized and, unlike ELISA, that no universal optimal operating parameters exist. The authors went on to detail the first report of an aptamer-based displacement assay. Based on the observation that the TBA had a lower affinity for a modified form of thrombin than for the native form, microtitre plates were coated with the aptamer to retain horse peroxidise (HRP) modified thrombin. Following the addition of the native thrombin, HRP-thrombin was displaced, achieving a LoD of below 10 nM. Although, displacement assays with antibodies have been a long-time goal, very few reports of successful assays have appeared (Barry and Soloviev 2004; Gerdes, Meusel and Spener 1997, 1999; Ngo and Narinesingh 2005, Ngo 2005). This first report of an aptamer in a displacement assay not only highlighted the flexibility to different assay formats of aptamers, but also demonstrated the long-term stability of immobilized aptamers. Demonstrating significant advantages of aptamers over their antibody counterparts, Cruz-Aguado and Penner (2008) also showed that the displacement assay is possible with other aptamers, obtaining nanomolar detection limits of ochratoxin A (OTA) through a displacement assay by means of anti-OTA aptamer and a complementary fluorescein-labelled oligo with fluorescent polarization detection. The group exploited this in a displacement assay where immobilized avidin aptamer was incubated with the streptavidin-HRP followed by a 15-min incubation with the target avidin, achieving nanomolar detection limits. This displacement phenomenon has been widely exploited in aptasensors, as detailed later in this chapter.

Magnetic beads (Mb) have also been exploited as supports for RLAA formats, using either immobilized or labelled aptamers as capture or detection reagents. Bruno and Kiel (2002) used magnetic beads to develop a RLAA sandwich able to detect non-pathogenic Sterne strain *Bacillus anthracis* spores, cholera whole toxin and *Staphylococcal enterotoxin B* at nanogram to low picogram levels. The approach used relies on a selected pool of DNA aptamer immobilized on tosyl-activated magnetic beads used to capture the analyte in solution. The detection was accomplished by exposing the beads to a second biotinylated aptamer pool followed by addition of either streptavidin-conjugated ruthenium-trisbypyridine or avidin-HRP conjugate. Rye and Nustad (2001) developed a hybrid immuno-beads assay based on 5'-biotinylated DNA thrombin aptamer and anti-thrombin antibody. In this case, sheep anti-mouse IgG was conjugated to magnetic beads and used to bind the IgG anti-thrombin monoclonal antibody. The modified beads were then exposed to a pre-incubated mixture of thrombin and biotinylated thrombin aptamer. Finally, europium (Eu)-labelled streptavidin was added for detection. The results revealed that the TBA could bind the target under both stringent conditions and physiological concentrations, again highlighting the enhanced flexibility of aptamers for analytical applications.

Demonstrating the impressive robustness of their use, aptamers have been tested in depth in many affinity assay formats: in addition to RLAAs, aptamers have been exploited in flow cytometry, affinity chromatography and capillary electrophoresis, among others. In flow cytometry analysis, labelled aptamers were used for detecting (and isolating) analytes immobilized on beads (K.A. Davis et al. 1996), as well as biomarkers expressed on cell surfaces (K.A. Davis et al. 1998; Shangguan et al. 2006; Herr et al. 2006). Fluorescent aptamers have also been used in capillary electrophoresis to detect IgE in buffer and serum samples (German, Buchanan and Kennedy 1998; Buchanan et al. 2003) or the reverse transcriptase of the human immunodeficiency virus type 1 (HIV-1) (Pavski and Le 2001; Fu, Guthrie and Le 2006). Furthermore, the high affinity, stability and the small size of the aptamers have facilitated their use in affinity chromatography. Aptamers immobilized in the stationary phase were demonstrated to obtain a very high selectivity in the retention of targets, as demonstrated by the efficient separation of arginine enantiomers (Geiger et al. 1996), the purification of fusion protein from cell lysate (Romig, Bell and Drolet 1999) and the separation of adenosine at different phosphorylation levels (Deng et al. 2001).

Affinity amplification assays

Nucleic acid aptamers are naturally predisposed to an assay format that combines the selectivity of the aptamers with the efficiency of nucleic acid amplification techniques, producing an impressive signal enhancement with considerably lower detection limits. The concept of integrating the sensitivity of nucleic acid amplification with an immunoassay, in the technique immunopolymerase chain reaction (immuno-PCR), precedes the exploitation of aptamers in bioanalysis. Immuno-PCR, first reported in 1992, is a method for the ultrasensitive detection of analytes (Sano, Smith and Cantor 1992). The reporting antibody is labelled with DNA, either directly or via a biotin-streptavidin bridge, improving the sensitivity of a conventional immunoassay and enhancing the detection limits up to

100 000-fold (Nam, Stoeva and Mirkin 2004; Mweene *et al.* 1996; Niemeyer, Adler and Wacker 2005; Adler 2005). The technique, however, does have some important drawbacks, such as difficulties in labelling the antibody with nucleic acids and, furthermore, this linkage – either directly to antibodies or via biotin-streptavidin linker bridges – is prone to a lack of precision often resulting in uneven numbers of oligos per antibody, giving high rates of error and affecting sensitivity (McKie *et al.* 2002; Niemeyer *et al.* 1999). Additionally, following the immunorecognition step, the DNA needs to be separated from the antibody for subsequent amplification. Aptamers do not require to be conjugated to a label as they can inherently act both as detecting and reporting molecule, simply by flanking the aptamer with two primer sequences, bypassing the problems with immuno-PCR. In aptamer affinity amplification methods, the antibody labelled with an oligonucleotide used in immuno-PCR is replaced by an aptamer ready to be amplified. This technique already has many different formats described in literature, where different sandwich formats and a variety of amplification methods have been reported.

Zhang *et al.* (2006) reported on the use of capillary electrophoresis (CE) for separation of the aptamer–target complex followed by downstream quantitative PCR, achieving the detection of just 180 molecules of HIV-1 reverse transcriptase. Fisher, Tarasow and Tok (2008) demonstrated detection of thrombin via rolling cycle amplification (RCA) and real-time quantitative PCR (qPCR). By using different concentrations of thrombin-modified magnetic microparticles to bind the TBA, flanked by two primer regions for amplification, the group achieved a 2 nM detection limit using RCA and as low as a few hundred femtomolar using qPCR. Exploiting two different thrombin-binding aptamers (TBAs), Pinto *et al.* (2009) detailed a sandwich format where qPCR was used. A detection limit of just a few hundreds femtomolar of thrombin was obtained where the biotinylated TBA for the heparin-binding site was immobilized onto a streptavidin-coated plate, which was used to capture the analyte in solution. Subsequently the TBA–thrombin complex was exposed to the TBA specific for the fibrinogen-binding site flanked by two primer regions, which was finally eluted and detected using qPCR, achieving a femtomolar detection limit. H.J. Lee and co-workers (2009) described an antibody/aptamer mixed sandwich to detect up to 10 *E. coli* cells/ml using qPCR.

In an interesting report, Waga's team achieved the selection of an RNA aptamer against the constant region (Fc) of rabbit IgG (Yoshida *et al.* 2008) to be used as a reporter molecule that could be taken advantage of in a multitude of assays where rabbit IgG antibodies were used. Detection was achieved using qPCR (Yoshida *et al.* 2009), highlighting the immense potential of the simple combination of an aptamer with antibody in affinity amplification assays.

Another example of the tremendous flexibility of aptamers is represented by the aptamer nuclease protection developed by X. Wang *et al.* (2004), where an exonuclease was added to a solution containing thrombin and its cognate aptamers, to digest all the oligonucleotides free in solution while not digesting those that form a complex with the analyte. Subsequent amplification of the surviving TBA molecules facilitated a detection limit of just a few hundred molecules of thrombin, demonstrating the power of this technique for highly sensitive analyte detection. The proximity ligation assay (PLA), similiarly to the nuclease protection assay, does not rely on the immobilization of either the analyte or the probe. In a typical PLA, two aptamers that bind two closely located binding sites are ligated using a template probe and a ligase enzyme to form a unique amplicon that is detected by qPCR. Using this technique, Fredriksson *et al* (2002) have been able to detect zeptomoles of the cytokine platelet-derived growth factor (PDGF), and the same group went on to use the technique to detect microbial pathogens (Gustafsdottir *et al*. 2006) and the prostate specific membrane agent (PMSA) (Zhu *et al*. 2006). In a similar format, L. Yang and Ellington (2008) developed a setup exploiting the conformation-switching of the TBA, achieving detection limits in the picomolar range. This technique is highly aptamer dependent as it takes advantage of the secondary structures assumed by the aptamers following target binding, with the conformational change the aptamer undergoes upon binding, promoting ligation within a primer region for subsequent real-time amplification. This concept has been used by the same group to detect PDGF in the nanomolar range using RCA, where the conformation-switching aptamer was circularized on interaction with its target (L. Yang *et al*. 2007).

Aptasensors

According to the IUPAC definition (Thevenot *et al*. 1999):

> A biosensor is a self-contained integrated device, which is capable of providing specific quantitative or semi-quantitative analytical information using a biological recognition element (biochemical receptor) which is retained in direct spatial contact with an electrochemical transduction element.

The analytical power of biosensors relies on the molecular recognition element used. Aptamers offer many advantages as biocomponents, such as their size, ease of modification, easy immobilization compatible with developed microarray technologies, as well as flexible detection formats. Furthermore, aptamers offer unique chemical and physical proprieties for construction of re-usable biosensors.

Optical aptasensors

One of the first examples of aptasensors was reported by Kleinjung *et al*. (Kleinjung *et al*. 1998) for the detection of L-adenosine through the cognate RNA aptamer immobilized over optical fibres via a biotin-streptavidin bridge, where total internal reflection fluorescence from the fluorescein isothiocyanate (FITC) labelled L-adenosine in competition with the unlabelled L- and D- enantiomers was used. In parallel, Ellington's group designed an aptasensor in which a FITC modified thrombin-binding aptamer (TBA) was used as a signalling probe for thrombin detection (Potyrailo *et al.* 1998), monitoring the evanescent-wave-induced fluorescence anisotropy of a microscope slide-immobilized aptamer, detecting subnanomolar levels of analyte in a few picolitres in less than 5 min. Using the conformational switch properties of aptamers, Jhaveri *et al.* (2000) exploited the unique properties of aptamers to achieve detection introducing fluorescent labels on the aptamer at positions where the fluorescent moiety enhanced the intensity of the emission upon target binding. The authors described the successful modification of both RNA and DNA ATP-binding aptamers by using a moiety of acridine phosphoramidite, thus positioning the fluorophore outwards following target binding. Using the same principle, Katilius, Katiliene and Woodbury (2006) developed DNA aptamer detection systems for thrombin, IgE and PDGF using fluorescent nucleotide analogues. The advantage of this last approach is that little or no decrease in binding affinity is expected due to the introduction of the fluorescent moiety, but it is still critical to identify correctly the optimal position for introduction of the reporter moiety.

The use of a reporter pair rather than one single molecule has been intensively investigated to detect the conformational changes induced by target recognition. The first report of this type of strategy was from Tyagi and Kramer (1996) who detailed a nucleic acid molecular beacon for the detection of specific nucleic acid sequences. Here single stranded nucleic acid probes form a characteristic hairpin structure with each terminal labelled with a fluorophore and a quencher, respectively, which in close proximity result in fluorescent quenching. Following hybridization with its complementary sequences, the hairpin structure undergoes a spontaneous conformational change, breaking apart the hairpin structure, and thus generating the signal. Due to the nucleic acid nature of aptamers, they can easily be formatted into a molecular beacon structure, again demonstrating their superior flexibility to alternative assay formats for analytical applications as compared with antibodies.

Anti-cocaine (Stojanovic, De Prada and Landry 2001) and anti-PDGF (Fang *et al.* 2003) aptamers were engineered to form an aptabeacon involving the 3'- and 5'- ends when bound to the analyte. Using a fluorophore at one extremity and a quencher on the opposite one, the binding produced a measurable 'switch-off' of the signal. The anti-thrombin aptamer (TBA) was used in a 'off–on' mechanism (Hamaguchi, Ellington and Stanton 2001), where the original DNA aptamer was appended with a sequence to form a hairpin structure that was destabilized by target binding, which inherently promoted the formation of the particular G-quadruplex structure of the TBA thus distancing the quencher from the fluorophore, with a concomitant increase in fluorescence signal. A similar approach was used by Z. Tang *et al.* (2008) who modified the aptamer with a polyethyleneglycol (PEG) spacer followed by a short antisense oligonucleotide modified with a quencher in the 3'-terminus, demonstrating the viability of the method with TBA and the anti-ATP aptamer, achieving 90% of signal intensity in just 5 s.

The intermolecular displacement of an oligonucleotide from a complementary sequence has been exploited by Nutiu and Li (2003). The authors used a fluorophore-labelled DNA aptamer and an antisense small oligonucleotide conjugated to a quenching moiety, and was facilitated by the signal generated due to displacement of the antisense oligonucleotide, where an aptamer/antisense–oligonucleotide duplex was used with the antisense sequence being released following recognition of the target, and the fluorophore thus able to generate signal. Similarly, Li and Ho described the detection of ATP using a switch-off mechanism taking advantage of a duplex formed by the anti-ATP aptamer and the antisense oligonucleotide labelled at the extremities with a fluorophore-quenching pair. The signal is quenched when the analyte promotes the displacement of the antisense strand, which is free to form a beacon structure (N. Li and Ho 2008). Avoiding direct conjugation of the aptamer with reporter moieties, a chimera comprised of two active aptamers fused together has been detailed by Stojanovic and Kolpashchilov (2004). The authors used ATP, theophylline or flavin mononucleotide (FMN)-binding aptamers merged with the anti-malachite green aptamer, forming a two domain chimera. When the analyte domain was bound to the target, an allosteric activation occurs allowing the binding of the malachite green to the second domain, promoting the enhancement of the dye fluorescence.

As well as organic dyes, metal complexes (Jiang, Fang and Bai 2004) and water-soluble cationic polymers have been used as reporters (Ho *et al.* 2002). An easy and universally applicable setup was described by Ho and Leclerc (2004) to detect thrombin using polythiophene, a cationic water-soluble polymer that binds to the negatively charged backbone of the oligonucleotide. Upon binding to thrombin, the TBA undergoes a conformational change, inducing the polymer to wrap the folded G-quadruplex structure rather than form a planar and highly conjugated structure with unfolded ssDNA, producing a colorimetric signal used to detect human thrombin at levels as low as 2×10^{-15} mol.

Gold nanoparticles (AuNPs) have also been used as reporters, exploiting their unique property to change color upon aggregation (Rao *et al.* 2000). This property has been used by Mirkin *et al.* to develop oligonucleotide-modified AuNP probes (Elghanian *et al.* 1997), where hybridization with the promoted nanoparticle aggregation inducing a color change. Using a similar principle Liu and Lu designed ATP and cocaine biosensors, by means of two oligonucleotides each conjugated to a AuNP and complementary to two different regions of the anti-ATP aptamer; in the absence of analyte a duplex is formed, cross-linking the nanoparticle and resulting in aggregation. However, when the aptamer folds up, binding the analyte, the duplex is broken causing disaggregation (J. Liu and Lu 2006). In a similar approach, Zhao *et al.* (2007) used an anti-ATP aptamer, which was hybridized to a short oligo attached to a AuNP, where the presence of the duplex prevented aggregation due to the repulsion of the negative charges of the aptamer backbone, with duplex destabilization due to the presence of the target resulting in aggregation.

Quantum dot nanoparticles (QDs), nanocrystals offering greater photostability, longer fluorescence lifetime and sharper emission bands than traditional dyes, have also been used as reporters. Levy, Cater and Ellington (2005) reported on a Fluorescence Resonance Energy Transfer (FRET)-based aptasensor where a QD nanoparticle was conjugated to TBA, which formed a duplex with an antisense oligonucleotide labelled with quencher, dampening the QD signal. Following addition of thrombin the aptamer underwent a conformational change, releasing the complementary antisense and the QD signal was restored.

Surface plasmon resonance (SPR) has also been used to monitor aptamer-target binding, for example for detection of the retinol-binding protein 4 (RBP4) in serum (S.J. Lee *et al.* 2008), as well as the detection of the $2'-5'$ oligoadenylate synthase (Potyrailo *et al.* 1998) and others (Tombelli, Minunni and Mascini 2005b; Ikebukuro, Kiyohara and Sode 2005; James and Barclay 1998a; Brody 1999; Van Ryk and Venkatesan 1999). SPR is also routinely used to monitor SELEX and characterize aptamer affinity and selectivity. In a further report exploiting surface plasmon phenomena, the use of nanoparticle plasmon resonance (NPPR) has also been reported, where Hernandez *et al.* (2009) detailed an aptananosensor consisting of a noble metal nanoparticle coated with an anti-avidin aptamer used to detect avidin levels as low as 20 nM.

Mass aptasensors   Piezoelectric transduction using a quartz crystal microbalance (QCM) has been used to detect thrombin and the HIV-1 Tat protein with low detection limits (Tombelli *et al.* 2005a). In addition, the catalytic activity of immobilized aptazymes has been monitored in real time looking at the changes in QCM frequency, where Knudsen *et al.* (2006) adapted the catalytic ligation activity of the anti-HIV-1 Rev peptide aptazyme to detect the HIV-1 Rev peptide as well as the cleavage activity of the theophylline aptazyme for the detection of theophylline. Another example of mass-based aptasensing exploits the use of a microcantilever functionalized with aptamers to detect Taq DNA polymerase and the hepatitis C virus (Savran *et al.* 2004; K.S. Hwang *et al.* 2007). Furthermore, Schlensong *et al.* (2004) immobilized aptamers on a surface acoustic wave (SAW) device to detect thrombin and HIV-1 Rev peptide, using the changes in the propagation of the acoustic waves for transduction.

Electrochemical aptasensors   The first example of an electrochemical aptasensor was reported by Ikebukuro *et al.* (2005), who reported a sandwich assay for detecting thrombin using chronoamperometry, where the TBA was immobilized via a thiol terminal on a gold electrode to capture the target analyte. Following exposure to the analyte, the electrode was exposed to a second aptamer against a different epitope of the thrombin and conjugated to pyrrole quinoline quinone glucose dehydrogenase (PQQGDH), which generated an electrochemical signal using glucose substrate and methoxyphenemethosulfonate mediator, achieving a LoD of 10 nM. A similar detection limit was obtained by Bang, Cho and Kim (2005) who used a thrombin aptabeacon immobilized over the electrode surface with the redox indicator methylene blue, which was released when the stem loop was opened upon binding with thrombin, with a concomitant decrease in electrochemical signal. A similar format was reported by Xiao *et al.* (2005a) who conjugated the methylene blue moiety directly to an extremity of the TBA, attaching this electrochemical aptabeacon to an electrode surface, achieving a detection limit of 10 nM LoD for the analysis of thrombin in real serum samples. In this case, the authors used the longer TBA and in the absence of target, the aptabeacon has enough elasticity to contact the electrode surface, but upon binding the quadruplex is induced, forming a rigid structure with the methylene blue label located far from the electrode surface in a 'switch-off' mechanism. To achieve a 'signal-on' mechanism, Xiao *et al.* (2005b) immobilized the TBA aptamer on a gold electrode surface with an oligonucleotide linker, and then hybridized the probe to an oligonucleotide conjugated at one extremity with the methylene blue. Upon the addition of the analyte, the aptamer favoured target binding with the thrombin and dehybridized from its complementary, rendering flexibility to the probe and facilitating accessibility of the methylene blue moiety to the electrode surface.

Using an alternative format, where the folding of TBA into a G-quadruplex structure upon binding with the thrombin was exploited, Radi *et al.* (2006) reported an electrochemical aptabeacon with a 'switch-on' mechanism. The TBA was immobilized on a gold electrode at one terminus and ferrocene labelled at the other terminus, which, in the absence of the target, was located too far from the electrode surface to facilitate electron transfer. However, the conformational change of the aptamer from a random coil to a G-quadruplex upon target binding brought the ferrocene label in close proximity to the gold surface, facilitating electron transfer and signal generation with a sub-nanomolar detection limit. Further reports of electrochemical aptabeacons also exploited the conformational changes undergone by the aptamer upon binding, for example the use of methylene blue to monitor the formation of the thrombin–TBA complex via chronoamperometry and DPV with a LoD of 10 nM (Hianik *et al.* 2005) as well as the cocaine–cognate aptamer complex via DPV with LoD of 500 μM (Baker *et al.* 2006).

In another report, an ion-selective field-effect transistor (ISFET) was used to detect adenosine in a particular displacement method where no label was required (Zayats *et al.* 2006). An anti-adenosine aptamer was immobilized on an electrode surface and hybridized to a shorter complementary oligonucleotide, and when adenosine interacted with the aptamer, it displaced the shorter oligonucleotide, and a change of charge was detected achieving a detection limit of a few micromoles. This detection limit has been improved to a few hundred nanomoles by L. Shen *et al.* (2007) who implemented the system described using $[Ru(NH_3)_6]^{3+}$, which binds electrostatically to the DNA and the signal decreases when one strand is displaced. Similar results have been obtained by B.L. Li *et al.* (2007), who modified the original setup by immobilizing a probe complementary to the aptamer on the surface and the aptamer is then liberated from the surface on addition of the target.

For label-less detection, impedance is one of the most commonly reported transduction techniques for electrochemical aptasensors, as detection can be achieved simply upon target binding, which changes the impedance. The transduction principle is generally based on monitoring the electron-transfer of the electrochemical redox marker $[Fe(CN)_6]^{4-/3-}$, whose access to the surface is impeded or enhanced, depending on the specific experimental setup. Furthermore, as the aptamer is negatively charged and is also relatively small compared to an antibody, much better sensitivity than for immunosensors exploiting impedance transduction can be expected. D.K. Xu and colleagues (2005) first reported the detection of human IgE using electrochemical impedance spectroscopy (EIS) with the cognate aptamer as recognition element, immobilizing the aptamer on a gold film electrode via self-assembly, achieving a detection limit of 0.1 nM. With an analogous setup, C.F. Pan et al. (2009) described the detection of a few thousand cells/ml of the T-leukemia cell line CCRF-CEM, while Radi et al. (2005) achieved a detection limit of 2 nM thrombin exploiting the 15-mer TBA. Interestingly, using the same approach but replacing the gold electrode with a microfabricated gold thin film, Cai, Lee and Hsing (2006) further decreased the detection limits of thrombin by about 20 times and J.L. Lee et al. (2008) achieved a detection limit as low as 0.5 nM thrombin with pyrolyzed carbon film.

Similarly, but using the opposite format, Rodriguez, Kawde and Wang (2005) immobilized the biotinylated anti-lysozyme aptamer onto a streptavidin-modified indium tin oxide surface, exploiting changes of the surface-modified charges upon the binding event through Faradaic impedance spectroscopy, achieving a detection limit of 10 nM of lysozyme. Before the binding event, the negatively charged aptamer on the surface creates an electrostatic barrier to the electron transfer of the negatively charged ferricyanide couple and this barrier is disrupted when the aptamer binds the positively charged cognate target. To enhance the detection limit of a system relying on impedance spectroscopy transduction, Y. Xu and co-workers (2006) realized an interesting approach, where following binding with its cognate aptamer immobilized on a gold surface, thrombin was denatured by the addition of the chaotropic agent guanidine hydrochloride, resulting in an increase in impedance and achieving a dramatically reduced detection limit of just 10 fM.

Electrochemical stripping analysis has also been used in aptasensors. Hansen et al. (2006) demonstrated a multidisplacement assay of thrombin and lysozyme labelled with quantum dots. Following deposition of thiolated anti-thrombin and anti-lysozyme aptamers on gold electrode surfaces, quantum dot labelled thrombin and lysozyme were captured by the aptamer and electrochemically stripped at a coated glassy carbon electrode. The sensitivity and selectivity of the system was tested with BSA and IgG as controls, giving a detection limit of 0.5 pM for thrombin in a signal displacement setup.

Magnetic beads have also been used in combination with electrochemical detection. One example cites the use of an anti-lysozyme aptamer labelled with magnetic beads for the separation, concentration and detection of lysozyme (Kawde et al. 2005), where the captured protein was released from the aptamer for its electrochemical detection by chronopotentiometric stripping, achieving a detection limit of 7 nM.

An indirect square wave voltammetry detection method of aptamer–thrombin interaction was reported by Le Foch, Ho and Leclerc (2006) using the biochemical properties of the nucleic acid aptamers. This strategy was based on the specific enzymatic hydrolysis of ssDNA molecules by a nuclease, which did not degrade aptamer attached to the target. Thus, following enzymatic degradation, thrombin was denatured to release the aptamer, which was then hybridized with a DNA capture probe for detection of the aptamer, achieving a detection limit of 75 nM.

An alternative approach was based on the direct adsorption of thrombin on a modified gold electrode surface, with subsequent specific interaction with anti-thrombin labelled aptamer (Mir, Vreeke and Katakis 2006), detected using chronoamperometry with a detection limit of 3 nM. The same group also reported another thrombin aptasensor, based on the thrombin-based catalysis of the chromogenic substrate, β-Ala–Gly–Arg–p-nitroaniline, producing p-nitroaniline, where the rate of p-nitroaniline formation was followed by UV adsorption at 405 nm, or electrochemically by the reduction of its nitro group. Electrochemical detection was carried out by DPV.

Finally carbon nanotubes have been exploited as elements in an electrochemical aptasensor architecture, where a single-walled carbon nanotube field-effect transistor was reported for thrombin sensing (So *et al.* 2005). The change of the electrical double layer before and after interaction with the target was detected and the system achieved a detection limit of 10 nM, with a linear range of 0–100 nM. Furthermore, Maehashi *et al.* (2007) produced label-free protein biosensors based on aptamer-modified carbon nanotube field-effect transistors for the real-time detection of IgE, demonstrating a better performance than monoclonal antibodies under the same assay conditions.

In conclusion, a wide range of aptasensor formats have been reported, with a large increase in the number of publications in the period 2005–2010. Exploiting formats not possible with immunosensors, impressive detection limits have been achieved with easy to use formats, which can be easily applied to mass-producible sensors. There has also been a move away from the 'model system' of the thrombin binding aptamer, demonstrating the flexibility and potential widespread application of aptasensors for highly sensitive, specific and rapid detection of targets, potentially at the site of analysis, such as at the point-of-care, for example at a physician's office.

Imaging with aptamers

The advantages of aptamers over others recognition elements, such as their flexibility to be chemically modified and their high affinity to their targets, have attracted attention from medical biologists for imaging and treating cancers and other diseases (Medintz *et al.* 2003; Chu *et al.* 2006; Young and Rozengurt 2006). An added advantage in the continuous search for earlier diagnosis and improved therapeutic solutions is the possibility of faster approval of aptamers than other candidates by authorities such as the FDA (Missailidis *et al.* 2008). As previously mentioned, aptamers offer unique benefits compared to other targeting agents: they bind with high affinity and selectivity; are easily and rapidly synthesized using *in vitro* techniques; are stable and consistent (Jayasena 1999), demonstrating their possibility use as powerful alternatives to antibodies and peptides for imaging (Ferreira *et al.* 2008; Hesselberth *et al.* 2000; Borbas *et al.* 2007; Missailidis and Perkins 2007). The first aptamer used for imaging – an anti-human neutrophil elastase aptamer labelled with ^{99m}Tc for the *in vivo* imaging of inflammatory sites – was reported by Charlton, Sennello and Smith (1997). ^{99m}Tc is a gamma-emitting radionuclide that has a 6-h half-life and an energy emission spectrum which is ideal for imaging applications. The isotopic labelling of the aptamer facilitated the capture of an image with a γ-camera where the aptamer showed significantly higher target-to-background (T/B) ratio, in less time, than its IgG counterpart (Famulok and Mayer 1999). Additionally, further reports detailed the higher rate of clearance of the aptamer from the peripheral circulation compared to the IgG, which effectively permitted superior T/B ratios (Charlton *et al.* 1997). Positron emission tomography (PET) is a sensitive, functional nuclear imaging technique, requiring radionuclide molecular imaging agents and which, combined with image contrast and the rapid pharmacokinetics of the aptamers, permits non-invasive assessment and quantification of specific biological and pharmacological processes at the molecular levels. However, photonic technologies have developed in the last decade aiming to evaluate the biodistribution of fluorescent aptamers, avoiding the complication of working with radioisotopes, such as the two complementary photonic technologies, which had been developed for small animal research – namely, whole body imaging and *in vivo* confocal fibre microscopy. Whole body imaging is a rapid semi-quantitative technique that allows comparisons of the tissue biodistribution of aptamers in superficial tissues such as xenografts over a very long time (Tavitian *et al.* 2009). This technique facilitates the analysis of many compounds at costs and throughputs that are not achievable by, for example, PET. However, the field of view is smaller than a square millimetre, the images are two-dimensional and quantification is not as precise as that for PET. Fibred confocal microscopes achieve micrometric spatial resolutions, facilitating the quantitative documentation of the uptake, penetration and distribution of aptamers tagged with a fluorescent dye at the cellular level, and *in vivo* confocal fibre microscopy is an important complement for images obtained with PET or whole body fluorescence imaging which have low spatial resolution (ca. 1–2 mm at best) (Hicke *et al.* 2006; D.W. Hwang *et al.* 2010).

Aptamers in cytometry

Aptamers coupled to fluorescence reporters are increasingly reported in cytometry research, as well as cytomics in which protein interactions are studied in a living cell. Several methods have been described since K.A. Davis *et al.* (1996) analysed the affinity of a FITC-labelled DNA aptamer and a complex of mouse antihuman neutrophil elastase (HNE) antibody and FITC-labeled rat anti-mouse antibody to HNE-labelled beads. The results obtained with flow cytometry showed equally efficient detection of HNE. The same group went on to report on the high specificity of fluorescein-labelled aptamers in the target recognition of human recombinant CD4-stained mouse T-cells expressing human CD4, using as negative control mouse T-cells lacking human CD4 (K.A. Davis *et al.* 1998). The most common reporter molecule used to conjugate aptamers for flow cytometry applications is FITC (Shangguan *et al.* 2006), despite the fact that it is known to generate weak fluorescence emission and others studies have been carried evaluating the efficiency of others reporter molecules with promising results, such as phycoerythrin (K.A. Davis *et al.* 1998), tetramethylrhodamine anhydride (TAMRA) (Herr *et al.* 2006) or nanoparticles like Au-Ag nanorods (Huang, Chang and Tan 2008). Recently several papers have been published demonstrating that imaging with quantum dot-conjugated aptamers can be used to simultaneously evaluate the expression of different cancer markers in a single cancer cell. As previously mentioned, quantum dots (QDs) are semiconductor nanocrystals which are being increasingly utilized as biological imaging and labeling probes because of their unique optical properties, including broad absorption with narrow photoluminescence spectra, high quantum yield, low photobleaching, and resistance to chemical degradation (Kang *et al.* 2009). In some cases, these unique properties have conferred advantages to QD over traditional fluorophores such as organic dyes (Gopalakrishnan *et al.* 2006; Medintz *et al.* 2005; Michalet *et al.* 2005; J. Yao *et al.* 2005). Once the aptamer is conjugated with the corresponding QD it can be applied to the sample being analysed, for example using a confocal laser scanning microscope to facilitate the differentiation of the target carcinogenic cell from the other non-carcinogenic cells (Bagalkot *et al.* 2007; Kang *et al.* 2009; D.W. Hwang *et al.* 2010). However, this imaging approach must solve problems associated with the biosafety of QDs and biostability of aptamers prior to clinical application (Kang *et al.* 2009).

Magnetic Resonance Imaging techniques

Novel techniques and the development of new imaging agents allow the *in vivo* imaging of some biological events at the cellular and subcellular level by Magnetic

Resonance Imaging (MRI). MRI is a non-invasive imaging technique capable of high spatial resolution, excellent soft tissue contrast, as well as an ability to simultaneously image tissue anatomy, physiology, and molecular events (Sosnovik, Nahrendorf and Weissleder 2007). Molecular MRI is already playing an important role in pre-clinical investigation and has the potential to play a major role in clinical diagnostics in the near future (Sosnovik 2008). In contrast to other molecular imaging techniques like PET, the main advantage of MRI lies in its lower sensitivity as the conventional gadolinium chelates used have a sensitivity in the micromolar range, which is significantly better than iodinated contrast agents (millimolar range) but significantly worse than radiotracer and fluorescence techniques (picomolar or better) (Sosnovik *et al.* 2007; Sosnovik 2008). As with the radioisotopic and fluorescent labeling, aptamers can be targeted by attachment to molecular magnetic resonance agents that have significantly higher relaxivities than conventional gadolinium chelates (Sosnovik *et al.* 2007), for example superparamagnetic iron oxide magnetic nanoparticles (MNPs) (T. Shen *et al.* 1993; C.H. Liu *et al.* 2007; Sosnovik 2008). These agents contain thousands of gadolinium or iron atoms for each targeting ligand attached to their surface, producing a high level of detection efficiency for each binding event (Sosnovik 2008).

To summarise, although there are few reports of the analytical applications of aptamers for imaging, there has been significant progress in their use, which, coupled with advances in detection technologies, show great promise for advanced image analysis.

Conclusions, outlooks and perspectives

As outlined in this chapter, the use of aptamers for analysis has found increasing application, with a plethora of innovative methodologies exploiting the unique properties of aptamers reported. Aptamers demonstrate remarkable flexibility as compared to their antibody counterparts, and asides from an increased stability, often improved selectivity, ease of production via chemical synthesis, easy labelling/immobilization and applicability to platforms developed for genomic analysis, aptamers possess characteristics that allow them to address the ever increasing demands of clinical diagnostic analysis as well as food control and environmental monitoring. As detailed, one format that has been considerably exploited using aptamers is that of a displacement type assay – a format very difficult to achieve with antibodies. Different displacement approaches have been explored, based on the displacement of DNA strands complementary to the aptamer sequence, modified targets, or hybrid complementary sequences, with all approaches only requiring sample addition, with extremely rapid response times and excellent sensitivity and selectivity. It can be envisaged that these formats can be easily adapted to lateral flow devices as well as diagnostic microsytems, where the need for complex multi-step assays requiring several washes, is completely avoiding thus rendering a cost-effective and easy-to-use tool. In another innovative assay, which again exploits the specific properties of aptamers, they have been used both as recognition and reporter molecule in an approach akin to immuno-PCR, achieving incredibly low detection limits, but in a much more straightforward manner than immuno-PCR as there is no requirement for labelling or precipitation. In additional to these formats, electrochemical, electro-chemiluminescent and fluorescent molecular aptabeacons that gain exploit properties specific to aptamers and not possible with antibodies, have been reported. Again with this format, the only required end-user intervention is that of sample addition, with the response being extremely fast, sensitive and highly selective. Some reports have exploited the inherent properties of the aptamer to undergo a conformational change upon interaction with its target, while others have engineered the aptamer to fold/unfold into/out of a beacon format upon binding. In addition to these elegant approaches, which truly exploit the unique properties of aptamers for application in analysis, aptamers have also been used in 'traditional' formats, simply replacing antibodies, and often demonstrating superior detection limits, increased stability and compatibility with non-physiological matrices. As well as the significant developments reported for the application of aptamers, there have been considerable advances in the selection/isolation of aptamers, resulting in a noteworthy improvement in the number and diversity of aptamers selected, with an increasing tendency to select aptamers for specific applications, rather than using 'model systems' to demonstrate different detection platforms. The combination of innovative detection formats, addressing high selectivity, low detection limits and complex matrices, coupled with easy to implement selection methodologies, provides great promise for the future application of aptamers, and it can be expected to soon see commercial exploitation of these remarkable molecule tools for advanced analytical techniques.

References

Adler, M. (2005) Immuno-PCR as a clinical laboratory tool. *Adv Clin Chem* **39**: 239–292.

Avci-Adali, M., Paul, A., Wilhelm, N. *et al.* (2009) Upgrading SELEX technology by using lambda exonuclease digestion for single-stranded DNA generation. *Molecules* **15**: 1–11.

Bagalkot, V., Zhang, L., Levy-Nissenbaum, E. *et al.* (2007) Quantum dot–aptamer conjugates for synchronous cancer imaging, therapy, and sensing of drug delivery based on bi-fluorescence resonance energy transfer. *Nano Lett* **7**: 3065–3070.

Baker, B.R., Lai, R.Y., Wood, M.S. *et al.* (2006) An electronic, aptamer-based small-molecule sensor for the rapid, label-free detection of cocaine in adulterated samples and biological fluids. *J Am Chem Soc* **128**: 3138–3139.

Baldrich, E., Restrepo, A. and O'Sullivan, C.K. (2004) Aptasensor development: elucidation of critical parameters for optimal aptamer performance. *Anal Chem* **76**: 7053–7063.

Baldrich-Rubio, E., Campas, M. and O'Sullivan, C.K. (2004) *Aptamers. Powerful Molecular Tools for Therapeutics and Diagnostics*. John Wiley & Sons, Ltd, Chichester.

Bang, G.S., Cho, S. and Kim, B.G. (2005) A novel electrochemical detection method for aptamer biosensors. *Biosens Bioelectron* **21**: 863–870.

Barry, R. and Soloviev, M. (2004) Quantitative protein profiling using antibody arrays. *Proteomics* **4**: 3717–3726.

Berezovski, M., Drabovich, A. and Krylova, S.M. (2005) Nonequilibrium capillary electrophoresis of equilibrium mixtures: a universal tool for development of aptamers. *J Am Chem Soc* **127**: 3165–3171.

Berezovski, M., Musheev, M., Drabovich, A. and Krylov, N. (2006) Non-SELEX selection of aptamers. *J Am Chem Soc* **128**: 1410–1411.

Bianchini, M., Radrizzani, M. and Brocardo, M.G. (2001) Specific oligobodies against ERK-2 that recognize both the native and the denatured state of the protein. *J Immunol Methods* **252**: 191–197.

Biesecker, G., Dihel, L., Enney, K. and Bendele, R.A. (1999) Derivation of RNA aptamer inhibitors of human complement C5. *Immunopharmacology* **42**: 219–230.

Biroccio, A.J., Hamm, J., Incitti, I. *et al.* (2002) Selection of RNA aptamers that are specific and high-affinity ligands of the hepatitis C virus RNA-dependent RNA polymerase. *J. Virol*, **76**: 3688–3696.

Blank, M., Weinschenk, T., Priemer, M. and Schluesener, H. (2001) Systematic evolution of a DNA aptamer binding to rat brain tumor microvessels: selective targeting of endothelial regulatory protein pigpen. *J Biol Chem* **276**: 16464–16468.

Bock, L.C., Griffin, L.C., Latham, J.A. *et al.* (1992) Selection of single-stranded DNA molecules that bind and inhibit human thrombin. *Nature* **355**: 564–566.

Boisgard, R., Kuhnast, B., Vonhoff, S. *et al.* (2005) In vivo biodistribution and pharmacokinetics of 18F-labelled spiegelmers: a new class of oligonucleotidic radiopharmaceuticals. *Eur J Nucl Med Mol Imaging* **32**: 470–477.

Borbas, K.E., Ferreira, C.S.M., Perkins, A. *et al.* (2007) Design and synthesis of mono- and multimeric targeted radiopharmaceuticals based on novel cyclen ligands coupled to anti-MUC1 aptamers for the diagnostic imaging and targeted radiotherapy of cancer. *Bioconjugate Chem* **18**: 1205–1212.

Breaker, R.R. (1997) DNA aptamers and DNA enzymes. *Curr Opin Chem Biol.* **1**: 26–31.

Brody, E.N. (1999) The use of aptamers in large arrays for molecular diagnostics. *Mol Diagn* **4**: 381–388.

Bruno, J.G. and Kiel, J.L. (1999) In vitro selection of DNA aptamers to anthrax spores with electrochemiluminescence detection. *Biosens Bioelectron* **14**: 457–464.

Bruno, J.G. and Kiel, J.L. (2002) Use of magnetic beads in selection and detection of biotoxin aptamers by electrochemiluminescence and enzymatic methods. *BioTechniques* **32**: 178–183.

Buchanan, D. D., Jameson, E. E., Perlette, J. *et al.* (2003) Effect of buffer, electric field, and separation time on detection of aptamer–ligand complexes for affinity probe capillary electrophoresis. *Electrophoresis* **24**: 1375–1382.

Bunka, D.H. and Stockley, P.G. (2006) Aptamers come of age –at last. *Nat Rev Microbiol* **4**: 588–596.

Burmeister, P.E., Lewis, S.D., Silva, R.F. *et al.* (2005) Direct in vitro selection of a 2'O-methyl aptamer to VEGF. *Chem Biol* **12**: 25–33.

Burmeister, P.E., Wang, C., Killough, J.R. *et al.* (2006) 2'Deoxy purine, 2'-O-methyl pyrimidine (dRmY) aptamers as candidate therapeutics. *Oligonucleotides* **16**: 337–351.

Cai, H., Lee, T.M.H. and Hsing, I.M. (2006) Label-free protein recognition using an aptamer-based impedance measurement assay. *Sensors Actuators B Chem* **114**: 433–437.

Cerchia, L., Duconge, F., Pestourie, C. *et al.* (2005) Neutralizing aptamers from whole-cell SELEX inhibit the RET receptor tyrosine kinase. *PLoS Biol* **3**: 697–704.

Charlton, J., Sennello, J. and Smith, D. (1997) In vivo imaging of inflammation using an aptamer inhibitor of human neutrophil elastase. *Chem Biol* **4**: 809–816.

Chu, T.C., Shieh, F., Lavery, L.A. *et al.* (2006) Labeling tumor cells with fluorescent nanocrystal-aptamer bioconjugates. *Biosens Bioelectron* **21**: 1859–1866.

Ciesiolka, J., Gorski, J. and Yarus, M. (1995) Selection of an RNA domain that binds Zn^{2+}. *RNA* **1**: 538–550.

Conrad, R. and Ellington, A.D. (1996) Detecting immobilized protein kinase C isozymes with RNA aptamers. *Anal Biochem* **242**: 261–265.

Cox, J.C. and Ellington, A.D. (2001) Automated selection of anti-proteins aptamers. *Bioorg Med Chem* **9**: 2525–2531.

Cox, J.C., Hayhurst, A., Hesselberth, J. R. *et al.* (2002b) Automated selection of aptamers against protein targets translated in vitro:from gene to aptamer. *Nucleic Acids Research*, 30, e108.

Cox, J.C., Rajedran, M., Riedel, T. *et al.* (2002a) Automated adquisition of aptamer sequences. *Comb Chem High* **5**: 289–299.

Cox, J.C., Rudolph, P. and Ellington, A. (1998) Automated RNA selection. *Biotechno. Prog*, **14**: 845–850.

Cruz-Aguado, J.A. and Penner, G. (2008) Fluorescence polarization based displacement assay for the determination of small molecules with aptamers. *Anal Chem* **80**: 8853–8855.

Daniels, D.A., Chen, H., Hicke, B.J. *et al.* (2003) A tensacin-C aptamer identified by tumor cell SELEX: systematic evolution of ligands by exponential enrichment. *PNAS* **100**: 15416–15421.

Davis, J.P., Janji, N., Javornik, B.E. and Zichi, D. (1996) Identifying consensus patterns and secondary structure in SELEX sequence sets. *Methods Enzymol* **267**: 302–306.

Davis, K.A., Abrams, B., Lin, Y. and Jayasena, S.D. (1996) Use of a high affinity DNA ligand in flow cytometry. *Nucleic Acids Res* **24**: 702–706.

Davis, K.A., Lin, Y., Abrams, B. and Jayasena, S.D. (1998) Staining of cell surface human CD4 with 2'-F-pyrimidine-containing RNA aptamers for flow cytometry. *Nucleic Acids Res* **26**: 3915–3924.

Deng, Q., German, I., Buchanan, D. and Kennedy, R.T. (2001) Retention and separation of adenosine and analogues by affinity chromatography with an aptamer stationary phase. *Anal Chem* **73**: 5415–5421.

Dietz, T.M. and Koch, T.H. (1987) Photochemical coupling of 5-bromuracil to tryptophan, tyrosine and histidine, peptide-like derivatives in aqueous fluid solution. *Photochem Photobiol* **46**: 971–978.

Dietz, T.M. and Koch, T.H. (1989) Photochemical reduction of 5-brouracil by cystine derivatives and coupling of 5-bromuracil to cystine derivatives. *Photochem Photobiol* **49**: 121–129.

Drabovich, A., Berezovski, M. and Krylov, N. (2005) Selection of smart aptamers by equilibrium capillary electrophoresis of equilibrium mixtures (ECEEM). *J Am Chem Soc* **127**: 11224–11225.

Drolet, D., Moon-McDermott, L. and Romig, T. (1996) An enzyme-linked oligonucleotide assay. *Nat Biotechnol* **14**: 1021–1025.

Eaton, B.E. (1997) The joys of in vitro selection: chemically dressing oligonucleotides to satiate protein targets *Curr Opin Chem Biol* **1**: 10–16.

Eaton, B.E., Gold, L. and Zichi, D.A. (1995) Let's get specific: the relationship between specificity and affinity. *Chem Biol* **2**: 633–638.

Elghanian, R., Storhoff, J.J., Mucic, R.C. *et al.* (1997) Selective colorimetric detection of polynucleotides based on the distance-dependent optical properties of gold nanoparticles. *Science* **277**: 1078–1081.

Ellington, A.D. and Szostak, J.W. (1990) In vitro selection of RNA molecules that bind specific ligands. *Nature* **346**: 818.

Ellington, A.D. and Szostak, J.W. (1992) Selection in vitro of single stranded DNA molecules that fold into specific ligand-binding strutures. *Nature* **355**: 850.

Engvall, E. and Perlmann, P. (1971) Enzyme-linked immunosorbent assay (ELISA) quantitative assay of immunoglobulin G. *Immunochemistry* **8**: 871–874.

Espelund, M., Stacy, R.A. and Jakobsen, K.S. (1990) A simple method for generating single-strand DNA probes labeled to high activities. *Nucleic Acids Res* **18**: 6157–6158.

Eulberg, D., Buchner, K., Maasch, C. and Klussmann, S. (2005) Development of an automated in vitro selection protocol to obtain RNA-based aptamers: identification of a biostable substance P antagonist. *Nucleic Acids Res* **33**: e45.

Famulok, M. and Huttenhofer, A. (1996) In vitro selection analysis of neomycin binding RNAs with a mutagenized pool of variants of the 16S rRNA decoding region. *Biochemistry* **35**: 4265–4270.

Famulok, M. and Mayer, G. (1999) Aptamers as tools in molecular biology and immunology. *Comb Chem Biol* **243**: 123–136.

Fang, X., Sen, A., Vicens, M. and Tan, W. (2003) Synthetic DNA aptamers to detect protein molecular variants in a high-throughput fluorescence quenching assay. *ChemBioChem* **4**: 829–834.

Faulhammer, D., Eschgfaeller, B., Stark, S. *et al.* (2004) Biostable aptamers with antagonistic properties to the neuropeptide nociceptin/orphanin FQ. *RNA* **10**: 516–527.

Ferreira, C.S.M., Papamichael, K., Guilbault, G. *et al.* (2008) DNA aptamers against the MUC1 tumour marker: design of aptamer–antibody sandwich ELISA for the early diagnosis of epithelial tumours. *Anal BioAnal Chem* **390**: 1039–1050.

Fischer, N.O., Tarasow, T.M. and Tok, J.B.H. (2008) Protein detection via direct enzymatic amplification of short DNA aptamers. *Anal Biochem* **373**: 121–128.

Fitzwater, T. and Polisky, B. (1996) A SELEX primer. *Methods Enzymol* **267**: 275–301.

Fredriksson, S., Gullberg, M., Jarvius, J. *et al.* (2002) Protein detection using proximity-dependent DNA ligation assays. *Nat Biotechnol* **20**: 473–477.

Fu, H., Guthrie, J.W. and Le, X.C. (2006) Study of binding stoichiometries of the human immunodeficiency virus type 1 reverse transcriptase by capillary electrophoresis and laser-induced fluorescence polarization using aptamers as probes. *Electrophoresis* **27**: 433–441.

Gartner, Z. J., B. N. Tse, R. Grubina, J. B. Doyon, T. M. Snyder and D. R. Liu (2004) DNA-Templated Organic Synthesis and Selection of a Library of Macrocycles. *Science* 305, 1601–1605.

Geiger, A., Burgstaller, P., Von der Eltz, H. *et al.* (1996) RNA aptamers that bind L-arginine with sub-micromolar dissociation constants and high enantioselectivity. *Nucleic Acids Res* **24**: 1029–1036.

Gerdes, M., Meusel, M. and Spener, F. (1997) Development of a displacement immunoassay by exploiting cross- reactivity of a monoclonal antibody. *Anal Biochem* **252**: 198–204.

Gerdes, M., Meusel, M. and Spener, F. (1999) Influence of antibody valency in a displacement immunoassay for the quantitation of 2,4-dichlorophenoxyacetic acid. *J Immunol Methods* **223**: 217–226.

German, I., Buchanan, D.D. and Kennedy, R.T. (1998) Aptamers as ligands in affinity probe capillary electrophoresis. *Anal Chem* **70**: 4540–4545.

Golden, M.C., Collins, B.D., Willis, M.C. and Koch, T.H. (2000) Diagnostic potential of PhotoSELEX-evolved ssDNA aptamers. *J Biotechnol* **81**: 167–178.

Goodman, S.D., Velten, N.J., Gao, Q. *et al.* (1999) In vitro selection of integration host factor binding sites. *J Bateriol* **181**: 3246–3255.

Gopalakrishnan, G., Danelon, C., Izewska, P. *et al.* (2006) Multifunctional lipid/quantum dot hybrid nanocontainers for controlled targeting of live cells. *Angew Chem Int Edit* **45**: 5478–5483.

Green, L.S., Jellinek, D., Bell, C. *et al.* (1995) Nuclease-resistant nucleic-acid ligands to vascular-permeability factor vascular endothelial growth-factor. *Chem Biol* **2**: 683–695.

Gustafsdottir, S.M., Nordengrahn, A., Fredriksson, S. *et al.* (2006) Detection of individual microbial pathogens by proximity ligation. *Clin Chem* **52**: 1152–1160.

Gyllensten, U.B. and Erlich, H.A. (1988) Generation of single-stranded DNA by the polymerase chain reaction and its application to direct sequencing of the HLA-DQA locus. *Proc Natl Acad Sci USA* **85**: 7652–7656.

Hamaguchi, N., Ellington, A. and Stanton, M. (2001) Aptamer beacons for the direct detection of proteins. *Anal Biochem* **294**: 126–131.

Hamm, J., Alessi, D.R. and Biondi, R.M. (2002) Bi-functional, substrate mimicking RNA inhibits MSK1-mediated cAMP-response element-binding protein phosphorylation and reveals magnesium ion-dependent conformational changes of the kinase. *J Biol Chem* **277**: 45793–45802.

Hamm, J., Huber, J. and Lührmann, R. (1997) Anti-idiotype RNA selected with an anti-nuclear export signal antibody is actively transported in oocytes and inhibits Rev- and cap-dependent RNA export. *Proc Natl Acad Sci USA* **94**: 12839–12844.

Hamula, C.L.A., Gurthrie, J.W., Zhang, H. *et al.* (2006) Selection and analytical applications of aptamers. *Trends Anal Chem* **25**: 681–689.

Hansen, J.A., Wang, J., Kawde, A.N. *et al.* (2006) Quantum-dot/aptamer-based ultrasensitive multi-analyte electrochemical biosensor. *J Am Chem Soc* **128**: 2228–2229.

Helmling, S., Maasch, C., Eulberg, D. *et al.* (2004) Inhibition of ghrelin action in vitro and in vivo by an RNA-Spiegelmer. *Proc Natl Acad Sci USA* **101**: 13174–13179.

Hermann, T. and Patel, D.J. (2000) Adaptive recognition by nucleic acid aptamers. *Science* **287**: 820–825.

Hernandez, F.J., Dondapati, S.K., Ozalp, V.C. *et al.* (2009) Label free optical sensor for Avidin based on single gold nanoparticles functionalized with aptamers. *J Biophotonics* **2**: 227–231.

Herr, J.K., Smith, J.E., Medley, C.D. *et al.* (2006) Aptamer-conjugated nanoparticles for selective collection and detection of cancer cells. *Anal Chem* **78**: 2918–2924.

Hesselberth, J., Robertson, M.P., Jhaveri, S. and Ellington, A.D. (2000) In vitro selection of nucleic acids for diagnostic applications. *J Biotechnol* **74**: 15–25.

Hianik, T., Ostatná, V., Zajacová, Z. *et al.* (2005) Detection of aptamer–protein interactions using QCM and electrochemical indicator methods. *Bioorganic Med Chem Lett* **15**: 291–295.

Hicke, B.J., Marion, C., Chang, Y.F. *et al.* (2001) Tenascin-C aptamers are generated using tumor cells and purified protein. *J Biol Chem* **276**: 48644–48654.

Hicke, B.J., Stephens, A.W., Gould, T. *et al.* (2006) Tumor targeting by an aptamer. *J Nucl Med* **47**: 668–678.

Higuchi, R.G. and Ochman, H. (1989) Production of single-stranded DNA templates by exonuclease digestion following the polymerase chain reaction. *Nucleic Acids Res* **17**: 5865.

Ho, H.A., Boissinot, M., Bergeron, M.G. *et al.* (2002) Colorimetric and fluorometric detection of nucleic acids using cationic polythiophene derivatives. *Angew Chem Int Edit* **41**: 1548–1551.

Ho, H.A. and Leclerc, M. (2004) Optical sensors based on hybrid aptamer/conjugated polymer complexes. *J Am Chem Soc* **126**: 1384–1387.

Hofmann, H.P., Limmer, S., Hornung, V. and Sprinzl, M. (1997) Ni^{2+} binding RNA motifs with an asymmetric purine-rich internal loop a G-A base pair. *RNA* **3**: 1289–1300.

Homann, M. and Goringer, H.U. (1999) Combinatorial selection of high affinity RNA ligands to live African trypanosomes. *Nulceic Acids Res* **27**: 2006–2014.

Huang, Y., Eckstein, F., Padilla, R. and Sousa, R. (1997) Mechanism of ribose 2′-group discrimination by an RNA polymerase. *Biochemistry* **36**: 8231–8242.

Huang, Y.F., Chang, H.T. and Tan, W.H. (2008) Cancer cell targeting using multiple aptamers conjugated on nanorods. *Anal Chem* **80**: 567–572.

Huizenga, D.E. and Szostak, J.W. (1995) A DNA aptamer that binds adenosine and ATP. *Biochemistry* **34**: 656–665.

Hultman, T., Stahl, S., Hornes, E. and Uhlen, M. (1989) Direct solid phase sequencing of genomic and plasmid DNA using magnetic beads as solid support. *Nucleic Acids Res* **17**: 4937–4946.

Hwang, K.S., Lee, S.M., Eom, K. *et al.* (2007) Nanomechanical microcantilever operated in vibration modes with use of RNA aptamer as receptor molecules for label-free detection of HCV helicase. *Biosens Bioelectron* **23**: 459–465.

Hwang, D.W., Ko, H.Y., Lee, J.H. *et al.* (2010) A nucleolin-targeted multimodal nanoparticle imaging probe for tracking cancer cells using an aptamer. *J Nucl Med* **51**: 98–105.

Hybarger, G., Bynum, J., Wiliams, R.F. *et al.* (2006) A microfluidic SELEX prototype. *Anal Bioanal Chem* **384**: 191–198.

Ikebukuro, K., Kiyohara, C. and Sode, K. (2005) Novel electrochemical sensor system for protein using the aptamers in sandwich manner. *Biosens Bioelectron* **20**: 2168–2172.

Irvine, D., Tuerk, C. and Gold, L. (1991) SELEXION. Systematic Evolution of Ligands by Exponential Enrichment with Integrated Optimization by Non-linear Analysis. *J Mol Biol* **222**: 739–761.

Ito, S., Saito, I. and Matsuura, T. (1980) Acetone sensitized photocoupling of 5-bromuridine to tryptophan derivatives via electrontransfer process. *J Am Chem Soc* **102**: 7535–7541.

James, W. (2000) Aptamers, in *Encyclopedia of Analytical Chemistry* (ed. R.A. Meyers), Johm Wiley & Sons, Ltd, pp. 4848–4871.

James, W. (2001) Nucleic acid and polypeptide aptamers: a powerful approach to ligand discovery. *Curr Opin Pharmacol* **1**: 540–546.

Jayasena, S. D. (1999) Aptamers: an emerging class of molecules that rival antibodies in diagnostics. *Clin Chem* **45**: 1628–1650.

Jellinek, D., Green, L.S., Bell, C. *et al.* (1995) Potent 2′-amino -2′-deoxypyrimidine RNA inhibitors of basic fibroblast growth factor. *Biochemistry* **34**: 11363–11372.

Jenison, R.D., Gill, S.C., Pardi, A. and Polisky, B. (1994) High-resolution molecular discrimination by RNA. *Science* **263**: 1425–1429.

Jensen, K.B., Atkinson, B.L., Willis, M.C. *et al.* (1995) Using in vitro selection to direct the covalent attachment of human immunodeficiency virus type 1 Rev protein to high-affinity RNA ligands. *Proc Natl Acad Sci USA* **92**: 12220–12224.

Jeong, S., Eom, T.Y., Kim, S.J. *et al.* (2001) In vitro selection of the RNA aptamer against the Sialyl Lewis X and its inhibition of the cell adhesion. *Biochem Biophys Res Commun* **281**: 237–243.

Jhaveri, S.D., Kirby, R., Conrad, R. *et al.* (2000) Designed signaling aptamers that transduce molecular recognition to changes in fluorescence intensity. *J Am Chem Soc* **122**: 2469–2473.

Jiang, Y., Fang, X. and Bai, C. (2004) Signaling aptamer/protein binding by a molecular light switch complex. *Anal Chem* **76**: 5230–5235.

Jones, L.A., Clancy, L.E., Rawlinson, W.D. and White, P.A. (2006) High-affinity aptamers to subtype 3a hepatitis C virus polymerase display genotypic specificity. *Antimicrob Agents Chemother* **50**: 3019–3027.

Kang, W.J., Chae, J.R., Cho, Y.L. *et al.* (2009) Multiplex imaging of single tumor cells using quantum-dot-conjugated aptamers. *Small* **5**: 2519–2522.

Katilius, E., Katiliene, Z. and Woodbury, N.W. (2006) Signaling aptamers created using fluorescent nucleotide analogues. *Anal Chem* **78**: 6484–6489.

Kato, Y., Minakawa, N., Komatsu, Y. *et al.* (2005) New NTP analogs: the synthesis of 4'-thioUTP and 4'-thioCTP and their utility for SELEX. *Nucleic Acids Res* **33**: 2942–2951.

Kawde, A.N., Rodriguez, M.C., Lee, T.M.H. and Wang, J. (2005) Label-free bioelectronic detection of aptamer-protein interactions. *Electrochem Commun* **7**: 537–540.

Keefe, A.D. and Cload, S.T. (2008) SELEX with modified nucleotides. *Curr Opin Chem Biol* **12**: 448–456.

Kleinjung, F., Klussmann, S., Erdmann, V.A. *et al.* (1998) High-affinity RNA as a recognition element in a biosensor. *Anal Chem* **70**: 328–331.

Klussman, S. (ed.) (2006) *The Aptamer Handbook. Fuctional Oligonucleotides and Their Applications*. Wiley-VCH Verlag GmbH & Co., KGaA, Weinheim.

Klussman, S., Nolte, A., Bald, R. *et al.* (1996) Mirror-image RNA that binds D-adenosine. *Nat Biotechnol* **14**: 1112–1116.

Knudsen, S.M., Lee, J., Ellington, A.D. and Savran, C.A. (2006) Ribozyme-mediated signal augmentation on a mass-sensitive biosensor. *J Am Chem Soc* **128**: 15936–15937.

Koch, T.H., Smith, D., Tabacman, E. and Zichi, D.A. (2004) Kinetic analysis of site-specific photoaptamer–protein cross-linking. *J Mol Biol* **336**: 1159–1173.

Kraus, E., James, W. and Barclay, A.N. (1998b) Cutting edge: novel RNA ligands able to bind CD4 antigen and inhibit CD4(+) T lymphocyte function. *J Immunol* **160**: 5209–5212.

Krylov, N. (2006) Nonequilibrium capillary electrophoresis of equilibrium mixtures (NECEEM): a novel method for biomolecular screening. *J Biomol Screen* **11**: 115–122.

Krylov, N. and Berezovski, M. (2003) Non-equilibrium capillary electrophoresis of equilibrium mixtures –appreciation of kinetics in capillary electrophoresis. *Analyst* **128**: 571–575.

Kujau, M.J. and Wölfl, S. (1997) Efficient preparation of single-stranded DNA for in vitro selection. *Mol Biotechnol* **7**: 333–335.

Kulbachinskiy, A.V. (2007) Methods for selection of aptamers to protein targets. *Biochemistry (Mosc)* **72**: 1505–1518.

Kusser, W. (2000) Chemically modified nucleic acid aptamers for in vitro selections: evolving evolution. *J Biotechnol* **74**: 27–38.

Latham, J.A., Johnson, R. and Toole, J.J. (1994) The application of a modified nucleotide in aptamer selection:novel thrombin aptamers containing 5-(1-pentynyl)-2′-deoxyuridine. *Nucleic Acids Res* **22**: 2817–2822.

Le Floch, F., Ho, H.A. and Leclerc, M. (2006) Label-free electrochemical detection of protein based on a ferrocene-bearing cationic polythiophene and aptamer. *Anal Chem* **78**: 4727–4731.

Lee, H.J., Kim, B.C., Kim, K.W. *et al.* (2009) A sensitive method to detect *Escherichia coli* based on immunomagnetic separation and real-time PCR amplification of aptamers. *Biosens Bioelectron* **24**: 3550–3555.

Lee, J.A., Hwang, S., Kwak, J. *et al.* (2008) An electrochemical impedance biosensor with aptamer-modified pyrolyzed carbon electrode for label-free protein detection. *Sensors Actuators B Chem* **129**: 372–379.

Lee, J.F., Hesselberth, J.R., Meyers, L.A. and Ellington, A.D. (2004) Aptamer database. *Nucleic Acids Res* **32**: 95–100.

Lee, S.J., Youn, B.S., Park, J.W. *et al.* (2008) ssDNA aptamer-based surface plasmon resonance biosensor for the detection of retinol binding protein 4 for the early diagnosis of type 2 diabetes. *Anal Chem* **80**: 2867–2873.

Legiewicz, M., Lozupone, C., Knight, R. and Yarus, M. (2005) Size, constant sequences, and optimal selection. *RNA* **11**: 1701–1709.

Leva, S., Lichte, A., Burmeister, J. *et al.* (2002) GnRH binding RNA and DNA Spiegelmers a novel approach toward GnRH antagonism. *Chem Biol* **9**: 351–359.

Levine, H.A. and Nilsen-Hamilton, M. (2007) A mathematical analysis of SELEX. *Comput Biol Chem* **31**: 11–35.

Levy, M., Cater, S.F. and Ellington, A.D. (2005) Quantum-dot aptamer beacons for the detection of proteins. *ChemBioChem* **6**: 2163–2166.

Li, B.L., Du, Y., Wei, H. and Dong, S.J. (2007) Reusable, label-free electrochemical aptasensor for sensitive detection of small molecules. *Chem Commun* (36): 3780–3782.

Li, N. and Ho, C.M. (2008) Aptamer-based optical probes with separated molecular recognition and signal transduction modules. *J Am Chem Soc* **130**: 2380–2381.

Li, Y., Geyer, C.R. and Sen, D. (1996) Recognition of anionic porphyrins by DNA aptamers. *Biochemistry* **35**: 6911–6922.

Lin, Y., Qiu, Q., Gill, S.C. and Jayasena, S.D. (1994) Modified RNA sequence pools for in vitro selection. *Nucleic Acids Res* **22**: 5229–5234.

Liu, C.H., Kim, Y.R., Ren, J.Q. *et al.* (2007) Imaging cerebral gene transcripts in live animals. *J Neurosci* **27**: 713–722.

Liu, J. and Lu, Y. (2006) Fast colorimetric sensing of adenosine and cocaine based on a general sensor design involving aptamers and nanoparticles. *Angew. Chem Int. Edit* **45**: 90–94.

Liu, J.J. and Stormo, G.D. (2005) Combining SELEX with quantitative assays to rapidly obtain accurate models of protein-DNA interaction. *Nucleic Acids Res* **33**: e141.

Lupold, S.E., Hicke, B.J., Lin, Y. and Coffey, D.S. (2002) Identification and characterization of nuclease-stabilized RNA molecules that bind human prostate cancer cells via the prostate -specific membrane antigen. *Cancer Res* **62**: 4029–4033.

Maehashi, K., Katsura, T., Kerman, K. *et al.* (2007) Label-free protein biosensor based on aptamer-modified carbon nanotube field-effect transistors. *Anal Chem* **79**: 782–787.

Mairal, T., Ozalp, V.C., Sanchez, P.L. *et al.* (2008) Aptamers: molecular tools for analytical applications. *Anal Bioanal Chem* **390**: 989–1007.

Mann, D., Reinemann, C., Stoltenburg, R. and Strehlitz, B. (2005) In vitro selection of DNA aptamers binding ethanolamine. *Biochem Biophys Res Commun* **338**: 1928–1934.

Mannironi, C., Di Nardo, A., Fruscoloni, P. and Tocchini-Valentini, G.P. (1997) In vitro selection of Dopamine RNA ligands. *Biochemistry* **36**: 9726–9734.

Marshall, K.A. and Ellington, A.D. (2000) In vitro selection of RNA aptamers. *Methods Enzymol* **318**: 193–214.

Masud, M.M., Kuwahara, M., Ozaki, H. and Sawai, H. (2004) Sialyllactose-binding modified DNA aptamer bearing additional functionality by SELEX. *Bioorg Med Chem* **12**: 1111–1120.

McKie, A., Samuel, D., Cohen, B. and Saunders, N.A. (2002) Development of a quantitative immuno-PCR assay and its use to detect mumps-specific IgG in serum. *J Immunol Methods* **261**: 167–175.

Medintz, I.L., Clapp, A.R., Mattoussi, H. *et al.* (2003) Self-assembled nanoscale biosensors based on quantum dot FRET donors. *Nature Mat* **2**: 630–638.

Medintz, I.L., Uyeda, H.T., Goldman, E.R. and Mattoussi, H. (2005) Quantum dot bioconjugates for imaging, labelling and sensing. *Nature Mat* **4**: 435–446.

Meisenheimer, K.M. and Koch, T.H. (1997) Photocross-linking of nucleic acids to associated proteins. *Crit Rev Biochem Mol Biol* **32**: 101–140.

Mendosa, S.D. and Bowser, M.T. (2004a) In vitro selection of high-affinity DNA ligands for human IgE using capillary electrophoresis. *Anal Chem* **76**: 5387–5392.

Mendosa, S.D. and Bowser, M.T. (2004b) In vitro evolution of functional DNA using capillary electrophoresis. *J Am Chem Soc* **126**: 20–21.

Mendosa, S.D. and Bowser, M.T. (2005) In vitro selection of aptamers with affinity for neuropeptide Y using capillary electrophoresis. *J Am Chem Soc* **127**: 9382–9383.

Michalet, X., Pinaud, F.F., Bentolila, L.A. *et al.* (2005) Quantum dots for live cells, in vivo imaging, and diagnostics. *Science* **307**: 538–544.

Mir, M., Vreeke, M. and Katakis, I. (2006) Different strategies to develop an electrochemical thrombin aptasensor. *Electrochem Commun* **8**: 505–511.

Misono, T.S. and Kumar, P.K.R. (2005) Selection of RNA aptamers against human influenza virus hemagglutinin using surface plasmon resonance. *Anal Chem* **342**: 312–317.

Missailidis, S. and Perkins, A. (2007) Update: aptamers as novel radiopharmaceuticals: their applications and future prospects in diagnosis and therapy. *Cancer Biother Radiopharm* **22**: 453–468.

Missailidis, S., Perkins, A., Santos, S.D. *et al.* (2008) Aptamer-based radiopharmaceuticals for diagnostic imaging and targeted radiotherapy of epithelial tumors. *Braz Arch Biol Technol* **51**: 77–82.

Mosing, R.K. and Bowser, M.T. (2007) Microfluidic selection and applications of aptamers. *J Sep Sci* **30**: 1420–1426.

Muller, J., El-Maarri, O., Oldenburg, J. *et al.* (2008) Monitoring the progression of the in vitro selection of nucleic acid aptamers by denaturing high-performance liquid chromatogprahy. *Anal Bioanal Chem* **390**: 1033–1037.

Murphy, M.B., Fuller, S.T., Richardson, P.M. and Doyle, S.A. (2003) An improved method for the in vitro evolution of aptamers and applications in protein detection and purification. *Nucleic Acids Res* **31**: e110.

Mweene, A.S., Ito, T. Okazaki, K. *et al.* (1996) Development of immuno-PCR for diagnosis of bovine herpesvirus 1 infection. *J Clin Microbiol* **34**: 748–750.

Naimuddin, M., Kitamura, K., Kinoshita, Y. *et al.* (2007) Selection-by-function: efficient enrichment of cathepsin E inhibitors from a DNA library. *J Mol Recognit* **20**: 58–68.

Nam, J.M., Stoeva, S.I. and Mirkin, C.A. (2004) Bio-bar-code-based DNA detection with PCR-like Sensitivity. *J Am Chem Soc* **126**: 5932–5933.

Ngo, T.T. (2005) Ligand displacement immunoassay. *Analytical Letters*, 38, 1057–1069.

Ngo, T.T. and Narinesingh, D. (2005) Ligand displacement fluorescence immunoassay for gentamicin and human IgG. *Anal Lett* **38**: 803–813.

Niemeyer, C.M., Adler, M. Pignataro, B. *et al.* (1990) Self-assembly of DNA-streptavidin nanostructures and their use as reagents in immuno-PCR. *Nucleic Acids Res* **27**: 4553–4561.

Niemeyer, C.M., Adler, M. and Wacker, R. (2005) Immuno-PCR: high sensitivity detection of proteins by nucleic acid amplification. *Trends Biotechnol* **23**: 208–216.

Nieuwlandt, D., Wecker, M. and Gold, L. (1995) In vitro selection of RNA ligands to substance P. *Biochemistry* **34**: 5351–5359.

Nikiforov, T.T., Rendle, R.B., Kotewicz, M.L. and Rogers, Y.-H. (1994) The use of phospho-rothioate primers and exonuclease hydrolysis for the preparation of single-stranded PCR products and their detection by solid-phase hybridization. *PCR Methods Appl* **3**: 285–291.

Nolte, A., Klussmann, S., Bald, R. *et al.* (1996) Mirror-design of L-oligonucleotide ligands binding to L-arginine. *Nat Biotechnol* **14**: 1116–1119.

Nutiu, R. and Li, Y. (2003) Structure-switching signaling aptamers. *J Am Chem Soc* **125**: 4771–4778.

Nutiu, R. and Li, Y. (2004) Structure-switching signaling aptamers:transducing molecular recognition into fluorescence signaling. *Chemistry (Easton)* **10**: 1868–1876.

O'Sullivan, C.K. (2002) Aptasensors –the future of biosensing? *Fresenius' J Anal Chem* **372**: 44–48.

Ohuchi, S.P., Ohtsu, T. and Nakamura, Y. (2006) Selection of RNA aptamers against recombinant transforming growth factor-β type III receptor displayed on cell surface. *Biochimie* **88**: 897–904.

Padilla, R. and Sousa, R. (1999) Efficient synthesis of nucleic acids heavily modified with non-canonical ribose 2'-groups using a mutant T7 RNA polymerase (RNAP). *Nucleic Acids Res* **27**: 1561–1563.

Pagratis, N.C. (1996) Rapid preparation of single-strand DNA from PCR products by strep-tavididin induced electrophoretic mobility shift. *Nucleic Acids Res* **24**: 3645–3646.

Pan, C.F., Guo, M.L., Nie, Z. *et al.* (2009) Aptamer-based electrochemical sensor for label-free recognition and detection of cancer cells. *Electroanalysis* **21**: 1321–1326.

Pan, W.H. and Clawson, G.A. (2009) The shorter the better: reducing fixed primer regions of oligonucleotide libraries for aptamer selection. *Molecules* **14**: 1353–1369.

Pan, W.H., Xin, P. and Clawson, G.A. (2008) Minimal primer and primer-free SELEX protocols for selection of aptamers from random DNA libraries. *Biotechniques* **44**: 351–360.

Pavski, V. and Le, X.C. (2001) Detection of human immunodeficiency virus type 1 reverse transcriptase using aptamers as probes in affinity capillary electrophoresis. *Anal Chem* **73**: 6070–6076.

Pinto, A., Bermudo Redondo, M.C., Cengiz Ozalp, V. and O'Sullivan, C.K. (2009) Real-time apta-PCR for 20 000-fold improvement in detection limit. *Mol Biosyst* **5**: 548–553.

Potyrailo, R.A., Conrad, R.C., Ellington, A.D. and Hieftje, G.M. (1998) Adapting selected nucleic acid ligands (aptamers) to biosensors. *Anal Chem* **70**: 3419–3425.

Purschke, W.G., Radtke, F., Kleinjung, F. and Klussmann, S. (2003) A DNA Spiegelmer to staphylococcal enterotoxin B. *Nucleic Acids Res* **31**: 3027–3032.

Radi, A.E., Acero Sánchez, J.L., Baldrich, E. and O'Sullivan, C.K. (2006) Reagentless, reusable, ultrasensitive electrochemical molecular beacon aptasensor. *J Am Chem Soc* **128**: 117–124.

Radi, A.E., Sánchez, J.L.A., Baldrich, E. and O'Sullivan, C.K. (2005) Reusable impedimetric aptasensor. *Anal Chem* **77**: 6320–6323.

Radrizzani, M., Brocardo, M.G., Solveyra, C.G. *et al.* (1999) Oligobodies: bench made synthetic antibodies. *Medicina Buenos Aires* **59**: 753–758.

Rao, C.N.R., Kulkarni, G.U., Thomas, P.J. and Edwards, P.P. (2000) Metal nanoparticles and their assemblies. *Chem Soc Rev* **29**: 27–35.

Rhie, A., Kirby, L., Sayer, N. *et al.* (2003) Characterization of 2′-fluoro-RNA aptamers that bind preferentially to disease-associated conformations of prion protein and inhibit conversion. *J Biol Chem* **278**: 39697–39705.

Robertson, D.L. and Joyce, G.F. (1990) Selection in vitro of an RNA enzyme that specifically cleaves single-stranded DNA. *Nature* **344**: 467–468.

Rodriguez, M.C., Kawde, A.N. and Wang, J. (2005) Aptamer biosensor for label-free impedance spectroscopy detection of proteins based on recognition-induced switching of the surface charge. *Chem Commun* (34): 4267–4269.

Romig, T.S., Bell, C. and Drolet, D.W. (1999) Aptamer affinity chromatography: combinatorial chemistry applied to protein purification. *J Chromatogr B Biomed Sci Appl* **731**: 275–284.

Ruan, C.C. and Fuller, C.W. (1991) Using T7 gene 6 exonuclease to prepare single-stranded templates for sequencing. *USB Comments* **18**: 1–8.

Ruckman, J., Green, L.S., Beeson, J. and Waugh, S. (1998) 2′-Fluoropyrimidine RNA-based aptamers to the 165-amino acid form of vascular endothelial growth factor (VEGF$_{165}$). *J Biol Chem* **273**: 20556–20567.

Rusconi, C.P., Scardino, E., Layzer, J. *et al.* (2002) RNA aptamers as reversible antagonists of coagulation factor IXa. *Nature* **419**: 90–94.

Rye, P.D. and Nustad, K. (2001) Immunomagnetic DNA aptamer assay. *BioTechniques* **30**: 290–295.

Sakurai, K., Snyder, T.M. and Liu, D.R. (2005) DNA-templated functional group transformations enable sequence-programmed synthesis using small-molecule reagents. *J Am Chem Soc* **127**: 1660–1661.

Sano, T., Smith, C.L. and Cantor, C.R. (1992) Immuno-PCR: very sensitive antigen detection by means of specific antibody-DNA conjugates. *Science* **258**: 120–122.

Sassanfar, M. and Szostak, J.W. (1993) An RNA motif that binds ATP. *Nature* **364**: 550–553.

Savran, C.A., Knudsen, S.M., Ellington, A.D. and Manalis, S.R. (2004) Micromechanical detection of proteins using aptamer-based receptor molecules. *Anal Chem* **76**: 3194–3198.

Sayer, N., Ibrahim, J., Turner, K. *et al.* (2002) Structural characterization of a 2′F-RNA aptamer that binds a HIV-1 SU glycoprotein, gp120. *Biochem Biophys Res Commun* **293**: 924–931.

Schlensog, M.D., Gronewold, T.M.A., Tewes, M.M. *et al.* (2004) A Love-wave biosensor using nucleic acids as ligands. *Sensors Actuators B Chem* **101**: 308–315.

Schneider, D., Gold, L. and Platt, T. (1993) Selective enrichment of RNA species for tight binding to *Escherichia coli* rho factor. *FASEB J* **7**: 201–207.

Schurer, H., Buchynskyy, A., Korn, K. *et al.* (2001) Fluorescence correlation spectroscopy as a new method for the investigation of aptamer/target interactions. *Biol Chem* **382**: 479–481.

Shangguan, D., Li, Y. Tang, Z.W. *et al.* (2006) Aptamers evolved from live cells as effective molecular probes for cancer study. *Proc Natl Acad Sci USA* **103**: 11838–11843.

Shen, L., Chen, Z., Li, Y.H. *et al.* (2007) A chronocoulometric aptamer sensor for adenosine monophosphate. *Chem Commun* (21): 2169–2171.

Shen, T., Weissleder, R., Papisov, M. *et al.* (1993) Monocrystalline iron-oxide nanocompounds (mion) –physicochemical properties. *Magn Reson Med* **29**: 599–604.

Shimada, T., Fujita, N., Maeda, M. and Ishihama, A. (2005) Systematic search for the Cra-binding promoters using genomic SELEX system. *Genes Cells* **10**: 907–918.

Shtatland, T., Gill, S.C. Javornik, B. E. *et al.* (2000) Interactions of *Escherichia coli* RNA with bacteriophage MS2 coat protein: genomic SELEX. *Nucleic Acids Res* **28**: E93.

So, H.M., Won, K., Kim, Y.H. *et al.* (2005) Single-walled carbon nanotube biosensors using aptamers as molecular recognition elements. *J Am Chem Soc* **127**: 11906–11907.

Sooter, L.J. and Ellington, A.D. (2002) Reflections on a novel therapeutic candidate. *Chem Biol* **9**: 857–858.

Sosnovik, D.E. (2008) Molecular imaging in cardiovascular magnetic resonance imaging: current perspective and future potential. *Top Magn Reson Imag* **19**: 59–68.

Sosnovik, D.E., Nahrendorf, M. and Weissleder, R. (2007) Molecular magnetic resonance imaging in cardiovascular medicine. *Circulation* **115**: 2076–2086.

Sousa, R. and Padilla, R. (1995) A mutant T7 RNA polymerase as a DNA polymerase. *EMBO J* **14**: 4609–4621.

Srinivasan, J., Cload, S.T., Hamaguchi, N. *et al.* (2004) DP-specific sensors enable universal assay of protein kinase activity *Chem Biol* **11**: 499–508.

Stojanovic, M.N., De Prada, P. and Landry, D.W. (2001) Aptamer-based folding fluorescent sensor for cocaine. *J Am Chem Soc* **123**: 4928–4931.

Stojanovic, M.N. and Kolpashchikov, D.M. (2004) Modular aptameric sensors. *J Am Chem Soc* **126**: 9266–9270.

Stoltenburg, R., Reinemann, C. and Strehlitz, B. (2005) FluMag-SELEX as an advantageous method for DNA aptamer selection. *Anal Bioanal Chem* **383**: 83–91.

Stoltenburg, R., Reinemann, C. and Strehlitz, B. (2007) SELEX –a (r)evolutionary method to generate high-affinity nucleic acid ligands. *Biomol Eng* **24**: 381–403.

Tang, J.J., Xie, J.W., Shao, N.S. and Yan, Y. (2006) The DNA aptamers that specifically recognize ricin toxin are selected by two in vitro selection methods. *Electrophoresis* **27**: 1303–1311.

Tang, Z.W., Mallikaratchy, P., Yang, R.H. *et al.* (2008) Aptamer switch probe based on intramolecular displacement. *J Am Chem Soc* **130**: 11268–11269.

Tavitian, B., Duconge, F., Boisgard, R. and Dolle, F. (2009) In vivo imaging of oligonucleotidic aptamers. *Methods Mol Biol* **535**: 241–259.

Thevenot, D.R., Toth, K., Durst, R.A. and Wilson, G.S. (1999) Electrochemical biosensors: recommended definitions and classification. *Pure Appl Chem* **71**: 2333–2348.

Tombelli, S., Minunni, M., Luzi, E. and Mascini, M. (2005a) Aptamer-based biosensors for the detection of HIV-1 Tat protein. *Bioelectrochemistry*, 67, 135–141.

Tombelli, S., Minunni, M. and Mascini, M. (2005b) Analytical applications of aptamers. *Biosens Bioelectron* **20**: 2424–2434.

Tuerk, C. and Gold, L. (1990) Systematic evolution of ligands by exponentional enrichment: RNA ligands to bacteiophage T4 DNA polymerase. *Science* **249**: 505–510.

Tyagi, S. and Kramer, F.R. (1996) Molecular beacons: probes that fluoresce upon hybridization. *Nat Biotechnol* **14**: 303–308.

Van Ryk, D.I. and Venkatesan, S. (1999) Real-time kinetics of HIV-1 Rev-Rev response element interactions –definition of minimal, binding sites on RNA and protein and stoichiometric analysis. *J Biol Chem* **274**: 17452–17463.

Vant-Hull, B., Payano-Baez, A., Davis, R.H. and Gold, L. (1998) The mathematics of SELEX against complex targets. *J Mol Biol* **278**: 579–597.

Vater, A., Jarosch, F., Buchner, K. and Klussmann, S. (2003) Short bioactive Spiegelmers to migraine-associated calcitonin gene-related peptide rapidly identified by a novel approach: tailored-SELEX. *Nucleic Acids Res* **31**: e130.

Vivekananda, J and Kiel, J.L. (2006) Anti-Francisella tularensis DNA aptamers detect tularemia antigen from different subspecies by aptamer-linked immobilized sorbent assay. *Lab Invest* **86**: 610–618.

Wang, C.L., Zhang, M., Yang, G. *et al.* (2003) Single-stranded DNA aptamers that bind differentiated but not parental cells: subtractive systematic evolution of ligands by exponential enrichment. *J Biotechnol* **102**: 15–22.

Wang, W. and Jia, L.-Y. (2009) Progress in aptamer screening methods. *Chin J Anal Chem* **37**: 454–460.

Wang, X.L., Li, F., Su, Y.H. *et al.* (2004) Ultrasensitive detection of protein using an aptamer-based exonuclease protection assay. *Anal Chem* **76**: 5605–5610.

Wen, J.D. and Gray, D.M. (2004) Selection of genomic sequences that bind tightly to Ff gene 5 protein: primer-free genomic SELEX. *Nucleic Acids Res* **32**: e182.

White, R., Rusconi, C.P., Scardino, E. *et al.* (2001) Generation of species cross-reactive aptamers using 'toggle' SELEX. *Mol Ther* **4**: 567–573.

Wiliams, K.P. and Bartel, D.P. (1995) PCR product with strands of unequal length. *Nucleic Acids Res* **23**: 4220–4221.

Williams, K., Liu, X., Schumacher, T.N.M. *et al.* (1997) Bioactive and nuclease-resistant L-DNA ligand of vasopressin. *Proc Natl Acad Sci USA* **94**: 11285–11290.

Wilson, C., Nix, J. and Szostak, J.W. (1998) Functional requirements for specific ligand recognition by a biotin-binding RNA psedoknot. *Biochemistry* **37**: 14410–14419.

Wilson, C. and Szostak, J.W. (1998) Isolation of a fluorophore-specific DNA aptamer with weak redox activity. *Chem Biol* **5**: 609–617.

Wochner, A. and Glokler, J. (2007) Nonradioactive fluorescence microtiter plate assay monitoring aptamer selections. *BioTechniques* **42**: 578–582.

Wu, L. and Curran, J. (1999) An allosteric synthetic DNA. *Nucleic Acids Res* **27**: 1512–1516.

Xiao, Y., Lubin, A.A., Heeger, A.J. and Plaxco, K.W. (2005a) Label-free electronic detection of thrombin in blood serum by using an aptamer-based sensor. *Ang Chem Int Edit* **44**: 5456–5459.

Xiao, Y., Piorek, B.D., Plaxco, K.W. and Heeger, A.J. (2005b) A reagentless signal-on architecture for electronic, aptamer-based sensors via target-induced strand displacement. *J Am Chem Soc* **127**: 17990–17991.

Xu, D.K., Xu, D.W., Yu, X.B. *et al.* (2005) Label-free electrochemical detection for aptamer-based array electrodes. *Anal Chem* **77**: 5107–5113.

Xu, Y., Yang, L., Ye, X. *et al.* (2006) An aptamer-based protein biosensor by detecting the amplified impedance signal. *Electroanalysis* **18**: 1449–1456.

Yang, L. and Ellington, A.D. (2008) Real-time PCR detection of protein analytes with conformation-switching aptamers. *Anal Biochem* **380**: 164–173.

Yang, L., Fung, C.W., Eun, J.C. and Ellington, A.D. (2007) Real-time rolling circle amplification for protein detection. *Anal Chem* **79**: 3320–3329.

Yang, Q., Goldstein, I.J., Mei, H.Y. and Engelke, D.R. (1998) Ligands that bind tightly and selectively to cellobiose. *Proc Natl Acad Sci USA* **95**: 5462–5467.

Yang, X.B., Li, X., Prow, T.W. *et al.* (2003) Immunofluorescence assay and flow-cytometry selection of bead-bound aptamers. *Nulceic Acids Res* **31**: e54.

Yang, Y., Yang, D., Schluesener, H.J. and Zhang, Z. (2007) Advances in SELEX and application of aptamers in the central nervous system. *Biomol Eng* **24**: 583–592.

Yao, J., Larson, D.R., Vishwasrao, H.D. *et al.* (2005) Blinking and nonradiant dark fraction of water-soluble quantum dots in aqueous solution. *Proc Natl Acad Sci USA* **102**: 14284–14289.

Yao, W., Adelman, K. and Bruenn, J.A. (1997) In vitro selection of packaging sites in a double-stranded RNA virus. *J Virol* **71**: 2157–2162.

Yoshida, Y., Horii, K., Sakai, N. *et al.* (2009) Antibody-specific aptamer-based PCR analysis for sensitive protein detection. *Anal Bioanal Chem* **395**: 1089–1096.

Yoshida, Y., Sakai, N., Masuda, H. *et al.* (2008) Rabbit antibody detection with RNA aptamers. *Anal Biochem* **375**: 217–222.

Young, S.H. and Rozengurt, E. (2006) Qdot nanocrystal conjugates conjugated to bombesin or ANG II label the cognate G protein-coupled receptor in living cells. *Am J Physiol Cell Physiol* **290**: C728–C732.

Yunusov, D., So, M., Shayan, S. *et al.* (2009) Kinetic capillary electrophoresis-based affinity screening of aptamer clones. *Anal Chim Acta* **631**: 102–107.

Yusuke, M., Nobuaki, S., Chiaki, O. and Akihiko, K. (2010) Selection of DNA aptamers using atomic force microscopy. *Nucleic Acids Res* **38**: e21.

Zayats, M., Huang, Y., Gill, R. *et al.* (2006) Label-free and reagentless aptamer-based sensors for small molecules. *J Am Chem Soc* **128**: 13666–13667.

Zhang, H., Wang, Z., Li, X.F. and Le, X.C. (2006) Ultrasensitive detection of proteins by amplification of affinity aptamers. *Ang Chem Int Edit* **45**: 1576–1580.

Zhao, W., Chiuman, W., Brook, M.A. and Li, Y. (2007) Simple and rapid colorimetric biosensors based on DNA aptamer and noncrosslinking gold nanoparticle aggregation. *ChemBioChem* **8**: 727–731.

Zhu, L., Koistinen, H., Wu, P. *et al.* (2006) A sensitive proximity ligation assay for active PSA. *Biol Chem* **387**: 769–772.

8

Use of Nanotechnology for Enhancing of Cancer Biomarker Discovery and Analysis: A Molecular Approach

Farid E. Ahmed

Introduction

Nanotechnology, deriving its name from the Greek word for 'dwarf', deals with structures that range from 1 to 100 nm and that are derived from naturally occurring material (e.g. carbohydrates, proteins), or are specifically engineered or modified for a biomedical application. There are about 700 products on the market that use nanotechnology, from sunscreens to electronics to the first cancer drug – Abraxane, used for breast cancer treatment – that uses nanoparticles of paclitaxel attached to albumin to achieve effective delivery of the medicine.

Nanotechology has been applied to cancer in two broad areas: therapy, involving the development of nanovectors (e.g. nanoparticles), which can be loaded with drugs or imaging agents and then actively targeted to tumours; and diagnostics, involving the development of high-throughput nanosensor devices for detecting the biological signatures of cancer (Nie *et al.* 2007). It is anticipated that new imaging agents, new diagnostic particles and new targeted therapies will soon come together to facilitate a form of personalized medicine in which early and more accurate detection could

Molecular Analysis and Genome Discovery, Second Edition. Edited by Ralph Rapley and Stuart Harbron.
© 2012 John Wiley & Sons, Ltd. Published 2012 by John Wiley & Sons, Ltd.

lead to rapid initiation of treatment, followed by diagnostic tests to find out if the patient is responding to treatment, that is real-time therapeutic monitoring (Kawasaki and Player 2005).

One of the hallmarks of particles is that their behaviour in the nano range differs from their intrinsic physical properties, as their behaviour in the nano range is not the same as when they are larger (i.e. in the macro range). For example, nanosized particles of gold and carbon may be toxic at the nanoscale, whereas large particles of the same material may not be, and the biological activity of the particle seems to increase as the particle size decreases (Brower 2006). Nanomaterial may also be carcinogenic because of the metals used to make them, as some metals can lead to reactive oxygen radicals that cause oxidative stress and DNA damage, or induce carcinogenicity if they contain a metal such as cadmium (Carter 2008). There is a strong likelihood that the biological activity of the nanoparticle may depend on physicochemical parameters not considered in routine toxicity screening studies (Igarashi 2006). Thus, it has been wisely recommended that the physicochemical *in vivo* and *in vitro* testing be carried out on all nanomaterial considered for use in drugs and devices for *in vivo* human use (Carter 2008).

In this chapter, we will focus on using nanomaterial for cancer diagnosis, for example by improving cancer protein biomarker capture and detection.

Proteomics and nanotechnology

Protein folding, protein–protein interactions and post-translational modifications (PTMS) are critical for protein function, but they also impede attempts at exponential protein amplification (Ahmed 2008a). A significant challenge in proteomics is that many of the clinically significant proteins are present in relatively low native concentrations in biological systems (Hortin 2006). Two dimension electrophoresis (2DE) in conjunction with mass spectrometry (MS) has sensitivity for protein detection in the $10^{-9}-10^{-12}$ M concentration range, whereas radioimmunoassay and ELISA techniques are sensitive to 10^{-15} M. Analysis of the low-abundance sets of the proteome is further complicated by the wide dynamic concentration range of proteins in biological systems that are not related to the expression of human disease and which spans an estimated 15 orders of magnitude (Anderson and Anderson 2002). Two general strategies have been developed to address these challenges: nanoscale multicomponent separation and nanoscale protein detection strategies.

Nanoscale multicomponent separation

Separation of proteins of interest from multicomponent mixtures has historically been carried out using 2DE methods that resolve proteins by mass and charge in isoelectric focused gels (O'Farrell 1975). However, these techniques are cumbersome and labour intensive. The resolution of protein fractions could be complicated by technical limitations and contamination by proteins with similar electrophoretic properties. In addition, their ability to resolve low-abundance proteins is limited and reproducibility is difficult (Ahmed 2008a).

Matrix-assisted laser desorption/ionization mass spectrometry (MALDI-MS) has been widely used for protein identification using organic mixtures (Ahmed 2008b). There are several improvements that need to be made for these matrices to improve detection. The most obvious problem is the interference of matrix peaks in the low molecular weight region, leading to an inability to collect peptide data below 800 m/z. Another problem is finding a suitable matrix for nanostructured surfaces. To solve these problems, use of several types of material have been attempted such as: metal and metal oxide particle such as Al, Mn, Mo, Si, Sn, TiO_2, W, WO_3, Zn and ZnO (Okuno *et al.* 2005); soluble-gel-deposited TiO_2 (Chen and Chen 2004); ordered mesoporous WO_3-TiO_2, Au-NPS (McLean, Stumpo and Russell 2005); self-assembling Ge nanorods (Seino *et al.* 2007); and silica-based nanoporous surfaces fabricated by coating silicon chips with a 500 nm thick nanoporous film of silicon oxide for capturing low molecular weight peptides from human plasma (Gaspari *et al.* 2006).

High-performance liquid chromatography (HPLC), particularly reverse-phase (RP) that uses a non-polar stationary phase to isolate non-polar proteins from complex mixtures has fewer technical limitations and can be coupled with downstream 2DE or MS (Ahmed 2009a). However, problems with protein denaturation under RP conditions have restricted its widespread use (Wu and Yates 2003). The advent of nanospray ionization (NSI) devices and their use in conjunction with tandem quadrupole time-of-flight mass spectrometers, resulted in faster sensitive systems capable of separating low-abundance proteins from multicomponent mixtures (Marko, Weil and Toms 2007).

Capillary electrophoresis (CE) offers a low running cost, a short separation time, various operation modes, a low sample volume requirement and a high-resolution technique that requires only a small amount of analyte and can be interfaced with MS for proteomic biomarker analysis. It is, however, not suited for peptides >20 kDa and peptides precipitate in capillaries in acidic conditions (Ahmed 2009b).

Protein microarray technology has been adapted from cDNA array technology; however, a fundamental difference lies in the concentration and dynamic range between genomic and proteomic arrays. While genomic variability occurs within a dynamic concentration of about 1–2 orders of magnitude, the dynamic range for protein concentration and differential expression is five- to tenfold greater (Ahmed 2006). Thus, while genome arrays require microscale volumes and spot sizes, protein arrays require larger sample volume and large spot sizes. Nanotechnology methods such as nanolithography (Lynch *et al.* 2004), atomic force microscopy (Lee *et al.* 2006) and other nanoscale techniques have addressed this problem as reduction in array spot size and increases in spot concentration afford acceptable dynamic range sensitivity. At the same time, sample requirement for array analysis is minimized, making the assays more practical and clinically acceptable for disease detection and characterization (Marko, Weil and Toms 2007).

RP protein microarrays using nanostructured surfaces for proteomics enable high-throughput screening of PTMs of signalling proteins (Johnson *et al.* 2008). Microarrays that used silicon instead of nitrocellulose as a surface, enabling creation of a large surface area by etching techniques for protein binding, have also used quantum dots as reporter agents to increase microarray sensitivity (Geho *et al.* 2005). The advantages and disadvantages of the various proteomic methods for biomarker analysis are presented in Table 8.1.

Use of nanoporous material surfaces as a fractionation tool for serum-based biomarker discovery used nanoporous silicon wafer to selectively deplete a fraction of proteins from serum. By controlling the pore size of nanoporous glass beads, distinct subsets of proteins were harvested and compared by surface-enhanced laser desorption/ionization (SELDI)-mass spectrometry (MS) (Geho *et al.* 2006a). Nanoporous materials that used monolith support were also used in proteomics technology because of their combinations of high flow rates and surface area, as well as the ability to tailor pore sizes and distributions to match a particular application (Josic 2007). Monolithic capillary columns for fractionation of serum proteins and peptides were reported using immobilized metal ion affinity chromatography, and eluted peptides were identified by matrix-assisted laser desorption/ionization (MALDI)-time-of-flight (TOF)-MS (Rainer *et al.* 2007).

Nanoscale protein detection strategies

Inorganic nanoparticles (nanobeads) are a crystallized aggregate of molecules with a diameter of generally less than 100 nm. When constructed for nanoscale detection of proteins, they are usually synthesized as oxides of iron, magnesium or silicone, although other metals (e.g. CdSe, ZnS) have been used (Ivanov *et al.* 2006). Nanoparticles could be fabricated by a variety of methods, most commonly those using superheating of raw material followed by self assembly of the particles during cooling, eventually maintaining them as emulsions in an inert liquid phase to prevent their aggregation and finally coating them with biological or chemical material such as phospholipids, specific ligand, antibodies or other agents that modify the physical properties of the beads and/or promote selective binding of proteins from solutions (Marko, Weil and Toms 2007).

The physicochemical characteristics and high surface-to-volume ratio of nanoparticles make them ideal candidates for developing biomarker harvesting platforms. Given the variety of nanoparticle technologies that are now available, it is feasible to tailor nanoparticle surfaces to selectively bind a subset of biomarkers and sequester them for label-free detection by sensitive analytical proteomic approaches (Johnson *et al.* 2008).

Table 8.1 Common advantages and limitations of proteomic methods for analysis

Ionization source/ analytical method	Advantages	Disadvantages
Surface-enhanced SELDI chips	Affinity capture on MALDI chips with chromatographic functionality, ease of use, automation, convenience, low sample volume, raw samples analyzed, various chip surfaces, does not require sophisticated bioinformatics tools, detection with a broad molecular mass region in a single analysis	Loss of important information, problems at pre-, analytical and post-analytical steps, bias towards high abundant proteins particularly in the low mass range, performance could change over time
Matrix-assisted laser/desorption ionization (MALDI)	Simple sample application and acquisition, high throughput, low downtime, can carry out sample reanalysis, tolerant towards contaminants such as detergents or common buffers	Variable sample preparation for different analytes, signal suppression, different discriminatory peaks for similar samples, biased towards highly abundant proteins, a limited mass window range, potential for data over fitting
Surface-derivatized magnetic beads	Can be automated, employs wide range of derivatized beads with different functional groups, sensitive, compatible with MS	Lack of reproducibility between commercial batches of the same beads
Nonporous substrates (silicon wafers, silica particles & glass beads	Allows for harvesting of distinct subsets of the proteome	Not a mature technology, needs standardization/validation
2DE chromatography	Applicable to large molecules, high resolution, allows visualizing changes in molecular mass (Mr), isoelectric point (pI) or PTMs	Not applicable to peptides <10 kDa, time consuming, low throughput, labor intensive, hydrophobic or LAPs, or those with extreme pIs or Mrs poorly represented
Capillary electrophoresis	Automation, relatively sensitive, low sample volume needed, low cost, MS/MS compatibility	Not well suited for peptides >20 kDa, precipitation of peptides in capillaries when acidic running buffers are used
Liquid chromatography	Automation, highly sensitive, accurate, multidimensional, versatile, HT potential, MS/MS compatibility	Time consuming, sensitive towards interfering compounds, limited mass range, often unsuitable for analysis of intact proteins
Protein microarrays	HT, low sample volume, chips have potential for assaying a wide range of biochemical activity, various platforms and detection methods are available	Antibodies are not availed for all screened proteins, no standardization is available for biomarker discovery, low sensitivity, qualitative

Magnetic nanoparticles

Nanoparticles measuring 6–12 nm diameter constructed with magnesium-, iron-, manganese- or cobalt-based cores and coated with proteins are susceptible to the effects of magnetic fields. Subsequent chemical desorption generates a concentrated fraction of the proteins (Osaka *et al.* 2006). Identification of protein biomarkers from multicomponent solutions from micolitre plasma samples using magnetic particles has been assessed with MALDI TOF-MS (Zhang *et al.* 2004), showing the potential of this technology to screen for human disease.

Quantum dots (QDs)

QDs are semiconductor light-emitting nanocrystals composed of an inert core of inorganic compounds (e.g. cadmium selenium (CdSe) nanoparticles) surrounded with a ZnS capping shell. Variation of the material used for nanocrystal and variation of the size of the nanocrystal afford a spectral range from the UV (ZnS) to near IR, 300–2500 nm (PbSe/TE) in peak emission. In comparison with organic dyes and fluorescent proteins, QDs have unique optical and electronic properties. For example, QDs have molar extinction coefficients that are 10–50 times larger than that of organic dyes, which cause them to be brighter in photon-limited *in vivo* conditions, making them suitable for *in vivo* biomolecular and cell imaging (Smith, Gao and Nie 2004). Because of their large Stokes shift, multiple coloured QDs can be simultaneously excited with a single wavelength, leading to their use in multiplexed immunoassays (Johnson *et al.* 2008).

QDs have widespread applications in *in vitro* cell biology fluorescent imaging techniques such as immunohistochemistry (IHS) and fluorescence resonance energy transfer (FRET) analysis of subcellular localization and interaction of proteins, which include single-particle tracking, fluorescence correlation spectroscopy and photomarking methods (Giepmans *et al.* 2006). However, because QDs are most commonly localized to endosomes, this makes it difficult to observe cytoplasmic trafficking of proteins when using microinjection or other subcellular localization methods. Moreover, the intermittent photoluminescence of QDs (blinking) results in challenges in tracking tagged protein movements in cells with automatic imaging techniques (Heuff, Swift and Cramb 2007).

QDs have an important application as immunohistochemical labels because the photo-bleaching organic fluorophores are a problem in both confocal and wide-field fluorescence microscopy samples, thereby allowing human prostate cancer cells to be stained with five different coloured labels, one for a house-keeping gene reference standard and the others against four different tumour biomarkers (Xing *et al.* 2007). QDs can be conjugated with molecules such as antibodies or nucleic acids to function as *in vivo* fluorescent biological detection probes; however, they must be made hydrophilic to ensure aqueous solubility (Bruchez *et al.* 1998).

QDs enhance the range of protein detection in western blots due to their ability to multiplex and detect a variety of protein samples with one blot (Bakalova *et al.* 2005). Luminescent QDs have been proposed as alternative to organic dyes or nanoporous silicon, rather than organic mixtures, as a MALDI matrix (Johnson *et al.* 2008). However, to lower the protein detection threshold beyond ELISA to detect antigens in the femtomolar range in biological samples, more complex magnetic nanoparticle detection systems combined with high resolution QDs were often needed (Agrawal, Sathe and Nie 2007).

Carbon nanotubes (CNTs)

CNTs are other nanodevices for biomarker detection that use single-walled carbon nanotubes. They belong to the family of fullerenes and are formed of coaxial graphite sheets (<100 nm) rolled into cylinders. These structures can be obtained either as single- (one graphite sheet), or multi-walled (several concentric graphite sheets) nanotubes. They exhibit excellent strength and electrical properties, and are efficient heat conductors. CNTs can be rendered water soluble by surface functionalization; therefore, they can be used as drug carriers and tissue-repair scaffolds (Polizu *et al.* 2006). Because of their metallic or semiconductor nature, CNTs have often been used as biosensors (Sanvicens and Marco 2008). They have been used for DNA haplotyping (Wooley *et al.* 2000), protein sensing, as well as carriers for imaging and therapeutic agent delivery (Wang, Liu and Jan 2004).

Single-wall CNTs are sensitive to single-protein binding events, and can be massively multiplexed with millions of tubes per chip for proteomic profiling. The tubes have extraordinary strength, unique electronic properties, and the ability to tag cancer-specific proteins to the surface. With a diameter of 1 nm and a length of 1 μm, these tubes are smaller than a single strand of DNA, and each tube becomes an atomic arrangement of one layer carbon atom on the surface. Protein binding events occurring on the surface of these nanotubes produce a measurable change in the mechanical and electrical properties of the sensor. By coating the surfaces of nanotubes with monoclonal antibodies, it is possible to detect cancer cells circulating in the blood (Jain 2008). Combination of electrochemical immunosensors using single-wall CNTs with multi-label secondary antibody-nanotube biconjugates could be used for the sensitive detection of cancer biomarkers in serum and other body fluids (Yu *et al.* 2006).

Bio-barcode assay

This assay provides a method of detecting proteins down to attomolar concentrations by amplifying protein targets with hundreds of copies of protein coding barcodes coated with unique nucleic acid oligonucleotides that could be released, amplified and detected by nucleic acids hybridization techniques (Nam, Thaxton and Mirkin 2003). Bifunctional nanoparticle probes could be coated with antibodies that bind a specific target protein and the bio-barcode of a unique nucleic acid sequence. The nanoprobes are mixed with a protein lysate and magnetic beads coated with antibody to the same target protein forming a nanoprobe-target, protein-magnetic bead sandwich. Gold nanoparticles (Au-NPs) contain hundreds of copies of protein coding bar-code strands. After reaction with the analyte, a magnetic field is used to localize and collect the sandwich structure, and a dithritol (DTT) solution at an elevated temperature is used to release the barcode strands. The barcode strands can then be identified microscopically using scantometric detection, which measures light scattering from nanostructured silver using a standard desktop scanner, or *in situ* if the barcodes have been pre-labelled with a detectable marker (Hill and Mirkin 2006). The barcode typically comprises 15- to 20-mer oligonucleotides, allowing pairing of a unique barcode to every conceivable recognition agent, for example for a 20-mer oligo, there are 420 unique combinations; thus, the potential for multiplexing is immense. The beads are magnetically separated from the reaction mixture and the nucleic acid probes are dehybridized from the multifunctional nanoprobes. The released nucleic acids may be amplified by PCR for nucleic acid chip-based detection of the nucleic acid bio-barcode of interest, revealing the protein identified from the lysate solution. When using this method for detection of prostate specific antigen (PSA), a linear dose–response over more than four orders of magnitude in both target concentration and concomitant signal and a 1000-fold improvement in detection limit compared to ELISA was observed, that is attomolar detection range of several hundred molecules of proteins in a microlitre of lysate (Bao *et al.* 2006).

Au-NPs containing protein-coding bar-code strands is considered to be the industry standard in nanotechnology. First introduced to the point-of-care (PoC) diagnostic market in the form of a pregnancy test, they are now used in the field of biomarker measurement because of their high diffusion rates, which results in improved mixing of the analyte and the capture particle with extremely high sensitivity. Their small size and large molar absorption coefficient also leads to the formation of a dense line of highly visible particles forming at the capture line, thus providing clear readouts (Johnson *et al.* 2008).

Surface-enhanced Raman scattering (SERS)

SERS depends on Raman spectroscopy to detect unique resonance frequencies generated by chemical bonds following irradiation, usually infrared. SERS uses gold nanoparticles modified to contain a monolayer of an intrinsically Raman scatterer. Two SERS formats are generally used as signal transducers in diagnosis: label-free assays where the SERS spectrum of the analyte is measured (Grow *et al.* 2003); and SERS reporter tags in sandwich immunoassay formats to capture the protein of interest from a lysate (Grubisha *et al.* 2003).

Nanocantelievers

Nanoelectromechanical systems are nanoscale detectors associated with molecular binding events that have a wide application as biosensor transducers, which can measure changes as low as 1 pg and can detect a single bacterium. They are example of a low-cost silicon microfabrication technology producing nanostructures whose size-dependent mechanical properties form the basis of a label-free biosensing platform with a multiplexing capability (Johnson *et al.* 2008). Nanocantelivers are most commonly constructed using focused ion-beam technology to fabricate cantilever arms from a single crystal silicon, which may be coated with gold or other material and subsequently modified with self-assembled monolayers (Pinnaduwage *et al.* 2004), polymeric films or antigen–antibody layers. The nanocantilevers could be combined with an electromechanical detection system to form a sensor capable of measuring binding events that occur on the surface of the modified cantilever arm (Mukhopadhyay *et al.* 2005).

Analyte binding to nanocantilevers results in measurable deflection due to change in the surface stress, detected either by measuring changes in resonance frequency associated with the binding event, or by detecting binding-induced deformation of the cantilever arm. The sensitivity of cantilever detectors that measure binding-induced bending is inversely proportional to the spring constant of the cantilever; thus, sensitivity improves as cantilever arm length increases. However, nanoscale cantilevers are constructed with short arms, resulting in high spring constant and low sensitivity for detecting binding-induced deformation. Conversely, the high spring constants increase the resonance frequency of the cantilever and decrease the threshold of mass detection. As the size of nanocantilever arms decrease, the resonance frequency increases allowing mass detection approaching the theoretical potential for single molecule detection (Marko, Weil and Toms 2007).

Proteomic application of nanocantilevers is challenging because of the difficulty in mobilizing ligands to the cantilever surface (Cheng *et al.* 2006). However, for prostate cancer PSA, a sensitivity and dynamic range of detection was shown to be in the range of 0.2 to 60 μg/ml (Wu *et al.* 2001). Advances in development of nanocantilever sensors will require improvements in detector fabrication, electron transfer for efficient signal transduction and adaptation of these devices for improved functions in aqueous biological systems (Cheng *et al.* 2006).

Nanowires

Nanowires are metallic or semiconducting particles having a high aspect ratio, with cross-sectional diameters <1 μm and lengths as long as tens of microns. Nanowires have been constructed from metallic elements (Au, Pt and Ni), semiconducting compounds (Si and GaN) and oxidized metals with insulating properties (SiO_2 and TiO_2) (Akutagawa *et al.* 2002). An alternate form of nanowire, sometimes called molecular nanowire, is composed of organic or inorganic complex molecules. The ideal nanowire is 10–20 nm in diameter, as this size minimizes noise without sacrificing sensitivity, and has a length-to-width radio (aspect ratio) larger than 1000 (Cheng *et al.* 2006). Neither conventional metallurgic techniques, nor the more complex nanolithographic methods have been successful at producing nanowires to these specifications; thus, nanowires are often grown and assembled using various microfluidic or materials engineering methods (Lu and Lieber 2006).

Nanowires show promise in several sensing strategies including optical, electrical, electrochemical and mass-based approaches (He, Morrow and Keating 2008). They are attractive because of their small size, high surface-to-volume ratios, and/ or electronic, optical and magnetic properties, which can differ markedly from those observed for bulk or thin film materials as the nanowire cross-sectional diameter decreases (Wanekaya *et al.* 2006). It is possible to incorporate large numbers of nanowires into large-scale arrays and complex hierarchical structures for high-density biosensors, electronics and optoelectronics (Lieber and Wang 2007).

Nanowires have been used for the simultaneous detection of multiple biomolecules (i.e. multiplex detection of markers), which is desirable for cancer diagnosis of multiple biomolecular targets such as nucleic acids or proteins (He, Morrow and Keating, 2008). Several nanowire-based detection strategies have shown promise for multiplexed bioanalysis, most of which can be classified as either optical (as in optically encoded nanowire suspension arrays) or electrical (as in semiconductor nanowire field-effect transistors, FETs), which have been demonstrated for biological multiplexing (as many as 30 and 3 targets, respectively).

Optical detection of barcoded nanowires

Striped nanowires can be used as encoded supports for fluorescence-based bioassays. Multiplexing is achieved by preparing optically distinguishable nanowires, by altering the sequence of different metal segments in the nanowires or by changing the diameter along metallic wires or silica tubes, or by on-wire lithography in which sacrificial Ni segments are etched to leave behind more noble metals (He, Morrow and Keating 2008). Nanowire patterns can be read out optically and different assays can be performed simultaneously on the different patterned particles in the same sample. Compared with encoded bead-based suspension arrays that use dyes for both identification and quantification (LaFratta and Walt 2008), barcoded nanowire platforms carry out multiplexed detection with only one dye to circumvent spectral overlapping. Conventional fluorescence optical microscopes are used for readout and quantification (He, Morrow and Keating 2008).

Barcoded metallic nanowires synthesized by templated electrodeposition of multiple metal segments are flexible suspension arrays for multiplexed protein and nucleic acids detection (Brunker, Cederquist and Keating 2007). They are ~6 μm long and ~300 nm in cross-sectional diameter, with segments ≥500 nm, and are stable indefinitely when stored under reducing conditions (Stoermer, Sioss and Keating 2005). Because of their relatively large size, these wires have bulk-like reflectivities for the adjacent metal segments, enabling the barcode striping pattern to be identified by optical microscopy (He, Morrow and Keating 2008).

The metallic surface of the nanowires can modulate the emission intensity of bound fluorescent dyes, which can be minimized by coating the wires with SiO_2 layers (Sloss *et al.* 2007), or the wire itself can act as a quencher for fluorescence in molecular beacon-style assays in the absence of bound target DNA sequence (Stoermer *et al.* 2006). These devices could be automated for imaging and readout and show promise for use in point-of-care clinical settings.

Electric detection in field-effect transistors (FETs)

Semiconductor nanowire-based FETs represent another route to ultrasensitive, label-free, real-time electrical detection of biomolecular interactions, in which conductance is monitored to detect binding events occurring on the nanowire surface (Patolsky, Zheng and Lieber 2006). The small diameter of the nanowire FETs provides extremely high sensitivity because the binding of target molecules causes accumulation/depletion of carriers throughout the wire cross-section for detecting analytes, including proteins (Cui *et al.* 2001). Because FETs respond to changes in surface charge, attention must be given to the buffer in which measurements are performed. Physiologically reasonable ionic strength negatively impacts sensitivity by compressing the electrical double layer around the wire. This effect has been overcome by desalting samples before analysis (Zheng *et al.* 2005).

Two complementary strategies are used to fabricate nanowire FETs: top-down (Stern *et al.* 2007), in which the wires themselves as well as the chip and electronic circuitry are all fabricated from a bulk silicon wafer using advanced microelectronics technologies (i.e. lithography, etching and deposition), but the incorporation of biological probe molecules is limited by the harsh manufacturing conditions commonly used in fabrication (high temperature, solvents or reactive-ion etching); and bottom-up (Patolsky, Zheng and Lieber 2006), in which nanowire building blocks are synthesized before assembly onto the chip surface, providing much greater flexibility in nanowire material properties and surface functionalization, although not offering the reliability of the top-down approach (He, Morrow and Keating 2008).

More recently, using the top-down approach, it was possible to overcome the problems of biomarker detection in blood such as biofouling and non-specific binding, and the need to use purified buffers that greatly reduces the clinical relevance of these sensors, by employing a microfluidic purification chip (MPC) that simultaneously captures multiple biomarkers from blood samples and releases them after washing into purified buffer for sensing by a silicon nanoribbon detector. The two-stage approach isolates the detector from the complex environment of the blood and

Table 8.2 Application of nanotechnology to biomarker discovery

Nanotechnology	Functions	Features
Magnetic nanoparticles	Detection of proteins, DNA and cells	Sensitive detection of proteins when combined with TOF-MS
Quantum dots (QDs)	Detection of proteins, DNA and immunohistochemistry labelling	Increased sensitivity, multiplexed immunoassays
Carbon nanotubes (CNTs)	Detection of proteins and DNA	Sensitive detection immunosensors
Bio-barcode assay	Detection of protein and DNA	Rapid tests with clear optical readout, high sensitivity
Surface-Enhanced Raman Scattering (SERS)	Detection of proteins, small molecules and DNA	Label-free detection or SERS active tags
Nanocantelievers	Detection of proteins, DNA, RNA, bacteria and viruses	Sensitive mass detector, can be functionalized with biologic molecules
Nanowires (NWs)	Detection of proteins, DNA and cells	Can be functionalized with appropriate antibodies

reduces its minimum required sensitivity by effectively pre-concentrating of two cancer biomarkers (prostate PSA and breast cancer carbohydrate antigen 15.3, CA15.3), allowing immuno-affinity detection from a 10 μl sample of whole blood, at ng/ml concentration, in less than 20 min, without the challenges associated with tailoring sensor operation for the medium of interest or engineering nanosensors that can withstand complex fluid media (Stern *et al.* 2010). Moreover, this system is capable of stand-alone use, or use in tandem with more demanding sensing methodologies, such as those used for rare circulating tumour cells (Nagrath *et al.* 2007).

Table 8.2 illustrates the application of nanotechnology to biomarker discovery.

Biomarker harvesting

Biomarker research revealed that low-abundance (picomole/l), and low molecular weight (LMW) circulating proteins and peptides represent a rich source of information regarding the state of the organism as a whole. However, two major hurdles have prevented these discoveries from achieving clinical benefits: (a) disease-relevant biomarkers in blood exists in very low concentrations within a complex mixture of biomolecules and are masked by highly abundant species such as high molecular weight (HMW) albumin, and immunoglobulins that account for ∼90% of total blood protein content covering a concentration range of ∼ 12 orders of magnitude

(Hortin, 2006); and (b) degradation of protein biomarkers can occur immediately following the minimally invasive collection of blood (e.g. by venipuncture) as a result of endogenous or exogenous proteinases (e.g. proteases associated with the blood clotting process, enzymes shed from blood cells or those associated with bacterial contamination), and continue acting during transportation or storage, especially if samples are not frozen (Ahmed 2009c).

Conventional proteomic methods such as 2DE do not have the sensitivity or resolution to quantify low abundance, LMW proteins (Ahmed 2008a). Quantitative modern mass spectrometers have a working range that spans at most five orders of magnitude of concentration (Ahmed 2009a) and thus cannot detect the less abundant protein markers that are masked by the more abundant proteins. It is possible to deplete the HMW proteins from plasma, although a recent study showed that it is difficult to completely remove albumin from convenient commercial immunoaffinity reusable spin column devices (Gundry et al. 2009). Nevertheless, removal of these abundant, HMW proteins can significantly reduce the yield of candidate biomarkers because most LMW proteins have been shown to be non-covalently and endogenously associated with the HMW carrier proteins that are being removed (Ahmed 2008b).

Two fundamental nanoparticle harvesting platforms have been envisioned: (a) particles with properties such as varying surface porosity, surface charge and functional groups that serve as bait molecules to provide binding sites for LMW proteins that circulate in the blood bound to carrier proteins such as albumin; and (b) particles conjugated with functional affinity groups such as antibodies added to their surfaces providing binding sites for circulating LMW proteins (Geho et al. 2006b). The first platform is simpler in concept and less daunting. The latter approach, although it may theoretically seem to provide more binding specificity, is complicated, depends on using many antibodies that may not be specific enough or even available, requires multiplexing that is not easy to perform, or is expensive in execution.

A simple configuration for a biomarker harvesting platform is to design a set on nanoparticles with distinct physicochemical properties. For example, relative surface area, presence or absence of surface chemistries of nanoparticles that can be tailored to provide a physicochemically based fractionation tool for blood proteins. Examples of nanoparticles to which proteins can be conjugated are QDs. In one instance they have been conjugated to streptavidin creating a reporter complex for molecular studies such as protein microarrays (Geho et al. 2005). Composite organic–inorganic nanoparticles (COINs) is another nanoparticle class that has been used as labels for immunoassays (Su et al. 2005), demonstrating that conjugation of an antibody to the COIN platform can be achieved without destroying the immunoaffinity characteristics of the antibody. Silica nanoparticles can be loaded or doped with metallo-organic luminophores (Santra et al. 2001). The surface of the luminophore-doped particles can then act as an intermediate substrate for bioconjugate formation with biomolecules, such as antibodies. Other potential particle labels include metallic barcodes (Nam, Thaxton and Mirkin 2003), silica nanoparticles with decoratable surfaces and striped metallic encoded nanoparticles (Nicewarner-Pena *et al.* 2001).

At the other end of the spectrum, hydrogel core-sheet nanoparticles have been fabricated to overcome barriers to biomarker harvesting. A hydrogel particle is a cross-linked particle of sub-micrometer size composed of hydrophilic polymers capable of swelling and contracting as a result of the application of an environmental trigger (e.g. temperature, pH, ionic strength or electric field) (Luchini *et al.* 2008). The nanoparticles simultaneously conduct both molecular sieve chromatography and affinity chromatography in one step in solution. The molecules captured and bound within the affinity matrix of the particles are protected from degradation by exogenous or endogenous proteases. A highly labile and low abundance biomarker such as platelet-derived growth factor (PDGF) when spiked in human serum was found to be completely sequestered from its carrier protein albumin, concentrated and fully preserved within minutes by the particles (Longo *et al.* 2009).

References

Agrawal, A., Sathe, T. and Nie, S. (2007) Single-bead immunoassay using magnetic nanoparticles and spectral-shifting quantum dots. *J Agric Food Chem* **55**: 3778–3782.

Ahmed, F. (2006) Expression microarray proteins and the dearch for cancer biomarkers. *Curr Genomics* **7**: 399–426.

Ahmed, F.E. (2008a) Mining the oncoproteome and studying molecular interactions for biomarker development by 2DE, ChIP and SPR technologies. *Exp Rev Proteomics* **5**: 469–496.

Ahmed, F.E. (2008b) Application of MALDI/SELDI mass spectrometry to cancer biomarker discovery and validation. *CurrProteomics* **5**: 224–252.

Ahmed, F.E. (2009a) Liquid Chromatography-mass spectrometry: a tool for proteome analysis and biomarker discovery and validation. *Exp Opinin Med Diagn* **3**: 429–444.

Ahmed, F.E. (2009b) The role of capillary electrophoresis-mass spectrometry to proteome analysis and biomarker discovery. *J Chromatogr B* **877**: 1963–1981.

Ahmed, F.E. (2009c) Sample preparation and fractionation for proteome analysis and cancer biomarker discovery by mass spectrometry. *J Sep Sci* **32**: 771–798.

Akutagawa, T., Ohta, T., Hasegawa, T. *et al.* (2002) Formation of oriented molecular nanowires on mica surfaces. *Proc Natl Acad Sci USA* **90**: 5028–5033.

Anderson, N.L. and Anderson, N.G. (2002) The human plasma proteome: history, character, and diagnostic prospects. *Mol Cell Proteomics* **1**: 845–867.

Bao, Y.P., Wei, T.F., Lefebver, P.A. *et al.* (2006) Detection of protein analytes via nanoparticle-based bio bar code technology. *Anal Chem* **78**: 2055–2059.

Bakalova, R., Zhelev, Z., Ohba, H. and Baba, Y. (2005) Quantum dot-based western blot technology for ultrasensitive detection of tracer proteins. *J Am Chem Soc* **127**: 9328–9329.

Brower, V.C. (2006) Is nanotechnology ready for primetime. *J Natl Cancer Inst* **98**: 9–11.

Bruchez, M., Morrone, M., Gin, P. *et al.* (1998) Semiconductor nanocrystals as fluorescent biological labels. *Science* **281**: 2013–2016.

Brunker, S.E., Cederquist, K.B. and Keating, C.D. (2007) Metallic barcodes for multiplexed bioassays. *Nanomedicine* **2**: 695–710.

Carter, A. (2008) Learning from history: understanding the carcinogenesis of nanotechnology. *J Natl Cancer Inst* **100**: 1664–1665.

Chen, C.T. and Chen, Y.C. (2004) Molecularly imprinted TiO_2-matrix-assisted laser desorption/ionization mass spectrometry for selectively detecting alpha-cyclodextrin. *Anal Chem* **76**: 1453–1457.

Cheng, M.M, Cuda, G., Bunimovich, Y.L. *et al.* (2006) Nanotechnologies for biomolecular detection and medical diagnostics. *Curr Opin Chem Biol* **10**: 11–19.

Cui, Y., Wei, Q., Park, H. and Lieber, C.M. (2001) Nanowire nanosensors for highly sensitive and selective detection of biological and chemical species. *Science* **293**: 1289–1292.

Gaspari, M., Ming-Cheng, C.M., Terracciano, R. *et al.* (2006) Nanoporous surfaces as harvesting agents for mass spectrometric analysis of peptides in human plasma. *J Proteome Res* **5**: 1261–1266.

Geho, D., Lahar, N., Gurnani, P. *et al.* (2005) Pegylated, streptavidin-conjugated quantum dots are effective detection elements for reverse-phase protein microarrays. *Bioconjug Chem* **16**: 559–566.

Geho, D., Cheng, M.M., Killian, K. *et al.* (2006a) Fractionation of serum components using nonporous substrates. *Bioconjug Chem* **16**: 654–661.

Geho, D.H., Jones, C.D., Petricoin, E.F. and Liotta L.A. (2006b) Nanoparticles: potential biomarker harvesters. *Curr Opin Chem Biol* **10**: 56–61.

Giepmans B.N., Adams, S.R., Ellisman, M.H. and Tsien, R.Y. (2006) The fluorescent toolbox for assessing protein location and function. *Science* **312**: 217–224.

Grow, A.E., Wood, L.L., Claycomb, J.L. and Thompson, P.A. (2003) New biochip technology for label-free detection of pathogens and their toxins. *J Microbiol Methods* **53**: 221–233.

Grubisha, D.S., Lipert, R.J., Park, H.Y. *et al.* (2003) Femtomolar detection of prostate-specific antigen: an immunoassay based on surface-enhanced Raman scattering and immunogold labels. *Anal Chem* **75**: 5936–5943.

Gundry, R.L., White, M.Y., Nogee, J. *et al.* (2009) Assessment of albumin removal from an immunoaffinity spin column: clinical implications for proteomic examination of the albuminome and albumin-depleted samples. *Proteomics* **9**: 2021–2028.

He, B., Morrow, T.J. and Keating, C.D. (2008) Nanowire sensors for multiplexed detection of biomolecules. *Curr Opin Chem Biol* **12**: 1–7.

Heuff, R.F., Swift, J.L. and Cramb, D.T. (2007) Fluorescence correlation spectroscopy using quantum dots: advances, challenges and opportunities. *Physl Chem Chem Phys* **9**: 1870–1880.

Hill, H.D. and Mirkin, C.A. (2006) The bio-barcode assay for the detection of protein and nucleic acid targets using DTT-induced ligand exchange. *Nat Protocols* **1**: 324–336.

Hortin, G.L. (2006) The MALDI-TOF mass spectrometric view of the plasma proteome and peptidome. *Clin Chem* **52**: 1223–1237.

Igarashi, E. (2008) Factors affecting toxicity and efficacy of polymeric nanomedicine. *Toxicol Appl Pharmacol* **229**: 121–134.

Issaq, H.J., Xiao, Z. and Veenstra, T.D. (2007) Serum and plasma proteomics. *Chem Rev* **107**: 3601–3620.

Ivanov, Y.D., Govorun, V.M., Bykov, V.A. and Archakov, A.I. (2006) Nanotechnologies in proteomics. *Proteomics* **6**, 1399–1414.

Jain, K.K. (2008) Recent advances in nanotechnology. *Technol Cancer Res Treat* **7**: 1–12.

Johnson, C.J., Zhukovsky, N., Cass, A.E.G. and Nagy, J.M. (2008) Proteomics, nanotechnology and molecular diagnostics. *Proteomics* **8**: 715–730.

Josic, D. (2007) Use of monolithic supports in proteomics technology. *J Chromatogr A* **1144**: 2–13.

Kawasaki, E.S. and Player, A. (2005) Nanotechnology, nanomedicine, and the development of new, effective therapies for cancer. *Nanomed Nanotechnol Biol Med* **1**: 101–109.

Kinumi, T, Saisu, T., Takayama, M. and Niwa, H. (2000) Matrix-assisted laser desorption/ionization time-of-flight mass spectrometry using n inorganic particle matrix for small molecule analysis. *J Mass Spectrom* **35**: 417–422.

LaFratta, C.N. and Walt, D.R. (2008) Very high density sensing arrays. *Chem Rev* **108**: 614–637.

Lee, M., Kang, D.K., Yang, H.K. *et al.* (2006) Protein nanoarray on ProLinker surface constructed by atomic force microscopy dip-pen nanolithography for analysis of protein interaction. *Proteomics* **6**: 1094–1103.

Lieber, C.M. and Wang, Z.L. (2007) Functional nanowires. *MRS Bull* **32**: 99–108.

Longo, C., Patanarut, A., George, T. *et al.* (2009) Core-shell hydrogel particles harvest, concentrate and preserve labile low abundance biomarkers. *PLoS ONE* **4**: e4763.

Lu, W. and Lieber, C.M. (2006) Semiconductor nanowires. *J Phys D (Appl Phys)* **39**: R387–R406.

Luchini, A., Geho, D.H., Bishop, B. *et al.* (2008) Smart hydrogel particles: biomarker harvesting: one-step affinity purification, size exclusion, and protection against degradation. *Nanotechnol Lett* **8**: 350–361.

Lynch, M., Mosher, C., Huff, J. *et al.* (2004) Functional protein nanoarrays for biomarker profiling. *Proteomics* **6**: 1695–1702.

Marko, N.F., Weil, R.J. and Toms, S.A. (2007) Nanotechnology in proteomics. *Exp Rev Proteomics* **4**: 617–626.

McLean, J.A., Stumpo, K.A. and Russell, D.H. (2005) Size-selected (2–10 nm) gold nanoparticles for matrix-assisted laser desorption/ionization of peptides. *J Am Chem Soc* **127**: 5304–5305.

Mukhopadhyay, R., Lorentzen, M., Kjems, J. and Besenbacher, F. (2005) Nanomechanical sensing of DNA sequences using piezoresistive cantilevers. *Langmuir* **21**: 8400–8408.

Nagrath, S., Sequist, L.V., Matheswaran, S. *et al.* (2007) Isolation of rare circulating tumor cells in cancer patients by microchip technology. *Nature* **450**: 1235–1239.

Nam, J.M., Thaxton, C.S. and Mirkin, C.A. (2003) Nanoparticle-based bio-bar codes for the ultrasensitive detection of proteins. *Science* **301**: 1884–1886.

Nicewarner-Pena, S.R., Freeman, R.G., Reiss, B.D. *et al.* (2001) Submicrometer metallic barcodes. *Science* **294**: 137–141.

Nie, S., Xing, Y., Kim, G.J. and Simons, J.W. (2007) Nanotechnology applications in cancer. *Annu Rev Biomed Eng* **9**: 257–288.

O'Farrell, H.P. (1975) High resolution two-dimensional electrophoresis for proteins. *J Biol Chem* **250**: 4007–4021.

Okuno, S., Arakawa, R., Okamoto, K. *et al.* (2005) Requirements for laser-induced desorption/ionization on submicrometer structures. *Anal Chem* **77**: 5364–5369.

Osaka, T., Matsunaga, T., Nakanishi, T. *et al.* (2006) Synthesis of magnetic nanoparticles and their application to bioassays. *Anal Bioanal Chem* **384**: 593–600.

Patolsky, F., Zheng, G. and Lieber, C.M. (2006) Nanowire-based biosensors. *Anal Chem* **78**: 4260–4269.

Pinnaduwage, L.A., Yi, D., Tian, F. *et al.* (2004) Adsorption of trinitrotoluene on uncoated silicon microcantilever surfaces. *Langmuir* **20**: 2690–2694.

Polizu, S., Savadogo, O., Poulin, P. and Yahia, L. (2006) Application of carbon nanotubes-based biomaterials in biomedical nanotechnology. *J Nanosci Nanotechnol* **6**: 1883–1904.

Rainer, M., Najam-ul-Haq, M., Bakry, R. *et al.* (2007) Mass spectrometric identification of serum peptides employing derivatized poly(glycidyl methacrylate/divinyl benzene) particles and mu-HPLC. *J Proteome Res* **6**: 382–386.

Ren, S.F., Zhang, L., Cheng, Z.H. and Guo, Y.L. (2005) Immobilized carbon nanotubes as matrix for MALDI-TOF-MS analysis: applications to neutral small carbohydrates. *J Am Soc Mass Spectr* **16**: 333–339.

Santra, S., Zhang, P., Wang, K. *et al.* (2001) Conjugation of biomolecules with luminophore-doped silica nanoparticles for photostable biomarkers. *Anal Chem* **73**: 4988–4993.

Sanvicens, N. and Marco, P. (2008) Multifunctional nanoparticles –properties and prospects for their use in human medicine. *Trends Biotechnol* **26**: 425–433.

Seino, T., Sato, H., Yamamoto, A. *et al.* (2007) Matrix-free laser desorption/ionization mass spectrometry using self-assembled germanium nanodots. *Anal Chem* **79**: 4827–4832.

Sloss, J.A., Stoermer, R.L., Sha, M.Y. and Keating, C.D. (2007) Silica-coated, Au/Ag striped nanowires for bioanalysis. *Langmuir* **23**: 11334–11241.

Smith, A.M., Gao, X. and Nie, S. (2004) Quantum dot nanocrystals for *in vivo* molecular and cellular imaging. *Photochem Photobiol* **80**: 377–385.

Stern, E., Klemic, J.F., Routenberg, D.A. *et al.* (2007) Label-free immunodetection with CMOS-compatible semiconducting nanowires. *Nature* **445**: 519–522.

Stern, E., Vacic, A., Rajan, N.K. *et al.* (2010) Label-free biomarker detection from whole blood. *Nat Nanotechnol* **5**: 138–142.

Stoermer, R.L., Sioss, J.A. and Keating, C.D. (2005) Stabilization of silver metal in citrate buffer: barcoded nanowires and their bioconjugates. *Chem Matter* **17**: 4356–4361.

Stoermer, R.L., Cederquist, K.B., McFarland, S.K. *et al.* (2006) Coupling molecular beacons to barcoded metal nanowires for multiplexed, sealed chamber DNA bioassays. *J Am Chem Soc* **128**: 16892–16903.

Su, X., Zhang, J., Sun, L. *et al.* (2005) Composite organic-inorganic nanoparticles (COINS) with chemically encoded optical signatures. *Nanotechnol Lett* **5**: 49–54.

Wang, J., Liu, G.D. and Jan, M.R. (2004) Ultrasensitive electrical biosensing of proteins and DNA: carbon-nanotube derived amplification of the recognition and transduction events. *J Am Chem Soc* **126**: 3010–3011.

Wanekaya, A.K., Chen, W., Myung, N.V. and Mulchandani, A. (2006) Nanowire-based electrochemical biosensors. *Electroanalysis* **18**: 533–550.

Wooley, A.T., Guillemette, C., Cheung, C.L. *et al.* (2000) Direct haplotyping of kilobase-size DNA using carbon nanotube probes. *Nat Biotechnol* **18**: 760–763.

Wu, C.C. and Yates, J.R. (2003) The application of mass spectrometry to membrane proteomics. *Nat Biotechnol* **21**: 262–267.

Wu, G., Datar, R.H., Hansen, K.M. *et al.* (2001) Bioassay of prostate-specific antigen (PSA) using microcantilevers. *Nat Nanotechnol* **19**: 856–860.

Xing, Y., Chaudry, Q., Shen, C. *et al.* (2007) Bioconjugated quantum dots for multiplexed and quantitative immunohistochemistry. *Nat Protocols* **2**: 1152–1165.

Yu, X., Munge, B., Patel, V. *et al.* (2006) Carbon nanotube amplification strategies for highly sensitive immunodetection of cancer biomarkers. *J Am Chem Soc* **128**: 11199–11205.

Zhang, X., Leung, S-M., Morris, C.R. and Shigenaga, M.K. (2004) Evaluation of a novel, integrated approach using functionalized magnetic beads, bench-top MALDI-TOF-MS with prestructured sample supports and pattern recognition software for profiling potential biomarkers in human plasma. *J Biomol Technol* **15**: 167–175.

Zheng, G., Patolsky, F., Cui, Y. *et al.* (2005) Multiplexed electrical detection of cancer markers with nanowire sensor array. *Nat Nanotechnol* **23**: 1294–1301.

9

Chip-Based Proteomics

Julian Bailes, Andrew Milnthorpe, Sandra Smieszek
and Mikhail Soloviev

Introduction

Characterization of the complement of expressed proteins from a single genome is
a central focus of the evolving field of proteomics. Monitoring the expression and
properties of a large number of proteins provides important information about the
physiological state of a cell and an organism. A cell can express a large number
of different proteins and the expression profile (the number of proteins expressed
and their expression levels) varies in different cell types, explaining why different
cells perform different functions. The central concept of modern proteomics is
'multiplexing', or a simultaneous analysis of all proteins in a defined protein
population. However, since one genome produces many proteomes and the number
of expressed genes in a single cell may exceed 10 000, the characterization of
thousands of proteins to evaluate proteomes ideally requires a high-throughput,
automated process. This is why new improved methods for high-throughput protein
identification and quantitation were needed. Many such technologies have been
developed recently, led by advances in the electronics industry and more recently
by the successes of the DNA chip-based technologies.

In the past, the majority of separation techniques used in protein and peptide analysis relied on their physical properties, such as protein or peptide size, shape, polarity,
pI, the distribution of ionizable polar and non-polar groups on the molecule surface,
and their affinity towards specific or non-specific affinity capture reagents. Modern
separation techniques rely on a combination of isoelectric focusing, electrophoretic
separation and a great variety of liquid chromatography techniques, often linked

Molecular Analysis and Genome Discovery, Second Edition. Edited by Ralph Rapley and Stuart Harbron.
© 2012 John Wiley & Sons, Ltd. Published 2012 by John Wiley & Sons, Ltd.

together to yield two- or three-dimensional separation approaches and frequently backed up by serious automation. Highly parallel analysis is often attempted through miniaturization (Marko-Varga, Nilsson and Laurell 2003) and the use of chip-based techniques (Hoa, Kirk and Tabrizian 2007; Lion *et al.* 2003, 2004) of which the Agilent 2100 Bioanalyzer is a good example (www.chem.agilent.com). The inherent heterogeneity of the proteins' and to a lesser degree peptides' physical properties which underlies all of the above separation options is, at the same time, the inherent problem of any highly parallel protein analysis.

Lab-on-a-chip

Most of the modern developments in the area of proteomics, including chip-based proteomics, are often based on miniaturized versions of traditional and established protein separation and purification techniques, such as ELISA, electrophoresis, chromatography or matrix-assisted laser desorption/ionization time-of-flight (MALDI-TOF) mass spectrometry. While traditional methods of protein analysis are generally time-consuming, labour-intensive and lack high-throughput capacity, lab-on-a-chip (LOC) approaches attempt to provide solutions to these problems by offering analytical devices that are easier to use, more cost effective, require smaller sample volumes and can rapidly deliver a level of results comparable to classical approaches. Smaller volumes also mean reduced reagent costs and less chemical waste, while a smaller physical footprint increases portability, permitting point-of-care testing away from centralized laboratories. High reproducibility is achieved due to the standardization and automation of experimental procedures involved with LOC analysis. As well as being able to perform operations such as sample handling, mixing, dilution, electrophoresis and chromatographic separation, staining and detection on a single chip, microfluidic devices can exploit the behaviour of liquids such as laminar flow and electro-osmotic flow that are unique to the microscale. These properties allow for greater control of molecular concentrations and interactions.

LOC devices are miniaturized technologies that integrate one or several laboratory operations on a very small scale, typically millimetres to a few square centimetres in size. These analytical systems essentially divide into two types of device: those that are microarray based and those that are microfluidic based.

Microarrays provide high-throughput parallel protein profile studies capable of assaying thousands of proteins simultaneously on a single chip. Selected probes are robotically arrayed and used to bind potential targets within an experimental sample. In the simplest of protein microarrays, fluorescent or otherwise labelled sample is incubated with immobilized antibodies in a direct assay (Knezevic *et al.* 2001; Haab, Dunhuam and Brown 2001), but the format is also compatible with competitive displacement assays where labelled and unlabelled sample compete for immobilized probes (Barry *et al.* 2003a), and also sandwich type (ELISA) assays. Microfluidic LOC technologies incorporate a network of channels and wells etched onto glass or polymer substrates with capillary, pressure or electrokinetic forces moving picolitre volumes in a finely controlled manner through the channels. These emerging technologies and a continuous trend of miniaturization are increasingly becoming commonplace across a wide range of disciplines, from research to point-of-care testing. Reduction in size often results in significant changes, with most being extremely useful and desirable in proteomic research. There are a few obvious and easily acceptable reasons for miniaturization. Miniaturization allows one to pack more material into the same volume. For example, hundreds of different protein affinity reagents may be spotted on a chip instead of using Western blots or Sepharose affinity columns. Other advantages include a potential to increase the reaction kinetics due to much smaller reaction volumes (and therefore faster reagent diffusion times) and significantly increased surface-to-volume ratios. This feature becomes especially useful for studying proteins since one of the two major approaches to protein separation, chromatography, relies on surface-mediated interactions. Truly miniature chromatographic applications do not require the use of porous resins and a careful selection of pore sizes, since a high surface-to-volume ratio of a small capillary channel, which is also easier to control, may be sufficient. In addition to these, most of the microfluidic applications allow capillary forces to manipulate the reagent solution. This often simplifies the design of the experiment (i.e. of the chip) and eliminates the need for additional equipment (i.e. for high precision pumping and mixing of the reagents, etc.). When capillary force is not sufficient for manipulating the reagents, spinning of a whole CD-based laboratory (e.g. www.gyros.com) could be used to finish the job. In addition, the lower power consumption of miniaturized devices and the possibility of portable applications (i.e. small chip-based devices with disposable chips) is also more attractive for routine diagnostic applications when compared to a desk-top sized HPLC system.

Another important implication of working at micro- or nanoscales is that at these dimensions liquid flow becomes laminar and even the simple task of mixing two samples becomes a problem. This is applicable to both nucleic acid and protein chips. Especially vulnerable are microfluidic applications, but even simple spotted arrays (whether DNA or proteins) require sample mixing, for example using surface acoustic waves (Advalytix AG, Germany; www.advalytix.de) or sophisticated sample re-circulation (e.g. Memorec Biotec GmbH, now a subsidary of Miltenyi Biotec GmbH; www.miltenyi.com) to improve the efficiency of hybridization. In the microworld, capillary forces, viscosity and surface tension become major factors determining liquid behaviour. These properties could be taken advantage of, but could also take over the whole microfluidic chip, for example by preventing liquid penetration through narrow capillaries with hydrophobic surfaces. Miniaturization therefore requires careful control of the chemical and physical properties of the material-to-liquid interface surface (i.e. capillary effects). Capillary channels the size of a single molecule (such as DNAs or large proteins) can now be manufactured. This may allow for single molecule applications to be carried out. As yet, there is little data available on how DNA and large proteins would behave under such conditions. However, because of physical limitations (i.e. prevalence of surface tension and viscosity over gravitational forces and absence of turbulence) miniaturization of liquid handling seems unreasonable below the low micrometre range.

Arrays

Increasingly sort after are multiplexed tests that increase throughput further still by performing multiple assays simultaneously, whether it be multiple tests on a single sample or multiple samples within a single run. Microarrays are a fine example of how LOC satisfy such needs, with multiplexity achievable through spatial separation of the probes, spectrally by using multiple fluorescent labels (separation by wavelengths, as in Voura et al. 2004) or by the use of mass-spectrometry based detection (separation by antigen masses, as in Scrivener et al. 2003; Barry et al. 2002a, 2002b, 2003b) or through serial detection (as in FACS-analysed bead arrays; Fulton et al. 1997).

DNA microarrays, pioneered by Ed Southern and Pat Brown and first developed commercially by Affimetrix (Southern, 1995, 1996, 2001; Eisen and Brown 1999; Schena et al. 1995; Pease et al. 1994; Fodor et al. 1993) preceded protein microarrays by some 10 years. A single array can be used to probe many thousands of spots of various cDNAs or oligonucleotides simultaneously. Transferring this technology to study proteins instead of DNA would seem like a logical step in creating new highly multiplexed proteomic tools to replace the traditional techniques that have been used for so long (Templin et al. 2002; Zhu and Snyder 2003; Barry et al. 2002c; Soloviev 2001). Closed systems (such as antibody arrays) are most suitable for applications where key priorities are the speed of analysis and sensitivity and where the number of analytes may not necessarily be large.

Most antibody microarrays rely on fluorescence detection to deduce protein expression levels in the samples tested. There are two main types of protein chip assays, direct binding and sandwich-type (ELISA-type) assays. In a direct binding assay the unknowns (i.e. proteins being assayed) are labelled directly with a detection reagent such as a fluorophore. This enables the labelled proteins to be bound by antibodies immobilized on the array and their relative abundance to be measured directly by a fluorescent scanner, following a brief wash step, without further processing of the chip. The advantages of this method lie with its simplicity, speed and capacity for quantitation. However, its disadvantage relates to the actual labelling process in that it may interfere with recognition epitopes on the protein thus preventing its antibody binding. Direct binding assays are suitable for both antibody and antigen screenings as well as for studying protein–protein interactions.

A sandwich assay negates the necessity for labelling the proteins of interest with a detection reagent. In a sandwich-based assay the array also comprises capture proteins (e.g. antibodies), which bind proteins in their unmodified native form. The detection step requires a second antibody possessing specificity to a second epitope on the captured protein, which is distinct from the capture antibody. The second antibody is conjugated to a detection molecule, that is either a fluorophore for primary detection, or an enzyme (e.g. horse radish peroxidase) or ligand (e.g. biotin) for secondary detection through a chemiluminescent molecule or DNA (Schweitzer *et al.* 2002). An advantage of this technique is in the degree of sensitivity achieved through amplification of the signal thus allowing lower abundance proteins to be identified. However, this is paralleled by a loss in the quantitative nature of the assay and requires double the amount of antibodies.

Direct and sandwich binding assays are often difficult to interpret (Barry and Soloviev 2004), while comparative binding or competitive binding assays allow easier interpretation and relative quantification options (Barry *et al.* 2003a; Alhamdani, Schröder and Hoheisel 2009; Wingren and Borrebaeck 2006). A number of protein affinity arrays have established themselves firmly and a great variety of formats have been reported, including more traditionally miniaturized microwell plate format micro-ELISAs (Mendoza *et al.* 1999) or nitrocellulose 'dot blots' (e.g. on Whatman/Schleicher-Schuell FAST slides; www.whatman.com), directly imported from the DNA array-field glass-slide arrays (Schweitzer and Kingsmore 2002; Angenendt *et al.* 2003), silicon (Jenison *et al.* 2001), or 3D substrates such as hydrogels (Scrivener *et al.* 2003). Other miniaturized protein array technologies include, format wise, flow cytometric microbead assays (or suspension arrays, e.g. Fulton *et al.* 1997) and, target wise, antigen arrays (Ge 2000; MacBeath and Schreiber 2000; Zhu *et al.* 2001; Michaud *et al.* 2003; Madoz-Gúrpide *et al.* 2001), whole cell lysate arrays (Paweletz *et al.* 2001), or *in vitro* synthesized protein arrays (He and Taussig 2001; Weng *et al.* 2002). Protein microarrays are often generated by high-speed robotics on a wide range of substrates (Figure 9.1) for which probes (e.g. antibodies) with known identity are used to bind experimental samples (i.e. proteins being assayed) enabling parallel protein profile studies to be undertaken. An experiment with a single protein chip can therefore supply information on thousands of proteins simultaneously and this provides a considerable increase in throughput.

Holt *et al.* (2000) have developed a Matrix Screening™ platform for high-throughput screening of protein–protein interactions. In this system, grids of intersecting lines are used to bring different V_H V_L antibody fragments together on the surface of a 'combinatorial' array. This enables thousands of protein–protein interactions or antibody heavy and light chain pairings to be screened. This technique may be used in affinity maturation studies of antibodies, proteomics assays and combinatorial library screening. Büssow *et al.* (1998) and Holz *et al.* (2001) have reported a technique useful for establishing catalogues of protein products of arrayed cDNA clones and for antibody repertoire screenings by gridding bacterially expressed cDNA libraries onto high-density filters. Simone and co-workers (2000) have coupled a laser capture microdissection (LCM) with sensitive quantitative chemiluminescent immunoassays in an attempt to quantitate the number of prostate-specific antigen molecules. Individual cells (about 30 μm in diameter)

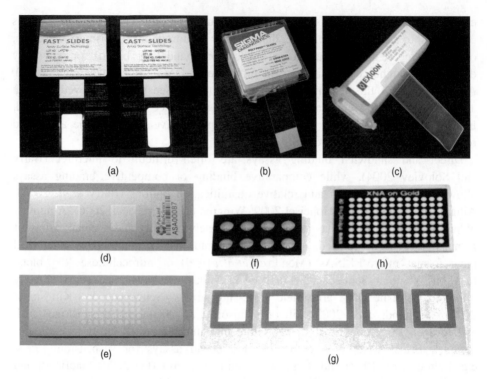

Figure 9.1 Examples of some commercially available microarray substrates (i.e. slides)

were captured under direct microscopic visualization from the heterogeneous tissue section onto a polymer transfer surface. The cellular macromolecules from the captured cells were solubilized in a microvolume of extraction buffer and directly assayed using an automated (1.5 h) sandwich chemiluminescent immunoassay. This technique is capable of measuring actual numbers of prostate-specific antigen molecules in microdissected tissue cells and has a broad applicability in the field of proteomics. Some other formats for functional proteomics on chips have also been published. Zhu *et al.* (2000) have expressed and immobilized almost the entire set of possible protein kinases (119 of 122) from *Saccharomyces cerevisiae*. These kinases were incubated with 17 different substrates to characterize their phosphorylation activity and this identified both known and previously uncharacterized activities of established and speculative kinases (homology predictions). Large arrays of immobilized proteins can be examined for protein–protein interactions in a system in some ways analogous to yeast two-hybrid experiments (Walter *et al.* 2000), and in a somewhat ironic assay, genome-wide protein–DNA interactions have been assessed using antibodies to immunoprecipitate DNA–protein complexes, which are then characterized on a DNA microarray (Iyer *et al.* 2001).

BD Biosciences (www.bdbiosciences.com) were one of the first Life Science providers to have produced an antibody array. Over 10 years later BD Biosciences moved the focus onto bead based arrays rather than 'chip' based. Randox Laboratories Ltd (www.randox.com) have developed a multianalyte immunoassay testing platform for the simultaneous array based biochip technology referred to as Evidence® platform. The results are achieved by combining a panel of tests on a single biochip with a single set of reagents, controls and calibrators. The company also provides a CCD based reader suitable for chemiluminescence detection and the necessary image-processing software. A range of specialized biochip panels are available; these are either DNA based (Colorectal cancer array) or antibody based (all other array products, including drugs of abuse arrays, cytokine arrays, adhesion molecules, cardiac, cerebral, tumour markers, fertility, thyroid, anti-microbial, growth promoters and synthetic steroids arrays). Randox arrays are becoming increasingly popular and a number of publications detailing their uses are available. We refer the reader to a few recent ones describing the use of Randox biochips for the analysis of cytokines, adhesion molecules, neurological disorders, reproduction and development hormone testing, thyroid function testing and for the monitoring and diagnosis of myocardial infarction (Badiou *et al.* 2008; Banfi *et al.* 2008; van Beem *et al.* 2008; Ermetici *et al.* 2008; Fabre *et al.* 2008; Horácek *et al.* 2008a, 2008b, 2008c, 2008d; Kavsak *et al.* 2008a, 2008b, 2008c; Lippi *et al.* 2008; Pezzilli *et al.* 2008; Åsberg *et al.* 2009; Berrahmoune *et al.* 2009; Kavsak *et al.* 2009; Roh *et al.* 2009a, 2009b; Zetterberg *et al.* 2009).

One of the recent developments in the field is a comprehensive range of antibody arrays for detecting and quantifying human cytokines developed by RayBiotech, Inc. (www.raybiotech.com). These are offered as a combination of five specialized anitbody arrays: (i) Inflammation Array, (ii) Growth Factor Array, (iii) Chemokine Array, (iv) Receptor Array and (v) Cytokine Array, totalling 200 different cytokines. These 'Quantibody(R)' arrays are used in a sandwich ELISA-based immunoassay format; for this the manufacturer provides a cocktail of biotinylated antibodies and fluorescently labelled streptavidin (suitable for detection with Cy-3 channel scanners). The kit is aimed at a wide range of users and contains all the buffers, standards and accessories to allow even less experienced users to use them. The user only has to provide their own fluorescent scanner and a bit of aluminium foil. Each of the glass slides is spotted with 16 grids of identical cytokine antibody arrays and each antibody is arrayed in quadruplicate. These and the standard mix of lyophilized cytokines allow users to determine the concentrations of 200 human cytokines in a single experiment within a day (Stechova *et al.* 2009; Willingham *et al.* 2009; El Karim *et al.* 2009; Souquière *et al.* 2009; Sharma *et al.* 2009; Altamirano-Dimas *et al.* 2009; Cheung *et al.* 2009; Du *et al.* 2009; Van Rossum *et al.* 2009; Zhai *et al.* 2008).

Other developments in the protein array filed include the emergence of pro-teomics/array service providers, the unique selling point of which are often a complete service package starting from the identification of novel biomarkers and targets to antibody testing and characterization. One such company, Kinexus Bioinformatics (www. kinexus.ca), developed Kinex™ microarray and Kinetworks™ immunoblotting services for the discovery and characterization of protein kinase biomarkers.

Custom services include the detection and quantitation of the expression levels and phosphorylation states of a large number of protein kinases using Kinex™ microarrays. Protein kinase Kinex™ arrays cover over 270 kinases and can be used for studying kinase inhibitors, for the identification of novel kinase substrates, establishing kinase antibody specificities, and for discovery and testing of protein kinase–protein interactions. The price of the service is currently $1.66 per protein and seems attractive until the user multiplies it by the number of spots – Kinexus microarrays include 650+ antibodies, ~230 cell lysates (a reverse microarray) and some ~100 tissues per microarray.

Other players in the 'Chip' field provide 'complete solutions' which use or are built around chip-based proteomics. One such complete solution incorporating custom-built microarrays and instrumentation is provided by Zeptosens (www.zeptosens.com). Their technology is based on thin film planar waveguides (PWG) for evanescent field detection and presents an alternative approach to both the structure of the array and the excitation light pathway. In this system, a thin film (Ta_2O_5) with a high refractive index is deposited on a transparent support with a lower refractive index. The excitation light is a parallel laser light beam integrated into the film by diffraction via a diffractive grating imprinted in the substrate, and this creates a strong evanescent field with highest intensity close to the surface. This effect can be exploited to selectively excite only those fluorophores situated in close proximity to the surface of the waveguide, that is at the chip surface where capture reagents are immobilized. The metal oxide layer can be modified to allow for the stable immobilization of DNAs and proteins for a wide range of bioassay applications. Therefore, when a fluorescently labelled analyte becomes bound to an immobilized capture reagent, excitation of the fluorophore by the evanescent field and detection of the analyte is limited only to the array surface. Since signals from unbound fluorophore molecules in solution are not detected there is no requirement for washing steps allowing the analysis to be carried out in real time. This also produces a significant improvement ($\times 10$- to $\times 100$-fold) in the signal-to-noise ratio and greatly increases the sensitivity of the assay in comparison to traditional fluorescence scanners. The ZeptoMARK™ array surfaces are modified to produce an accessible substrate, which is also stable and minimizes non-specific adsorption. This allows arrays to be applied in the analysis of antigen–antibody, enzyme–substrate and membrane receptor–ligand interactions. Arrays are imaged using the ZeptoREADER™, a high-throughput fluorescence imaging microarray readout system tailored to planar waveguide (PWG) technology and analysed using specialized ZeptoVIEW™ software. Zeptosens assays require a small amount of sample (analysis is performed in 12 μl chamber) and are capable of achieving higher signal-to-noise ratio compared to traditional scanners due to PWG technology, where fluorescent excitation is confined to the chip surface only.

ZeptoMARK glass slides were used by Ghatnekar-Nilsson *et al.* (2007) to generate a miniaturized custom attolitre-vial-based microarray for the sensitive detection of low abundant proteins from human sera. The atto-vial patterned substrates were manufactured using ZeptoMARK glass slides covered with positive polymer resist PMMA A4 and patterned using electron beam lithography achieving high-density arrays with at least 250 000 vials/mm^2. The authors showed sufficient sensitivity (in the pg/ml range) using only minute amounts of sample. The potential of such nanopatterned chips is great, but the main difficulty that remains is the targeted immobilization of the capture reagents inside the atto-vials.

A simpler way to manufacture antibody arrays is to spot the capture reagents on, for example, immobilized nitrocellulose. Such arrays are often referred to as reverse-phase protein arrays (RPPA). One such array containing some ~200 antibodies has been recently reported by Ruan and colleagues (2008) from the Institute Curie. Their RPPAs were made by spotting low 'nl' volumes of antibodies onto nitrocellulose-covered FastTM (Whatman) or NC-W (Schott Nexterion) slides. An assay would be similar to a Western blot or ELISA and would require little if any additional reagents or accessories. RPPAs are easy to manufacture using contact spotters (Q-array from Genetix, Flexys, Microsys, OmniGrid and MicroGrid from Genomics Solutions), non-contact spotters (Biochip Arrayer from Perkin-Elmer and GeSiM from NanoPlotter) or even hand spotters (Microcaster from Whatman). FAST Frame (Whatman) provides a convenient slide holder. Additional practical information on making and using DIY RPPAs can be found on the Whatman web pages (www.whatman.com and www.arraying.com). Due to their simplicity and affordability, RPPAs have been widely used for medium- to high-throughput proteomic studies for ~10 years, mostly in oncology for monitoring disease states and cancer progression (Paweletz et al. 2001; Omenn et al. 2005; Nishizuka et al. 2003; Scrivener et al. 2003). More recent examples include serum biomarker screening profiling (Grote et al. 2008) and analysis of primary acute myelogenous leukaemia samples (Tibes et al. 2006).

A common feature of all protein/antibody arrays (whether traditional or RPPA) is that they are useful only for assaying known proteins and require careful control of initial loading to maintain quantitative assay conditions. The importance of data analysis is difficult to underestimate when hundreds to thousands of data points require quantitative analysis. The recently reported analysis of leukaemia-affected cells from matched blood and marrow samples from 256 AML patients using RPPAs focused on determining protein expression levels with the aim of identifying cancer biomarkers. With the assumption being that protein function regulates the phenotypic characteristics of cancer, a functional proteomic classification system should allow one to predict response. Kornblau et al. (2009) reported distinct elevated expression levels of 24 proteins. Using principal component analysis, seven protein signature groups were distinguished. These profiles were found to correlate with known morphologic features such as remission attainment, relapse and overall survival (Kornblau et al. 2009). The RPPA approach allows signal amplification steps and can therefore be made to detect up to zeptomole levels of target with variance below 10%. Other solutions to the data analysis problem include variable slope normalization (Neeley et al. 2009) and competitive assays (Barry et al. 2003a). As is true of any antibody array, the BD protein array represents a 'closed' system, and is limited in its capacity to those antibodies present on the array.

Peptide microarrays are not dissimilar to protein arrays, except that peptides are used. The possibility of chemical synthesis of peptides provides additional more cost-effective opportunities for array manufacturing (Okochi *et al.* 2008). Sets of synthetic peptides can now be obtained from a number of providers at affordable prices (e.g. Peptide 2.0 Inc., www.peptide2.com). Another attractive opportunity is to have biologically active peptides for functional screenings. Parikh *et al.* (2009) used arrays containing 1176 peptides for determining recombinant kinase substrate preferences. Other examples include using high-density peptide arrays for mapping at the proteome level of the domain-mediated protein–protein interaction networks mediated by the SH3 domain (Li and Wu 2009), and using MHC-peptide arrays for identification of T cell epitopes (Stone *et al.* 2005). Peptide arrays, more so than protein arrays, require careful adjustments of the linker and spacers. Andresen *et al.* (2006) achieved picomolar level detection of monoclonal antibodies from diluted sera samples by carefully adjusting a spacer length. Binding of larger molecules, such as immunoglobulin Gs, depended on the spacer length but, surprisingly, the orientation of peptides, controlled through site-specific solution-phase coupling of biotinylated synthetic peptides to NeutrAvidin surface, was found to play only a minor role. Peptide arrays are suitable for a variety of functional screenings. A review by Uttamchandani and Yao (2008) provides ample examples of peptide arrays and their uses for wider drug discovery and point-of-care applications.

Protein array reviews are often limited to antibody arrays and peptide arrays, yet many other useful functional arrays exist and have been reported. For example, lectins are proteins derived from plants or bacteria and possess binding sites for particular single, or arrangements of, carbohydrate units, some of which display mitogenic activity. These represent another attractive functional assay, especially in array format. A few dozen different lectins are commercially available, ten or more as agarose conjugates (e.g. lectin, wheat germ, *Triticum vulgaris*, agarose conjugate from Calbiochem). The specificity and variety of lectins makes them useful tools for the selective affinity isolation of subsets of glycoproteins or glycopeptides; indeed arrays of lectins are becoming available for multiplex profiling of glycan structures (Zheng, Pelen and Smith 2005). Other types of arrays include tissue microarrays (Kononen *et al.* 1998) and microelectrode array (MEA) for the detection of signalling processes in cells and tissues (Egert *et al.* 1998; Stett *et al.* 2003). These are two powerful and widely used array technologies, but fall just outside the chip-based proteomics area. We therefore direct readers to relevant reviews.

There is no doubt that the potential of protein array based proteomics could have many advantages over 2D gels and chromatography, not least by way of higher throughput analysis, better reproducibility and more quantitative protein expression analysis. There is an ever increasing market for new, miniaturized, more accurate and cost-effective diagnostics that can be supplied in kit format for use in the field. The ability to multiplex such assays is also highly desirable, allowing for the simultaneous detection of more than one analyte in a given assay. Microarray techniques have provided such capabilities and revolutionized the fields of nucleic acid analysis and proteomics. For protein arrays, some of the downstream technologies are available but there is a huge shortage of affinity reagents, and even for proteins where antibodies are known to work for IHC, ELISA and western blotting, the conversion of these into a chip format for homogeneous assays of protein expression is not straightforward. Despite the considerable technical and financial hurdles, functional protein chips will undoubtedly have a dramatic impact on our ability to find disease-related protein changes, ascribe functionality to many proteins and elucidate pathways and interactions at an unprecedented pace.

Chip-based mass spectrometry

Microfluidic LOCs provide numerous advantages to bench-top systems when processing samples, but the choice of detection method afterwards is hugely important. Mass spectrometry (MS) has established itself as a revolutionary tool in proteomic analysis over the past two decades and so it is little wonder that it has been interfaced with LOC systems. Lee *et al.* (2009) provide an in-depth review of LOC fabrication considerations and the integration of microfluidic LOCs with MALDI-TOF mass spectrometry (MS) through both online and offline approaches. To increase accuracy of detection and identification of proteins by MS, sample proteins are digested (typically with trypsin) prior to MS analysis. The peptides obtained are typically identified using MALDI-TOF MS followed by database mass matching. Further confirmation can be obtained using MS/MS techniques with collision-induced dissociation (CID) to fragment the peptide enabling an amino acid sequence to be generated. MS-based detection has been used by Gygi *et al.* (1999) in their isotope-coded affinity tags (ICAT) approach, although this is not really a chip-based technology. MALDI-MS is used in addition to fluorescence detection of peptides in the peptidomics approach developed by Scrivener *et al.* (2003).

Other commercial systems that integrate LOC arrays and MS for protein profiling and biomarker identification include the ProteinChip SELDI-TOF-MS technology originally developed by Ciphergen Biosystems and later purchased by Bio-Rad for $20m in 2006 (www.bio-rad.com), which has been used to identify biomarkers of ovarian cancer with greater sensitivity and specificity than the CA125 test that is routinely used (Yu *et al.* 2004; Zhang *et al.* 2004). The ProteinChip® system developed by Ciphergen consists of the instrumentation, protein arrays and software to compare the presence and abundance of individual proteins in crude samples. The ProteinChip Array® comprises a series of non-selective affinity matrices (e.g. hydrophobic, normal phase, ion exchange, metal ion exchange), which are incubated with crude protein preparations. The bound samples are subjected to a series of wash steps of differing properties (e.g. water, organic, salt, pH, detergents, denaturants). This leads to the sample becoming split into sub-fractions. The captured proteins are then analysed by surface-enhanced laser desorption/ionization (SELDI) mass spectrometry (Chapman 2002). Such crude fractionation is suitable for taking 'snapshots' of the protein mixtures (e.g. quality control of cell culture media), but may not be very useful for studying differential protein expression due to the poor mass resolution of this technique, which is largely limited to lower MW proteins (i.e. <25 kDa). The ProteinChip® approach is often used for the identification of crude protein expression patterns in disease diagnostics. Ishida *et al.* (2008) used this approach to search for biomarkers in adult T cell leukaemia (ATL). Serum samples were obtained from controls, HTLV-1 virus carriers and ATL patients. ProteinChip® arrays coupled with SELDI-TOF-MS generated spectral maps with molecular weights and relative concentrations allowing detection of disease-specific proteins, including five upregulated ($P < 0.001$) and four downregulated ($P < 0.001$) proteins, when compared with non-ATL samples (1779, 1866, 2022, 4467, and 8930 m/z and 4067, 4151, 8130, and 8597 m/z, respectively). The 1779, 1866 and 2022 m/z peaks were identified with MS/MS ion search as being from C3f complement protein, which seems to be upregulated in ATL samples. Additionally, three pathways whose activation results in formation of C3f were investigated. Concentration of MBL/MASP-2, serum lectin pathway marker was found to be significantly increased ($P < 0.05$) in ATL patients indicating that activation takes place through the lectin pathway. These peaks were selected as biomarker candidates for ATL.

In another study, Aivado *et al.* (2007) used SELDI-TOF-MS to generate serum proteome profiles of patients with myelodysplastic syndrome (MDS) and non-MDS cytopenias. The aim of the investigation was to identify diagnostic molecular markers of MDS. Profiling of 218 patients yielded a predictive MDS serum proteome profile. Robustness of the proteome profile was validated by predicting MDS in a control set of sera samples with 81.9% accuracy, 83.3% sensitivity and 79.2% specificity ($P < 0.001$) (Aivado *et al.* 2007). In a multivariate logistic regression model, the identified serum proteome profile that is decreased serum levels of both chemokine ligand 4 (CXCL4) and chemokine ligand 7 (CXCL7) remained an independent predictor of MDS, disproving spurious associations. The levels were 4.0-fold lower than in controls and 2.9-fold lower than in non-MDS cytopenias. The two CXC chemokine ligands identified using SELDI-TOF-MS were both corroborated using immunoassays.

Another 'chip'-based proteomics development came from Agilent (www.agilent.com) in the form of integrated nanospray LC/MS Ion Source on a chip. Agilent's new bench-top 6520 Quadrupole Time-of-Flight LC/MS and its LC/MS chip-based ion source provide an easier to use alternative to traditional nanoflow LC separations and delivers increased sensitivity in a wide range of applications such as proteomics, drug discovery and development, biomarker discovery, metabolomics and metabolite ID. In the Agilent LCMS Chip, all LC fluidics (75 micron internal diameter channels), including the trapping column, valves controlled by the chromatograph software, connectors and nano-electrospray emitter are integrated into a single polyimide 'chip' with laser ablated channels, ports and frit structures, approximately double the size of a microscope slide (Yin *et al.* 2005; Yin and Killeen 2007). Some recent publications reporting the use of Agilent LCMS Chips include the study of protein–protein interactions by Swamy *et al.* (2009), Glycosaminoglycan Glycomics Profiling using hydrophilic interaction chromatography of heparin, heparan sulfate and chondroitin/dermatan sulfate glycosaminoglycans by Staples *et al.* (2009), and profiling of phosphoproteome of non-stimulated primary human leukocytes by Raijmakers *et al.* (2010). In the latter work, the authors identified 1012 unique phosphopeptides and 960 phosphorylation sites in circulating white blood cells. In all the above cases, the integrated chip-based approach allowed for increased sensitivity and eliminated the need to optimize connections between chromatograph, injector, trapping column, analytical column and spray needle.

Surface plasmon resonance (SPR) and quartz crystal microbalance (QCM) chip instruments

Other detection methods applicable to microarrays include surface plasmon resonance (SPR) and quartz crystal microbalance (QCM). These biosensors can monitor interactions by measuring the mass concentration of biomolecules close to a surface of the sensor. The SPR effect arises when light is reflected under certain conditions from a conducting film at the interface between two media of different refractive indices. The measured SPR response (which is typically the angle of minimum reflected light intensity) is directly proportional to the mass of molecules that bind to the surface. Therefore SPR enables discrimination between a sample in solution and molecules associated with the surface of the sensor and so measurements can be made in real time (Cullen, Brown and Lowe 1987–1988; Fagerstam *et al.* 1990). In addition, SPR allows detection of unlabelled molecules, which makes this technique very attractive for use in biomedical research and diagnostics. SPR-based detection is exploited commercially by GE Healthcare (www.biacore.com), XanTec bioanalytics GmbH (Germany) (www.xantech.com), Genoptics/Horiba Scientific (www.genoptics-spr.com), Reichert (www.reichertspr.com), Biosensing Instrument Inc. (biosensingusa.com), Cole-Parmer (www.coleparmer.com) SensíQ instruments (www.discoversensiq.com) and the SPRTM 100 module from Thermo (www.thermo.com). SPR-based BIAcore instruments (GE Healthcare) currently dominate the SPR market. BIAcore was the first to combine an SPR detection system with a microfluidics system and data handling software. Users are offered a number of protein immobilization techniques such as through amine groups, carboxyl groups, thiol groups, cis-diols on carbohydrate residues, histidine tags or biotin-mediated. Underivatized gold surfaces are also available for custom applications. The sensitivity of detection using BIAcore reaches ca. 1 pg of the surface-immobilized unlabelled protein. However, the exact conversion factor between the reflective angle and surface concentration is dependent on the properties of the sensor surface and buffer system as well as the nature of the molecules bound, thus limiting the method's capability of absolute quantitation of the bound protein. BIAcore instruments are capable of measuring protein binding in real-time (kinetic binding) as well as performing traditional end-point binding analysis. BIAcore chips are compatible for use with MALDI mass spectrometry thus allowing additional integration. BIAcore analysis can be applied to membrane proteins and such studies have been attempted (through immobilizing cell membrane preparations or proteoliposomes or by means of a lipid deposition and surface reconstitution approach), but such experiments are less reliable and result in less quantitative data. Unfortunately, SPR-based detection is not suitable for measuring small molecule binding, since an SPR signal is proportional to a total mass associated with the sensor surface, not the number of molecules.

Genoptics was one of the first to commercialize the SPR-based bio-microarray system. Such an array is capable of parallel detection of up to several hundreds of protein (antibody)–ligand interactions with no requirement for labelling. This permits a real-time analysis of differential protein profiles. Biosensing Instrument Inc. offers Hydrodynamic Isolation™ continuous flow, microfluidic sample delivery technology. Hydrodynamic Isolation™ combines hydrodynamic focusing with location-specific cell evacuation to create a high-performance, continuous-flow, sample addressing method that is robust, flexible and simple to multiplex. The use of hydrodynamic addressing, now also used in some GE Healthcare instruments, provides a simple and flexible sample addressing method. Biosensing Instrument's 16 spot Hydrodynamic Isolation™ flow cell is truly an 'Any Sample, Any Sensor, Any Time™' chip. Among other SPR chip instrument providers, Reichert (www.reichert.com) developed a highly flexible, affordable, modular SPR instrument capable of combining other measurements with SPR (e.g. electrochemistry, fluorescence and photochemistry measurements). Graffinity (www.graffinity.com) have developed a SPR-based fragment screening process for rapid identification and development of diverse novel quality leads and drug candidates. The immobilized compounds (small drug fragments) are incubated with a purified and solubilized target protein to generate an affinity fingerprint. This delivers information on the potential for small molecules as drug leads. The platform has been validated in a series of projects and collaborations with leading pharmaceutical and biotechnology companies and has been successfully applied to over 85 protein targets since 2003.

Developments of QCM instrumentations closely followed those in the SPRC field. The principle of QCM sensing is in detection and monitoring a self-resonance frequency of a quartz crystal. This depends on the mechanical properties of the crystal, the medium and any molecular interactions (e.g. molecular sorption) on the chip surface. QCM piezoelectric sensors are attractive alternatives to SPR-based chips, because in addition to the detection of a 'mass', QCM also allows the user to monitor viscoelasticity and micromechanics of surface processes. One of the earlier players in the field, Attana (www.attana.com) developed QCM biosensors for analysis of biomolecular interactions suitable for studying proteins, nucleic acids and carbohydrates and also binding moieties of vastly different sizes, ranging from peptides to cells. Nowadays basic QCM setups are produced (e.g. Elbatech, www.elbatech.com; or SRS, distributed by www.lambdaphoto.co.uk); these are priced at around $2 k–$3 k. At the other end of the scale is a fully automated, high-throughput, four-channel resonance acoustic profiling (RAP) RAPid4 (TTP Labtech, Royston, UK), costing well over $100 k. In addition to measuring viscoelastic prot-perties, QCM allows for novel derivatives of the classical QCM, such as rupture event scanning (Cooper *et al.* 2001) in which the principle of detection is in measuring the amplitude of crystal oscillations at which the molecular interaction is disrupted. This unusual detection principle, as well as the example of the Reichter SPR, com-bined with electrochemical and optical detection clearly reveal the advantages of chip-based miniaturized sensors with integrated multiple detection techniques.

Microfluidics

Further miniaturization and improved automation could be achieved by using microfluidics, a technique that is often realized in a chip and which allows rapid analysis of very low sample volumes. Microfluidic platforms are divided into capillary, pressure driven, centrifugal, electrokinetic and acoustic systems. While microfluidic devices may concede some ground in high-throughput capacity to microarrays, they require only minimal sample processing and handling and do not need expensive automated spotting robots, or strict reaction conditions such as humidity and temperature (Cesaro-Tadic *et al.* 2004; Golden *et al.* 2005; Delamarche, Juncker and Schmid 2005; Mukhopadhyay, 2009).

One of the forerunners in commercial LOC technology is Agilent Technologies (www.agilent.com) whose 2100 Bioanalyzer microfluidic systems are capable of protein detection as low as 1 pg/μl, as well as performing other protein analysis procedures such as checking antibody quality down to 10 ng/μl. Antes *et al.* (2010) provide a comprehensive comparison of Agilent's system with traditional SDS-PAGE for quality control of pharmaceutical products, considering long-term assay robust-ness as a key feature. The group found the microfluidic device to be notably more accurate, with a broader assay range and lower limit of quantification, while citing greatly reduced assay times (just 1 h compared to around 5 h for SDS-PAGE), and more convenient handling as additional benefits. The 2100 Bioanalyzer is designed to streamline the processes of protein expression, gene expression analysis and RNA isolation. The technology can therefore be applied to either electrophoretic chip applications or cell-based assays. Essentially the system is an automated capillary gel runner and still requires gels to be loaded, that is not a definitive dry-run tech-nique. However, the system is suitable for routine sizing and quantitation of proteins (as well as DNAs).

One platform that combines microfluidic chip and MALDI and demonstrates the commercial potential of LOC devices comes in the form of a novel compact disk (CD) device available from GyrosAB, Uppsala, Sweden (Gustafsson *et al.* 2004;

Hirschberg *et al.* 2004). The device harnesses the centrifugal force generated upon spinning the disk to move liquid through various microstructures containing components such as chromatography columns. The CD-based laboratory integrates multiple steps into a streamlined process. A flexibility of design enables CD products to be built to suit specific processes. Each CD-Lab can comprise hundreds of application-specific microstructures. As the CD spins, samples are processed in parallel in nanolitre amounts within the microstructures. Gustafsson *et al.* (2004) used CDs with reversed-phase chromatography columns to concentrate, desalt and elute peptides into target MALDI areas combined with a solvent containing the MALDI matrix. The CD was then inserted into the MALDI unit and peptide mass fingerprinting performed with proteolytic peptide detection achieved as low as 50-amol. Protein identification using such CD technology is reported to be twice as successful as with C18 ZipTips and standard MALDI steel targets. Gyros also market a variety of other CD-based protein chips that are based upon fluorescence detection (www.gyros.com).

Ko *et al.* (2008) have developed a microfluidic biochip capable of multiplexed detection of the cancer biomarkers alpha-fetoprotein (AFP), carcinoembryonic antigen (CEA) and prostate-specific antigen (PSA). The device comprises a biocompatible PDMS layer over a glass substrate that contains a platinum electrode to detect electrical signals. Pillar-type microfilters are used to immobilize antibody-conjugated microbeads, and an immunogold silver staining technique is used to amplify the electrical signal generated when an antibody–antigen complex forms. Assay times were reduced from between 3 and 8 h durations for conventional methods to under 1 h, with a working range of target protein concentrations between 10^{-3} and 10^{-1} µg/ml, demonstrating great promise for possible clinical application with future development.

An example of the application of microfluidics in antibody screening was an investigation in which an electrokinetic platform was used to separate 5×10^8 antibodies displayed on bacteria. This method relies on the induction of a charge in a liquid by a solid surface of opposite charge and the movement of the liquid towards the surface (electro-osmosis). Any ions dissolved in the liquid are also separated according to their charge (electrophoresis). The field is then rotated 90° to further separate the constituents in solution (this is known as dielectrophoresis). The results in the antibody investigation were comparable to those obtained using the latest commercial cell sorting instruments available at the time, and the method can also be used for gene transfection or the transport or separation of whole cells (Bessette *et al.* 2007)

Microfluidics monoclonal antibody-based analysis has also shown promise as a prognostic tool for patients because a monoclonal antibody against a cell surface receptor of interest can be immobilized in a microfluidic device to allow cells expressing the target antigen to bind. This has been used to quickly isolate and elucidate the quantity of circulating tumour cells, which can be produced during the early stages of tumourigenesis and from which metastases form. The number of these cells present in serum can indicate the probability of success of possible therapeutic regimes and the stage of the disease, thus providing a prognosis for the patient and helping clinicians decide the appropriate treatment (Adams *et al.* 2008; Bessette *et al.* 2007).

One criticism of LOC devices is that they are often more 'Chips-in-Lab' than 'Lab-on-Chips', highlighting the lab-based equipment that can also be required. LabNow (www.labnow.com) in conjunction with the University of Texas at Austin, are striving to produce truly portable testing devices and have developed a fully integrated analytical LOC system that incorporates advances in nanochemistry, microfluidics, imaging analysis and digital fluorescence microscopy to provide rapid turnaround of results from low sample volume for applications such as infectious disease diagnostics and cardiac and cancer biomarkers. The platform is described as a programmable Nano-Bio-Chip (NBC) Sensor, comprising various modular components and permitting the user to insert new assay types quickly and efficiently. In 2005, Christodoulides *et al.* (2005) reported its use for the simultaneous detection of two critical markers of coronary heart disease – C-reactive protein levels and leukocyte count. Simultaneous increase of the levels of both markers correlates with sevenfold increased risk of heart disease (Margolis *et al.* 2005). Previously these two markers were assayed separately, but by using a dual-function microchip incorporating a membrane-based assay for the capture and detection of blood cells, and a bead-based assay platform for the capture and measurement of blood proteins, the group produced a multiplexed assay capable of highly accurate diagnosis of an individual's risk of heart disease, the foremost health issue in the developed world. The system has also been shown to be a promising new diagnostic tool for early detection of oral cancer, identifying four key parameters separating both dysplastic and malignant lesions from healthy oral epithelium samples, including the differential expression of the biomarker, epidermal growth factor receptor (EGFR) (Weigum *et al.* 2010). A more detailed insight into this platform can be found in the *Analytical Chemistry* feature (Jokerst *et al.* 2010), complete with a link to an online podcast with its founder, Professor John T. McDevitt.

Conclusion

The development of protein expression profiling (proteomics) techniques based on a chip format has expanded rapidly in recent years. A wide range of companies have entered the market and an increasingly large range of products are becoming available to proteomics researchers. Although a whole proteome-based analysis, as is potentially attainable for genomics using DNA chips, still remains in the future for proteins, a number of methodologies and technologies to realize this objective are now in place. This type of analysis for proteins will undoubtedly emerge within the next few years as the number of capture agents (antibodies, aptamers, etc.), which are suitable for application to an array format continues to increase. In the interim, both specialized chips for a focused analysis of groups of proteins or peptides and diagnostic chips for patient sample-screening are available, are being used and their complexity is expanding rapidly. A combination of almost 15 years of experience of DNA microarray technology and the realization that proteins need to be analysed for expression and function to speed up advances in medicine has driven the automation, miniaturization, investment and technological advances necessary to begin making functioning protein chips. Will the development of protein chips go along the familiar way paved by DNA chip developers, through continuing improvements in surface immobilization techniques, further increases in affinity and density of the immobilized antibodies and steady growth in the public acceptance of the new technology, or will it take another route? Early evidence from conferences and published reports suggests much greater variability in the style of protein arrays in all areas including surfaces, arrayers, affinity reagents, assays, analyses and readers. In a recent survey by Select Biosciences (www.SelectBiosciences.com) into the future of multiplexed diagnostics, predictions have been made within the multiplexed diagnostics realm by the participants from a variety of disciplines and with experience in a broad range of methods within the field (see Figure 9.2). By their nature, proteins present a number of challenges that have prevented protein arrays from achieving quite the same widespread impact that nucleic acid arrays achieved in the mid 1990s. Instability, immobilization and a great variety of protein size, structure and concentration ranges are but a few of the hurdles that must be overcome if the number of probes available to protein chips is to rival the number of probes available to nucleic

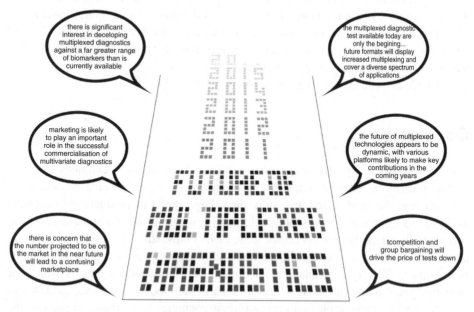

Figure 9.2   The future of multiplexed diagnostics (data source: Select Biosciences survey, *Multiplexed Diagnostics 2010,* available from www.SelectBiosciences.com)

acid arrays. Improvements in detection hardware are proving just as important to the advancement of these exciting devices as the novel biochemistry that underpins them, with ever more sensitive technology improving detection limits all the time. As is often the case, LOC technologies emerged initially in research laboratories where cost is less of a concern. However, what is encouraging is that in recent years much of this research has translated into microanalytical tools targeted towards applications that they were initially intended for those requiring small sample volumes and operated routinely at low cost by untrained individuals and used in settings such as medical diagnostics in developed and developing countries, counterterrorism, first responders (emergency services), environmental and food monitoring to name but a few. While their medical use in developing countries is in diagnosis of existing disease states, their use in developed countries is likely to be in turning healthcare from a treatment-orientated discipline to one that is anticipatory and where routine screening for biomarkers of disease becomes the norm. There already exists a large body of literature, particularly in cancer biomarker diagnostics, to support this outlook. As with all technology that promises so much, there are often equal doses of pessimism and optimism, but with continued innovation and careful application it seems that LOC technology will fulfil its much-hyped potential.

References

Adams, A.A., Okagbare, P.I., Feng, J. *et al.* (2008) Highly efficient circulating tumor cell isolation from whole blood and label-free enumeration using polymer-based microfluidics with an integrated conductivity sensor. *J Am Chem Soc* **130**: 8633–8641.

Aivado, M., Spentzos, D., Germing, U. *et al.* (2007) Serum proteome profiling detects myelodysplastic syndromes and identifies CXC chemokine ligands 4 and 7 as markers for advanced disease. *Proc Natl Acad Sci USA* **104**: 1307–1312.

Alhamdani, M.S., Schröder, C. and Hoheisel, J.D. (2009) Oncoproteomic profiling with antibody microarrays. *Genome Med* **1**: 68.

Altamirano-Dimas, M., Sharma, M. and Hudson, J.B. (2009) Echinacea and anti-inflammatory cytokine responses: Results of a gene and protein array analysis. *Pharm Biol* **47**: 500–508.

Andresen, H., Grötzinger, C., Zarse, K. *et al.* (2006) Functional peptide microarrays for specific and sensitive antibody diagnostics. *Proteomics* **6**: 1376–1384.

Angenendt, P., Glokler, J., Sobek, J. *et al.* (2003) Next generation of protein microarray support materials: evaluation for protein and antibody microarray applications. *J Chromatogr A* **1009**: 97–104.

Antes, B., Oberkleiner, P., Nechansky, A. and Szolar, O.H. (2010) Qualification of a microfluidics-based electrophoretic method for impurity testing of monoclonal antibodies. *J Pharm Biomed Anal* **51**: 743–749.

Åsberg, M., Nygren, Å., Leopardi, R. *et al.* (2009) Novel biochemical markers of psychosocial stress in women. *PloS One* **4**: e3590.

Badiou, S., Cristol, J.P., Jaussent, I. *et al.* (2008) Finetuning of the prediction of mortality in hemodialysis patients by use of cytokine proteomic determination. *Clin J Am Soc Nephrol* **3**: 423–430.

Banfi, G., Migliorini, S., Pedroni, F. *et al.* (2008) Strenuous exercise activates growth factors and chemokines over-expression in human serum of top-level triathlon athletes during a competitive season. *Clin Lab Med: CCLM / FESCC* **46**: 250–252.

Barry, R., Platt, A.E., Scrivener, E. *et al.* (2002a) *Detection of Peptides*. International Patent WO 02/25287.

Barry, R., Platt, A.E., Scrivener, E. *et al.* (2002b) *Detection of Peptides*. US Patent 2002055186.

Barry, R., Scrivener, E., Soloviev, M. and Terret J. (2002c) Chip-based proteomics technologies. *Int GenomicProteomic Technol*: 14–22.

Barry, R., Diggle, T., Terrett, J. and Soloviev, M. (2003a) Competitive assay formats for high-throughput affinity arrays. *J Biomolec Screen* **8**: 257–263.

Barry, R., Platt, A.E., Scrivener, E. *et al.* (2003b) *Detection of Peptides*. European Patent EP1320754.

Barry, R. and Soloviev, M. (2004) Quantitative protein profiling using antibody arrays. *Proteomics* **4**: 3717–3726.

Berrahmoune, H., Lamont, J.V., Herbeth, B. *et al.* (2009) Association between EGF and lipid concentrations: a benefit role in the atherosclerotic process? *Clin Chim Acta* **402**: 196–198.

Bessette, P.H., Hu, X., Soh, H.T. and Daugherty, P.S. (2007) Microfluidic library screening for mapping antibody epitopes. *Anal Chem* **79**: 2174–2178.

Büssow, K., Cahill, D., Nietfeld, W. *et al.* (1998) A method for global protein expression and antibody screening on high-density filters of an arrayed cDNA library. *Nucleic Acids Res* **26**: 5007–5008.

Cesaro-Tadic, S., Dernick, G., Juncker, D. *et al.* (2004) High-sensitivity miniaturized immunoassays for tumor necrosis factor alpha using microfluidic systems. *Lab Chip* **4**: 563–569.

Chapman, K. (2002) The ProteinChip Biomarker System from Ciphergen Biosystems: a novel proteomics platform for rapid biomarker discovery and validation. *Biochem Soc Trans* **30**: 82–87.

Cheung, J.K., Li, J., Cheung, A.W. *et al.* (2009) Cordysinocan, a polysaccharide isolated from cultured *Cordyceps*, activates immune responses in cultured T-lymphocytes and macrophages: signaling cascade and induction of cytokines. *J Ethnopharmacol* **124**: 61–68.

Christodoulides, N., Floriano, P.N., Acosta, S.A. *et al.* (2005) Toward the development of a lab-on-a-chip dual-function leukocyte and C-reactive protein analysis method for the assessment of inflammation and cardiac risk. *Clin Chem* **51**: 2391–2395.

Cooper, M.A., Dultsev, F.N., Minson, T. *et al.* (2001) Direct and sensitive detection of a human virus by rupture event scanning. *Nat Biotechnol* **19**: 833–837.

Cullen, D.C., Brown, R.G. and Lowe, C.R. (1987–1988) Detection of immuno-complex formation via surface plasmon resonance on gold-coated diffraction gratings. *Biosensors* **3**: 211–225.

Delamarche, E., Juncker, D. and Schmid, H. (2005) Microfluidics for processing surfaces and miniaturizing biological assays. *Adv Materials* **17**: 2911–2933.

Du, Y., Wei, X., He, Y. *et al* (2009) P2-380: Identification and characterization of human autoantibodies that may be used for the treatment of prion diseases. *Alzheimer's Dementia* **4**: T484.

Egert, U., Schlosshauer, B., Fennrich, S. *et al.* (1998) A novel organotypic long-term culture of the rat hippocampus on substrate-integrated multielectrode arrays. *Brain Res Brain Res Protocols* **2**: 229–242.

Eisen, M.B. and Brown, P.O. (1999) DNA arrays for analysis of gene expression. *Methods Enzymol* **303**: 179–205.

El Karim, I.A., Linden, G.J., Irwin, C.R. and Lundy, F.T. (2009) Neuropeptides regulate expression of angiogenic growth factors in human dental pulp fibroblasts. *J Endodontics* **35**: 829–833.

Ermetici, F., Malavazos, A.E., Corbetta, S. *et al.* (2008) Soluble adhesion molecules levels in patients with Cushing's syndrome before and after cure. *J Endocrinol Invest* **31**: 389–392.

Fabre, S., Dupuy, A.M., Dossat, N. *et al.* (2008) Protein biochip array technology for cytokine profiling predicts etanercept responsiveness in rheumatoid arthritis. *Clin Exp Immunol* **153**: 188–195.

Fagerstam, L.G., Frostell, A., Karlsson, R. *et al.* (1990) Detection of antigen–antibody interactions by surface plasmon resonance. Application to epitope mapping. *J Mol Recogn* **3**: 208–214.

Fodor, S.P., Rava, R.P., Huang, X.C. *et al.* (1993) Multiplexed biochemical assays with biological chips. *Nature* **364**: 555–556.

Fulton, R.J., McDade, R.L., Smith, P.L. *et al.* (1997) Advanced multiplexed analysis with the FlowMetrix system. *Clin Chem* **43**: 1749–1756.

Ge, H. (2000) UPA, a universal protein array system for quantitative detection of protein–protein, protein–DNA, protein–RNA and protein–ligand interactions. *Nucleic Acids Res* **28**: e3.

Ghatnekar-Nilsson, S., Dexlin, L., Wingren, C. *et al.* (2007) Design of atto-vial based recombinant antibody arrays combined with a planar wave-guide detection system. *Proteomics* **7**: 540–547.

Golden, J., Shriver-Lake, L., Sapsford, K. and Ligler F. (2005) A 'do-it-yourself' array biosensor. *Methods* **37**: 65–7.

Grote, T., Siwak, D.R., Fritsche, H.A. *et al.* (2008) Validation of reverse phase protein array for practical screening of potential biomarkers in serum and plasma: accurate detection of CA19-9 levels in pancreatic cancer. *Proteomics* **8**: 3051–3060.

Gustafsson, M., Hirschberg, D., Palmberg, C. *et al.* (2004) Integrated sample preparation and MALDI mass spectrometry on a microfluidic compact disk. *Anal Chem* **76**: 345–350.

Gygi, S.P., Rist, B., Gerber, S.A. *et al.* (1999) Quantitative analysis of complex protein mixtures using isotope-coded affinity tags. *Nat Biotechnol* **17**: 994–999.

Haab, B.B., Dunham, M.J. and Brown, P.O. (2001) Protein microarrays for highly parallel detection and quantitation of specific proteins and antibodies in complex solutions. *Genome Biol* **2**: RESEARCH0004.

He, M. and Taussig, M.J (2001), Single step generation of protein arrays from DNA by cell-free expression and *in situ* immobilisation (PISA method). *Nucleic Acids Res* **29**: E73.

Hirschberg, D., Tryggvason, S., Gustafsson, M. *et al.* (2004) Identification of endothelial proteins by MALDI-MS using a compact disc microfluidic system. *Protein J* **23**: 263–271.

Hoa, X.D., Kirk, A.G. and Tabrizian, M. (2007) Towards integrated and sensitive surface plasmon resonance biosensors: a review of recent progress. *Biosens Bioelectron* **23**: 151–160.

Holt, L.J., Bussow, K., Walter, G. and Tomlinson, I.M. (2000) By-passing selection: direct screening for antibody–antigen interactions using protein arrays. *Nucleic Acids Res* **28**: E72.

Holz, C., Lueking, A., Bovekamp, L. *et al.* (2001) A human cDNA expression library in yeast enriched for open reading frames. *Genome Res* **11**: 1730–1735.

Horácek, J.M., Tichy, M., Jebavy, L. *et al.* (2008a) Use of multiple biomarkers for evaluation of anthracycline-induced cardiotoxicity in patients with acute myeloid leukaemia. *Exp Oncol* **30**: 157–159.

Horácek, J.M., Tichy, M., Pudil, R. and Jebavy, L. (2008b) Glycogen phosphorylase BB could be a new circulating biomarker for detection of anthracycline cardiotoxicity. *Ann Oncology* **19**: 1656–1657.

Horácek, J.M., Tichy, M., Pudil, R. *et al.* (2008c) New biomarkers of myocardial injury and assessment of cardiac toxicity during preparative regimen and hematopoietic cell transplantation in acute leukaemia. *Clin Chem Lab Med CCLM / FESCC* **46**: 148–149.

Horácek, J.M., Tichy, M., Pudil, R. *et al.* (2008d) Multimarker approach to evaluation of cardiac toxicity during preparative regimen and hematopoietic cell transplantation. *Neoplasma* **55**: 532–537.

Ishida, Y., Yamashita, K., Sasaki, H. *et al.* (2008) Activation of complement system in adult T-cell leukemia (ATL) occurs mainly through lectin pathway: a serum proteomic approach using mass spectrometry. *Cancer Lett* **271**: 167–177.

Iyer, V.R., Horak, C.E., Scafe, C.S. *et al.* (2001) Genomic binding sites of the yeast cell-cycle transcription factors SBF and MBF. *Nature* **409**: 533–538.

Jenison, R., La, H., Haeberli, A. *et al.* (2001) Silicon-based biosensors for rapid detection of protein or nucleic acid targets. *Clin Chem* **47**: 1894–1900.

Jokerst, J.V., Jacobson, J.W., Bhagwandin, B.D. *et al.* (2010) Programmable nano-bio-chip sensors: analytical meets clinical. *Anal Chem* **82**: 1571–1579.

Kavsak, P.A., Ko, D.T., Newman, A.M. *et al.* (2008a) 'Upstream markers' provide for early identification of patients at high risk for myocardial necrosis and adverse outcomes. *Clin Chim Acta* **387**: 133–138.

Kavsak, P.A., Ko, D.T., Newman, A.M. *et al.* (2008b) Vascular versus myocardial dysfunction in acute coronary syndrome: Are the adhesion molecules as powerful as NT-proBNP for long-term risk stratification? *Clin Biochem* **41**: 436–439.

Kavsak, P.A., Lee, A., Hirte, H. *et al.* (2008c) Cytokine elevations in acute coronary syndrome and ovarian cancer: a mechanism for the up-regulation of the acute phase proteins in these different disease etiologies. *Clin Biochem* **41**: 607–610.

Kavsak, P.A., Henderson, M., Moretto, P. *et al.* (2009) Biochip arrays for the discovery of a biomarker surrogate in a phase I/II study assessing a novel anti-metastasis agent. *Clin Biochem* **42**: 1162–1165.

Knezevic, V., Leethanakul, C., Bichsel, V.E. *et al.* (2001) Proteomic profiling of the cancer microenvironment by antibody arrays. *Proteomics* **1**: 1271–1278.

Ko, Y.J., Maeng, J.H., Ahn, Y. *et al.* (2008) Microchip-based multiplex electro-immunosensing system for the detection of cancer biomarkers. *Electrophoresis* **29**: 3466–3476.

Kononen, J., Bubendorf, L., Kallioniemi, A. *et al.* (1998) Tissue microarrays for high-throughput molecular profiling of tumor specimens. *Nat Med* **4**: 844–847.

Kornblau, S., Tibes, R., Qiu, Y. *et al.* (2009) Functional proteomic profiling of AML predicts response and survival. *Blood* **113**: 154–164.

Lee, J., Soper, S.A. and Murray, K.K. (2009) Microfluidics with MALDI analysis for proteomics – a review. *Anal Chim Acta* **649**: 180–190.

Lion, N., Rohner, T.C., Dayon, L. *et al.* (2003) Microfluidic systems in proteomics. *Electrophoresis* **24**: 3533–3562.

Lion, N., Reymond, F., Girault, H.H. and Rossier, J.S. (2004) Why the move to microfluidics for protein analysis? *Curr Opin Biotechnol* **15**: 31–37.

Li, S.S. and Wu, C. (2009) Using peptide array to identify binding motifs and interaction networks for modular domains. *Methods Mol Biol* **570**: 67–76.

Lippi, G., Schena, F., Montagnana, M. *et al.* (2008) Influence of acute physical exercise on emerging muscular biomarkers. *Clin Chem Lab Med* **46**: 1313–1318.

MacBeath, G. and Schreiber, S.L. (2000) Printing proteins as microarrays for high-throughput function determination. *Science* **289**: 1760–1763.

Madoz-Gúrpide, J., Wang, H., Misek, D.E. *et al.* (2001) Protein based microarrays: a tool for probing the proteome of cancer cells and tissues. *Proteomics* **1**: 1279–1287.

Margolis, K.L., Manson, J.E., Greenland, P. *et al.* (2005) Leukocyte count as a predictor of cardiovascular events and mortality in postmenopausal women: the Women's Health Initiative Observational Study. *Arch Intern Med* **165**: 500–508.

Marko-Varga G., Nilsson, J. and Laurell, T. (2003) New directions of miniaturization within the proteomics research area. *Electrophoresis* **24**: 3521–3532.

Mendoza, L.G., McQuary, P., Mongan, A. *et al.* (1999) High-throughput microarray-based enzyme-linked immunosorbent assay (ELISA). *BioTechniques* **27**: 778–780, 782–786, 788.

Michaud, G.A., Salcius, M., Zhou, F. *et al.* (2003) Analyzing antibody specificity with whole proteome microarrays. *Nat Biotechnol* **21**: 1509–1512.

Mukhopadhyay, R. (2009) Microfluidics: on the slope of enlightenment. *Anal Chem* **81**: 4169–4173.

Neeley, S., Kornblau, S., Coombes, K. and Baggerly, K. (2009) Variable slope normalization of reverse phase protein arrays. *Bioinformatics* **25**: 1384–1389.

Nishizuka, S., Charboneau, L., Young, L. *et al.* (2003) Proteomic profiling of the NCI-60 cancer cell lines using new high-density reverse-phase lysate microarrays. *Proc Natl Acad Sci USA* **100**: 14229–14234.

Okochi, M., Nomura, S., Kaga, C. and Honda, H. (2008) Peptide array-based screening of human mesenchymal stem cell-adhesive peptides derived from fibronectin type III domain. *Biochem Biophys Res Commun* **371**: 85–89.

Omenn, G.S., States, D.J., Adamski, M. *et al.* (2005) Overview of the HUPO Plasma Proteome Project: results from the pilot phase with 35 collaborating laboratories and multiple analytical groups, generating a core dataset of 3020 proteins and a publicly-available database. *Proteomics* **5**: 3226–3245.

Parikh, K., Diks, S.H., Tuynman, J.H.B. *et al.* (2009) Comparison of peptide array substrate phosphorylation of c-raf and mitogen activated protein kinase kinase kinase 8. *PloS One*, **4**: e6440.

Paweletz, CP., Charboneau, L., Bichsel, V.E. *et al.* (2001) Reverse phase protein microarrays which capture disease progression show activation of pro-survival pathways at the cancer invasion front. *Oncogene* **20**: 1981–1989.

Pease, A.C., Solas, D., Sullivan, E.J. *et al.* (1994) Light-generated oligonucleotide arrays for rapid DNA sequence analysis. *Proc Natl Acad Sci USA* **91**: 5022–5026.

Pezzilli, R., Corsi, M.M., Barassi, A. *et al.* (2008) Serum adhesion molecules in acute pancreatitis: time course and early assessment of disease severity. *Pancreas* **37**: 36–41.

Raijmakers, R., Kraiczek, K., de Jong, A.P. *et al.* (2010) Exploring the human leukocyte phosphoproteome using a microfluidic reversed-phase-TiO2-reversed-phase high-performance liquid chromatography phosphochip coupled to a quadrupole time-of-flight mass spectrometer. *Anal Chem* **82**: 824–832.

Roh, M.I., Kim, H.S., Song, J.H. *et al.* (2009a) Effect of intravitreal Bevacizumab injection on aqueous humor cytokine levels in clinically significant macular edema. *Ophtalmology* **116**: 80–86.

Roh, M.I., Kim, H.S., Song, J.H. *et al.* (2009b) Concentration of cytokines in the aqueous humor of patients with naïve, recurrent and regressed CNV associated with amd after bevacizumab treatment. *Retina* **29**: 523–529.

Ruan, M., Lu, Y., Wulfkuhle, J. *et al.* (2008) An integrated image analysis tool for the systematic analysis of reverse phase protein array data. *Proc Am Assoc Cancer Res Annu Meet* **49**: 1220.

Schena, M., Shalon, D., Davis, R.W. and Brown, P.O. (1995) Quantitative monitoring of gene expression patterns with a complementary DNA microarray. *Science* **270**: 467–470.

Schweitzer, B. and Kingsmore, S.F. (2002) Measuring proteins on microarrays. *Curr Opin Biotechnol* **13**: 14–19.

Schweitzer, B., Roberts, S., Grimwade, B. *et al.* (2002) Multiplexed protein profiling on microarrays by rolling-circle amplification. *Nat Biotechnol* **20**: 359–365.

Scrivener, E., Barry, R., Platt, A. *et al.* (2003) Peptidomics: a new approach to affinity protein microarrays. *Proteomics* **3**: 122–128.

Sharma, M., Anderson, S.A., Schoop, R. and Hudson, J.B. (2009) Induction of multiple pro-inflammatory cytokines by respiratory viruses and reversal by standardized Echinacea, a potent antiviral herbal extract. *Antiviral Res* **83**: 165–170.

Simone, N.L., Remaley, A.T., Charboneau, L. *et al.* (2000) Sensitive immunoassay of tissue cell proteins procured by laser capture microdissection. *Am J Pathol* **156**: 445–452.

Soloviev, M. (2001) EuroBiochips: spot the difference! *Drug Discovery Today* **6**: 775–777.

Souquière, S., Mouinga-Ondeme, A., Makuwa, M. *et al.* (2009) T-cell tropism of simian T-cell leukaemia virus type 1 and cytokine profiles in relation to proviral load and immunological changes during chronic infection of naturally infected mandrills (Mandrillus sphinx). *J Med Primatol* **38**: 279–289.

Southern. E.M. (1995) DNA fingerprinting by hybridisation to oligonucleotide arrays. *Electrophoresis* **16**: 1539–1542.

Southern, E.M. (1996) High-density gridding: techniques and applications. *Curr Opin Biotechnol* **7**: 85–88.

Southern, E.M. (2001) DNA microarrays. History and overview. *Methods Mol Biol* **170**: 1–15.

Staples, G.O., Bowman, M.J., Costello, C.E. *et al.* (2009) A chip-based amide-HILIC LC/MS platform for glycosaminoglycan glycomics profiling. *Proteomics* **9**: 686–695.

Stechova, K., Spalova, I., Durilova, M. *et al.* (2009) Influence of maternal hyperglycaemia on cord blood mononuclear cells in response to diabetes-associated autoantigens. *Scand J Immunol* **70**: 149–158.

Stett, A., Egert, U., Guenther, E. *et al.* (2003) Biological application of microelectrode arrays in drug discovery and basic research. *Anal Bioanal Chem* **377**: 486–495.

Stone, J.D., Demkowicz, W.E. Jr and Stern, L.J. (2005) HLA-restricted epitope identification and detection of functional T cell responses by using MHC-peptide and costimulatory microarrays. *Proc Natl Acad Sci USA* **102**: 3744–3749.

Swamy, M., Molnar, E., Bock, T. *et al.* (2009) Detection of protein complex interactions via a Blue Native-PAGE retardation assay. *Anal Biochem* **392**: 177–179.

Templin, M.F., Stoll, D., Schrenk, M. *et al.* (2002) Protein microarray technology. *Trends Biotechnol* **20**: 160–166.

Tibes, R., Qiu, Y., Lu, Y. *et al.* (2006) Reverse phase protein array: validation of a novel proteomic technology and utility for analysis of primary leukemia specimens and hematopoietic stem cells. *Mol Cancer Ther* **5**: 2512–2521.

Uttamchandani, M. and Yao, S.Q. (2008) Peptide microarrays: next generation biochips for detection, diagnostics and high-throughput screening. *Curr Pharm Design* **14**: 2428–2438.

van Beem, R.T., Noort, W.A., Voermans, C. *et al.* (2008) The presence of activated CD4(+) T cells is essential for the formation of colony-forming unit-endothelial cells by CD14(+) cells. *J Immunol* **180**: 5141–5148.

van Rossum, A.P., Vlasveld, L.T., Vlasveld, I.N. *et al.* (2009) Granulocytosis and thrombocytosis in renal cell carcinoma: a pro-inflammatory cytokine response originating in the tumour. *Neth J Med* **67**: 191–194

Voura, E.B., Jaiswal, J.K., Mattoussi, H. and Simon, S.M. (2004) Tracking metastatic tumor cell extravasation with quantum dot nanocrystals and fluorescence emission-scanning microscopy. *Nat Med* **10**: 993–938.

Walter, G., Bussow, K., Cahill, D. *et al.* (2000) Protein arrays for gene expression and molecular interaction screening. *Curr Opin Microbiol* **3**: 298–302.

Weigum, S.E., Floriano, P.N., Redding, S.W. *et al.* (2010) Nano-bio-chip sensor platform for examination of oral exfoliative cytology. *Cancer Prev Res* **3**: 518–528.

Weng, S., Gu, K., Hammond, P.W. *et al.* (2002) Generating addressable protein microarrays with PROfusion™ covalent mRNA-protein fusion technology. *Proteomics* **2**: 48–57.

Willingham, S.B., Allen, I.C., Bergstralh, D.T. *et al.* (2009) NLRP3 (NALP3, Cryopyrin) facilitates in vivo caspase-1 activation, necrosis, and HMGB1 release via inflammasome-dependent and -independent pathways. *J Immunol* **183**: 2008–2015.

Wingren, C. and Borrebaeck, C.A. (2006) Antibody microarrays: current status and key technological advances. *OMICS: J Integr Biol* **10**: 411–427.

Yin, H. and Kileen, K. (2007) The fundamental aspects and applications of Agilent HPLC-Chip. *J Sep Sci* **30**: 1427–1434.

Yin, H., Killeen, K., Brennen, R. *et al.* (2005) Microfluidic chip for peptide analysis with an integrated HPLC column, sample enrichment column, and nanoelectrospray tip. *Anal Chem* **77**: 527–533.

Yu, J.K., Zheng, S., Tang, Y. and Li, L. (2005) An integrated approach utilizing proteomics and bioinformatics to detect ovarian cancer. *J Zhejiang Univ Sci B* **6**: 227–231.

Zetterberg, H., Tanriverdi, F., Unluhizarci, K. *et al.* (2009) Sustained release of neuron-specific enolase to serum in amateur boxers. *Brain Inj* **23**: 723–726.

Zhai, Y., Zhong, Z., Chen, C.Y. *et al.* (2008) Coordinated changes in mRNA turnover, translation, and RNA processing bodies in bronchial epithelial cells following inflammatory stimulation. *Mol Cell Biol* **28**: 7414–7426.

Zhang, Z., Bast, R.C. Jr, Yu, Y. *et al.* (2004) Three biomarkers identified from serum proteomic analysis for the detection of early stage ovarian cancer. *Cancer Res* **64**: 5882–5890.

Zheng, T., Peelen, D. and Smith, L.M. (2005) Lectin arrays for profiling cell surface carbohydrate expression. *J Am Chem Soc* **127**: 9982–9983.

Zhu, H. and Snyder, M. (2003) Protein chip technology. *Curr Opin Chem Biol* **7**: 55–63.

Zhu, H., Klemic, J.F., Chang, S. *et al.* (2000) Analysis of yeast protein kinases using protein chips. *Nat Genet* **26**: 283–289.

Zhu, H., Bilgin, M., Bangham, R. *et al.* (2001) Global analysis of protein activities using proteome chips. *Science* **293**: 2101–2105.

10

Antibody Microarrays in Proteome Profiling

Mohamed Sail Saeed Alhamdani and Jörg D. Hoheisel

Introduction

Following the completion of sequencing of the first human genome, high-throughput technologies such as DNA-microarrays (Hoheisel 2006) and – more recently – second-generation sequencing (Ansorge 2009) have become prominent and formidable analytical tools in biology. Adding to this instrumentation, the availability of powerful and user-friendly bioinformatics tools (Mychaleckyj 2007) has paved the way for researchers to deal with the massive amount of data generated by these technologies. In consequence, non-reductionist approaches have become possible and opened a new epoch in investigating biological phenomena, laying the foundations for a new biology. In extension to this tremendous development at the molecular-genetic level, there has been a surge in interest in the comprehensive analysis of proteins and protein networks (Wilkins *et al.* 2006). The field of proteomics, a term coined by Marc Wilkins and colleagues (Wilkins *et al.* 1996), evolved out of necessity, especially from a biomedical perspective. The intrinsic advantage of proteomics over genomics is that to a large extent the proteome is the biological end-product of the genome (Martin and Nelson 2001), although more and more information becomes available about the direct involvement of nucleic acids in regulation, structure building and functional activities. However, genomic sequences do not offer any apparent information on protein interaction, localization and post-translation modification. Also, transcript abundance levels do not necessarily predict the corresponding protein level (Schmidt *et al.* 2007), nor do

Molecular Analysis and Genome Discovery, Second Edition. Edited by Ralph Rapley and Stuart Harbron.
© 2012 John Wiley & Sons, Ltd. Published 2012 by John Wiley & Sons, Ltd.

they provide information about its regulatory status, which is usually governed by endogenous and exogenous factors. Most pharmacological interventions are directed at proteins rather than genes.

With this in mind, it is obvious that tools for an analysis of the proteome are of crucial importance. While for example genome-wide expression profiling by DNA-microarrays and sequencing have become routine, analogous tools are not yet available for proteins, although protein arrays and mass spectrometry strongly move in this direction. In addition to the technical shortcomings, proteins are immensely more diverse in structure and biophysical properties than nucleic acids. Post-translation modifications add even another level of complexity. In simple terms, an organism has only one genome but, at the same time, several proteomes.

So far, two-dimensional gel electrophoresis and mass spectrometry represent the most widely used analytical tools in proteomics and have evolved into indispensable tools for proteomic research (X. Han, Aslanian and Yates 2008; Wittmann-Liebold, Graack and Pohl 2006). Since the mid-1990s, the performance of these technologies in handling small sample sizes and analysing complex protein mixtures (Aebersold and Mann 2003) have been improved significantly. With the continuing need for more sensitive and robust techniques that can cope with very high complexity, however, these classical approaches still have limitations in resolution, sensitivity and cost (Bunai and Yamane 2005; Diamandis and van der Merwe 2005; Koomen *et al.* 2005; Beranova-Giorgianni 2003). Assays with affinity-reagents are a good and complementary alternative, since they are technically based on a different principle and additionally already well established in clinical diagnostics. Therefore, protein-array technology seems to be well suited to overcome some of the limitations inherent to other methodologies and to contribute substantially to the tool set used in global proteomic profiling.

Technical aspects

Antibody microarrays are miniaturized analytical systems generated by spatially arraying small amounts (volumes at a nanolitre scale or less) of individual capture molecules – mostly antibodies – at discrete positions on a solid support (Figure 10.1) (Lv and Liu 2007; Kusnezow *et al.* 2006a; Angenendt 2005; Pavlickova, Schneider and Hug 2004; Glokler and Angenendt 2003; Haab 2003; Kusnezow and Hoheisel 2003; Kusnezow and Hoheisel 2002). The feasibility of such miniaturized and multiplexed immunoassays was first discussed by Ekins in the late 1980s (Ekins and Chu 1991; Ekins 1989). To date, the number of antibodies used in an assay has varied from a few to several hundred, and has rarely been larger than one thousand. Upon incubation with a complex protein sample (in micro-liter quantities), the amount of bound antigen is determined for each spot. The acquired images contain signal intensities, which are converted to numerical values reflecting the protein profile within the biological sample. Assay sensitivities in the picomolar to femtomolar range have

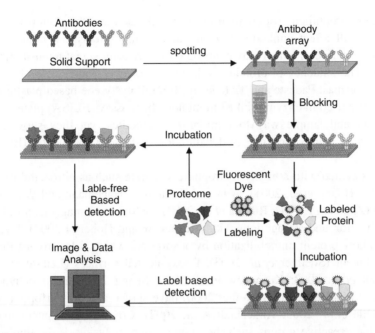

Figure 10.1   A schematic representation of the steps of an antibody array analysis

been reported (Kusnezow *et al.* 2006a). The technical factors that determine the performance of antibody microarrays are the array surface, the antibody quality, sample processing, incubation conditions, signal detection and data analysis.

The solid support

Unlike DNA, proteins have very divergent inhomogeneous structures and biophysical and chemical properties and are prone to perturbation by denaturation and/or modification (Doerr 2005). Hence, attaching proteins to a solid support is not without the risk of compromising functionality and, consequently, performance. There have been continuous efforts to generate support media that have a minimal detrimental effect on the arrayed proteins. The selection of the appropriate solid support is usually governed by several factors such as the nature of protein(s), the mode of attachment, the compatibility with available hardware (e.g. contact or non-contact printing), the array density, and the sample size or reagent consumption. Generally, microscopic glass slides are the most frequently used solid support for antibody microarray fabrication. The surface is usually coated with substrates that facilitate the attachment. A physical or chemical interaction between substrate and antibodies takes place by means of affinity, adsorption or covalent binding.

Choosing the appropriate solid support amongst the plethora of commercially available support media could be difficult. Care should be taken to consider factors such as surface hydrophilicity, orientation of immobilization and the chemistry of attachment, since all affect antibody–antigen interaction. Proteins can bind strongly to hydrophobic surfaces, which may contribute to increasing signal intensity and detection limit. The tendency of proteins to denature, however, increases on hydrophobic surfaces (Sorribas, Padeste and Tiefenauer 2002). Polystyrene-based plastic surfaces, such as those used in enzyme-linked immunosorbent assay (ELISA) plates, represent an efficient and simple way to immobilize antibodies, yet they show a significant degree of inactivation due to denaturation and steric obstruction (Butler *et al.* 1993, 1992).

Surfaces coated with a highly hydrophilic substrate such as nitrocellulose (Knight *et al.* 2004; Huang *et al.* 2001), agarose (Afanassiev, Hanemann and Wolfl 2000) or hydrogel (Zhou *et al.* 2004; Rubina *et al.* 2003) could be advantageous for preserving protein function and storage longevity (Kusnezow and Hoheisel 2003). Hydrophilic surfaces allow protein immobilization by adsorption, which is less robust than covalent attachment (Soellner *et al.* 2003). Epoxy-coated surfaces (Letarte *et al.* 2005; Angenendt *et al.* 2003; Kusnezow *et al.* 2003; Seong 2002) and aldehyde (Hahn *et al.* 2007; MacBeath and Schreiber 2000) have shown a better performance in this regard (Olle *et al.* 2005; Angenendt *et al.* 2002). Covalent attachment makes use of side-chain reactive groups such the amino-group of lysine or the thiol-group of cysteine, which are present in virtually all proteins. However, covalent reaction may occur at random, with the disadvantage that the protein may bind somewhere near to the active site. Attaching proteins in a uniform rather than random manner can provide a substantial advantage in assay sensitivity (Peluso *et al.* 2003). For antibodies, methods have been implemented that benefit from their unique structure and allow the establishment of an orientation that leaves the antigenic sites well exposed for capturing the target proteins. These methods depend on the affinity of certain molecules such as proteins A, G, A/G and L (Yuan, He and Lee 2009; Bonroy *et al.* 2006; Danczyk *et al.* 2003; Podlaski and Stern 2000) and synthetic peptides (Jung *et al.* 2008) to a certain antibody segment, usually the F_c part (constant region) (Ghose, Hubbard and Cramer 2007; Arora, Hammes and Oas 2006; H. Yang, Gurgel and Carbonell 2005; Guss *et al.* 1986). Other methods target the carbohydrate moiety (Nisnevitch *et al.* 2000) or the hinge sulfohydryl-group (Peluso *et al.* 2003) as site of attachment.

Affinity reagents

The quality and robustness of microarrays rely strongly on the affinity molecules used in their fabrication. Currently, access to well-validated affinity reagents is one of the challenging issues in profiling the human proteome (Uhlen 2008; Taussig *et al.* 2007). Antibody-based arrays have been fabricated using monoclonal antibodies (Barber *et al.* 2009; Chaga 2008; Ehrlich *et al.* 2008; Christopherson *et al.* 2006; Belov *et al.* 2005; Woolfson *et al.* 2005; Yeretssian *et al.* 2005; Hudelist *et al.* 2004) and polyclonal (Schröder *et al.* 2010; Rivas *et al.* 2008; M.K. Han *et al.* 2006), as well as recombinant antibody fragments such as $F(ab')_2$ and scFv (Ingvarsson *et al.* 2007; Song *et al.* 2007). Still, there are several other types of affinity reagents that can act as capture molecules (Plückthun 2009), such as affibodies (Nygren 2008; Tolmachev *et al.* 2007), small molecule scaffold binders (Xiao *et al.* 2009; Prakesch *et al.* 2008), aptamers (nucleic acid scaffolds) (Stoevesandt and Taussig 2007; Wilson, Keefe and Szostak 2001; Tuerk and Gold 1990), peptides (Nygren and Skerra 2004), proteins such as lipocalins (Beste *et al.* 1999), ankyrin repeat proteins (Stumpp and Amstutz 2007; Binz *et al.* 2004), fibronectin (Xu *et al.* 2002), Zn-finger (Bianchi *et al.* 1995) and other small chemical entities (Schuffenhauer *et al.* 2005; Roque, Taipa and Lowe 2004; Peczuh and Hamilton, 2000). Each molecule class has its advantages and disadvantages. Nevertheless, currently antibodies and antibody fragments are still the most attractive affinity probes. In the analysis, it seems to be advantageous to use at least two binders, recognizing two non-overlapping epitopes of the same target (Uhlen and Hober 2009; Uhlen 2008; Stoevesandt and Taussig 2007). Mono-specific polyclonal antibodies are therefore highly attractive probes because of the cooperative effect of a mixture of antibodies that bind to several epitopes of a particular target protein (Nilsson *et al.* 2005).

To date, most antibody arrays have been produced from monoclonal and polyclonal antibodies obtained from commercial sources. Commercial antibodies, however, are generally manufactured to serve particular applications such as ELISA, Western blot, immunoprecipitation (IP), immunohistochemistry (IHC), or immunocytochemistry (ICC). Some antibody manufacturers are now starting to include information about the suitability for microarray applications in their data sheets. However, as there are about 22 000 protein-encoding genes in the human genome (Birney *et al.* 2007; Clamp *et al.* 2007), there is a considerable imbalance in antibody coverage. It is possible to find hundreds of antibodies against particular targets – for instance, there are more than 900 antibodies against p53 – while none are available for many others. To fill this gap, several initiatives exist for creating a global resource of well-characterized affinity reagents for an analysis of the human proteome (Table 10.1). Among the main tasks of these efforts are the generation of highly specific antibodies using high-throughput technologies, such as recombinant-antibody phage display

Table 10.1 Initiatives for a global production of human protein affinity reagents

Initiative	Website	Reference
Human Proteome Atlas	http://www.proteinatlas.org/	(Berglund *et al.* 2008; Uhlen *et al.* 2005)
ProteomeBinders	http://www.proteomebinders.org/	(Taussig *et al.* 2007)
Clinical Proteomic Technologies Initiative	http://proteomics.cancer.gov/	
Antibody Factory	http://www.antibody-factory.de/	(Konthur, Hust and Dübel 2005)

(Winter *et al.* 1994), ribosome display (He and Taussig 2005), RNA display libraries (Lipovsek and Plückthun 2004), bacterial surface display (Jostock and Dübel 2005), and yeast surface display (Levy *et al.* 2007), validation of the established antibodies, and developing novel non-antibody affinity reagents.

Protein samples

Sample processing procedures that isolate an organism's proteome in a representative and reproducible manner are another crucial factor. The complexity and the enormous dynamic range of molecule concentrations in a proteome as well as the susceptibility of proteins to even small changes in their environment provide a formidable challenge. However, once a protein sample is adequately brought into solution, whether in a denatured or native form, it can be subjected to antibody array analysis. However, protein preparation under native conditions could be advantageous, since functionality may be preserved and the detection of protein isoforms may be possible (Alhamdani *et al.* 2010), as these are important to be identified for pharmacological reasons, for example. High sample complexity could cause unspecific binding and complicate uniform labelling in other proteomic approaches. For antibody microarrays, however, it has been shown that depletion of highly abundant proteins had no significant impact on the quality of the assay (Schröder *et al.* 2010) but, in contrast, introduced a considerable bias in proteome representation.

Since plasma is easy to collect in a nearly non-invasive process, the application of antibody microarrays for studying this type of specimen has created much interest (Bergsma *et al.* 2010; Schröder *et al.* 2010; Lal *et al.* 2009; C. Li *et al.* 2009; Rimini *et al.* 2009; Zeng *et al.* 2010; Hon *et al.* 2008; Lukesova *et al.* 2008; Sun *et al.* 2008; Loch *et al.* 2007). Conditioned media of cultured cells has received similarly wide attention (Cai *et al.* 2009; L.L. Chen *et al.* 2009; Grassel *et al.* 2009; Gruber *et al.* 2009; Chou *et al.* 2008; Inai *et al.* 2008; Y.C. Lee *et al.* 2008; Ohshima *et al.* 2008; Perera *et al.* 2008; Seeber *et al.* 2008; Ebihara *et al.* 2007; Lu *et al.* 2007; Neuhoff *et al.* 2007; Sze *et al.* 2007; Huang *et al.* 2001). However, other types of specimens have also been investigated, such as urine (Schröder *et al.* 2010; Hu *et al.* 2009; Liu *et al.* 2006), prostatic fluid (Fujita *et al.* 2008), cerebrospinal fluid (Dhungana, Sharrack and Woodroofe 2009; Tsai *et al.* 2008), tears (Leonardi *et al.* 2009), saliva (Lal *et al.* 2009), exhaled breath (Barta *et al.* 2010), cultured cells (Wong *et al.* 2009; Lin *et al.* 2003; Sreekumar *et al.* 2001) and tissue biopsies (Zander *et al.* 2009; Y. Hao *et al.* 2008; Moschos *et al.* 2007; Hudelist *et al.* 2004; Anderson *et al.* 2003). Owing to the differences in proteome composition and complexity, sample processing should be optimized empirically. The tissue proteome for example is more complex and gives rise to higher background noise as compared to plasma (Haab, 2003). Procedures have been established to overcome this obstacle (Alhamdani *et al.* 2010).

Labelling and detection strategies

Once a protein sample is isolated, the subsequent analysis may involve either label-based or label-free processing. Currently, labelling is predominantly used for the assays because of its simplicity, robustness and sensitivity (Kusnezow *et al.* 2003, 2006a, 2007; Wingren *et al.* 2007; Zhou *et al.* 2004). Sensitivities in the range of picomolar to femtomolar can be achieved, which is comparable to ELISA but without the necessity of signal amplification (Wingren *et al.* 2007; Kusnezow *et al.* 2006a). Proteins can be labelled either directly with fluorescent dyes (such as Cy3 or Cy5 dyes) or indirectly with a hapten such as biotin or biotin derivatives, followed by incubation with a fluorescently tagged molecule (such as streptavidin in the case of biotin) that has a high affinity to the hapten. There is a strong preference for indirect protein labelling with one colour, as this approach has enhanced sensitivity compared to direct labelling (Kusnezow *et al.* 2007; Wingren *et al.* 2007). Recently, direct labelling using a two-colour approach has been reported, substantially improving microarray performance for reproducibility and discriminative power (Schröder *et al.* 2010).

Despite the favourable virtues of sample-labelling approaches, there are concerns that the introduction of too many label molecules might affect the epitopes required for the formation of the antibody–antigen complex. Hence, label-free techniques are currently gaining more recognition. The simplest example of a label-free method is the sandwich assay (Ekins 1998). This approach is similar to ELISA, in which a specific protein is targeted by two antibodies. The first antibody is arrayed on a solid support and serves as a capture reagent. A second antibody is required to bind to the same target molecule. Usually it carries a label directly or is in turn identified by a third labelled antibody. However, the need for two highly specific antibodies and particularly the lack of scalability for a complex array, are severely limiting factors for such an approach (Templin *et al.* 2004). No more than about 30 antibody pairs can be used at the same time. Other more promising methods are emerging and may replace labelling once their practical feasibility and competitiveness have been demonstrated. Compared to labelling, these new methods currently still exhibit lower sensitivity, the need for expensive materials or equipment and restrictions to multiplexing. Table 10.2 lists some of these technologies, which were reviewed in more detail by Ray, Mehta and Srivastava (2010).

Bioinformatics

Since the principle is nearly identical, antibody array data analysis and interpretation can be carried out using the means developed for DNA-microarray studies. A major difference is data normalization, since reliable house-keeping proteins for data

Table 10.2 Label-free detection technologies

Technology	Sensitivity/ resolution	Array complexity	Reference
Surface plasmon resonance (SPR)	10 ng/ml	High	(Nedelkov, Tubbs and Nelson 2006; Usui-Aoki *et al.* 2005)
SPR-imaging (SPRi)	nM-zM	Very high	(Ladd *et al.* 2009; Lausted, Hu and Hood 2008; Suraniti *et al.* 2007; H.J. Lee, Nedelkov and Corn 2006; Kyo, Usui-Aoki and Koga 2005)
Oblique-incidence reflectivity difference-based (OI-RD)		Very high	(Fei *et al.* 2008; Zhu *et al.* 2007)
Scanning Kelvin nanoprobe (SKN)	<50 nm	High	(Sinensky and Belcher 2007; Thompson *et al.* 2005; Cheran *et al.* 2004)
Atomic force microscope (AFM)	Picolitre volume	High	(Kim *et al.* 2009; Huff *et al.* 2004)
Nanowires and nanotubes	nM-fM	Very high	(Drouvalakis *et al.* 2008; L. Yang *et al.* 2008; Okuno *et al.* 2007; Zheng *et al.* 2005; Cui *et al.* 2001)
Biological compact disk (BioCD)	30–70 pg/ml	High	(Morais *et al.* 2008; X. Wang, Zhao and Nolte 2008)
Ellipsometry	1 ng/ml	High	(Valsesia *et al.* 2006; Z.H. Wang *et al.* 2006; G. Jin *et al.* 2004; Z.H. Wang and Jin 2003)
Microcantilevers	0.2 ng/ml	High	(Backmann *et al.* 2005)

normalization cannot be picked easily and array complexity is frequently insufficient to use the majority of (unchanged) signals for normalization (Royce *et al.* 2006). To overcome the problem, other approaches have been used, such as an internally normalized ratio algorithm after dual-colour labelling (Andersson *et al.* 2005), spike-in protein control(s) of known concentration (Hamelinck *et al.* 2005) or relative normalization to particular probes, which are also assayed by another method (e.g. ELISA) (Hamelinck *et al.* 2005). With normalized data at hand, subsequent analyses such as the search for differentially expressed proteins, clustering or validation by Western blotting or, reversely, qRT-PCR, are performed as for the results from DNA-microarray.

Antibody array applications

With their rapid development during the past 10 years, antibody microarrays have seen a rise in applications, especially for clinically relevant analyses and

Table 10.3 Application of antibody arrays in human studies

Disease	Reference
Allergy	(Leonardi *et al.* 2009; Zander *et al.* 2009)
Bone Diseases	(Jarvinen *et al.* 2008)
Brain Diseases	(Tsai *et al.* 2008; Hergenroeder *et al.* 2010)
Cardiovascular Diseases	(Dhungana, Sharrack and Woodroofe 2009; Lal *et al.* 2009, 2004)
Chronic Fatigue Syndrome	(Cho *et al.* 2009)
Cystic Fibrosis	(Srivastava *et al.* 2006)
Dermatitis	(Harper *et al.* 2010; Hon *et al.* 2008)
HIV	(Wu *et al.* 2007, 2008)
Bowl Disease	(Kader *et al.* 2005)
Kidney Diseases	(Hu *et al.* 2009; Kaukinen *et al.* 2008; Liu *et al.* 2006)
Muscle Diseases	(Anderson *et al.* 2003)
Pulmonary Diseases	(Barreiro *et al.* 2008)
Transplantation	(Hu *et al.* 2009; Lal *et al.* 2004)
Cancer	
Bladder	(Sanchez-Carbayo *et al.* 2006)
Breast	(Carlsson *et al.* 2008; Smith *et al.* 2006; Hudelist *et al.* 2004)
Colorectal	(Toh *et al.* 2009; Madoz-Gurpide *et al.* 2007; Spisak *et al.* 2010)
Gastric	(Ellmark *et al.* 2006)
Leukemia	(Scupoli *et al.* 2008; Belov *et al.* 2003, 2006)
Liver	(Lausted, Hu and Hood 2008; Sun *et al.* 2008; Nonomura *et al.* 2007)
Lung	(Barta *et al.* 2010; M.K. Han *et al.* 2009; Kullmann *et al.* 2008; Gao *et al.* 2005)
Lymphoma	(Belov *et al.* 2006)
Melanoma	(Moschos *et al.* 2007)
Multiple Myeloma	(Kline *et al.* 2007)
Pancreas	(Schröder *et al.* 2010; C. Li *et al.* 2009; Ingvarsson *et al.* 2008; S. Chen *et al.* 2007; Hamelinck *et al.* 2005)
Prostate	(Fujita *et al.* 2008; Iiizumi *et al.* 2008; Shafer *et al.* 2007; Orchekowski *et al.* 2005)
Renal	(Lukesova *et al.* 2008)

investigations of the pathophysiology of human diseases. A broad range of human diseases has been investigated, predominantly in the field of oncology (Table 10.3). However, more basic biologically oriented investigations have also been performed (Table 10.4). Reviewing all applications in detail is beyond the scope of this chapter. In summary, however, the technique has demonstrated its potential and versatility in multiplexed proteomics profiling. Antibody arrays are promising tools for cancer diagnostics, biomarker discovery, therapy monitoring and the identification of new drug target leads (Alhamdani, Schröder and Hoheisel 2009; Chatterjee, Wojciechowski and Tainsky 2009; Loch *et al.* 2007; Borrebaeck, 2006; Sanchez-Carbayo 2006; Haab 2005).

Table 10.4   Biological fields of application

Angiogenesis	(Xi *et al.* 2009; Karaca *et al.* 2008; Ahn *et al.* 2006; Azizan *et al.* 2006)
Apoptosis	(Ray *et al.* 2010)
Biomarker discovery	(Schröder *et al.* 2010; Barber *et al.* 2009; M.K. Han *et al.* 2009; C. Li *et al.* 2009; Carlsson *et al.* 2008; Hao *et al.* 2008; Ingvarsson *et al.* 2008; Rivas *et al.* 2008; Sun *et al.* 2008; Loch *et al.* 2007; Shafer *et al.* 2007; Song *et al.* 2007; Wu *et al.* 2007; Christopherson *et al.* 2006; Sanchez-Carbayo *et al.* 2006; Spisak *et al.*)
Cell cycle analysis	(Uemura *et al.* 2009; Jamesdaniel *et al.* 2008; Tapias *et al.* 2008; X. Li *et al.* 2007);
Cell signaling	(Qi *et al.* 2009; Uemura *et al.* 2009; Wong *et al.* 2009; Zhong *et al.* 2009; Dewing *et al.* 2008; Iiizumi *et al.* 2008; Korf *et al.* 2008; Pelech *et al.* 2008; Skalnikova *et al.* 2008; Madoz-Gurpide *et al.* 2007)
Cytokine profiling	(L.L. Chen *et al.* 2009; Eads *et al.* 2009; Fenton *et al.* 2009; Gasparrini *et al.* 2009; Haddad and Belosevic 2009; Q. Hao, Wang and Tang 2009; Jiang *et al.* 2009; Takeda *et al.* 2009; Bandyopadhyay, Romero and Chattopadhyay 2008; Chang *et al.* 2008; Chou *et al.* 2008; Inai *et al.* 2008; K.B. Jin *et al.* 2008; Karaca *et al.* 2008; S. Li *et al.* 2008; Lu *et al.* 2008; Ohshima *et al.* 2008; Oscarsson *et al.* 2008; Huang 2007; Neuhoff *et al.* 2007; Watanabe *et al.* 2005; Foster *et al.* 2010; Yin *et al.* 2010)
Glycoproteomics	(C. Li *et al.* 2009; Zeng *et al.* 2010; Bereczki *et al.* 2007; S. Chen *et al.* 2007)
Monitoring of drug treatment and therapy	(Y.A. Lee, Cho and Yokozawa 2010; Cao *et al.* 2009; M.H. Chen *et al.* 2009; Cho *et al.* 2009; M.H. Lee *et al.* 2009; Qi *et al.* 2009; Zander *et al.* 2009; Jarvinen *et al.* 2008; K.B. Jin *et al.* 2008)
Phosphoproteomics	(Chung and Chin 2009; Fischer *et al.* 2009; Nystrom *et al.* 2009; Qi *et al.* 2009; Wong *et al.* 2009; Pelech *et al.* 2008; Rajala 2008; Gembitsky *et al.* 2004; Y.A. Lee, Cho and Yokozawa 2010)
Stem cell research	(Skalnikova *et al.* 2008; Tomchuck *et al.* 2008; Sze *et al.* 2007; Ko, Kato and Iwata 2005)

The microarray platforms used in these studies were either home-made or commercial. Currently, there are several manufacturers who provide arrays with a fixed antibody number (e.g. RayBiotech, www.raybiotech.com; Full Moon Biosystems, www.fullmoonbiosystems.com; Sigma, www.sigmaaldrich.com; R&D Systems, www.rndsystems.com) or offer customized arrays (e.g. Lampire Biological Laboratories, www.lampire.com; Kinexus, www.kinexus.ca). Generally, in studies involving home-made arrays, the focus was more on global proteomics profiling, while those with commercial arrays used them as a secondary tool among other methodologies. Interestingly, many home-made arrays have shown better functionality, performance and flexibility than those from commercial sources. Their superiority may result from the transparency of the system and its components to the operator, which has benefits for troubleshooting and tracking the sources of error. Also, the propriety nature and the price of commercial kits and arrays may restrict the performance of replicate experiments.

Summary

Antibody arrays are powerful, robust and sensitive proteomic tools that enable a broad spectrum of applications, to date particularly in biomedical sciences and clinical research. They provide expression and/or conformational data on a currently still limited set of proteins consuming only small amounts of both antibody and protein. Many technical issues have been addressed successfully, improving the performance of the best protocols to and beyond that of ELISA assays, the current standard format of immuno-based protein analysis. Further improvement can be expected from the increase in the number and quality of the affinity reagents. Access to more antibodies or equivalent binder molecules of enhanced specificity and affinity will add substantially to the power of the technology. Technically, there is no major obstacle to scale-up microarray analyses to several 10 000 probe molecules The production of solid support media that preserve protein functionality and allow for a well-oriented attachment could improve resolution and detection limit and strongly affect the ability to discriminate between protein isoforms. The assay format could also benefit from ongoing developments in detection methodologies. Surface Plasmon Resonance imaging methods, for example, are currently showing sensitivities comparable to those of fluorescence-based imaging and may become an alternative in future. However, fluorescence is likely to develop to single-molecule sensitivity, thus enabling real quantitative counting of bound protein molecules rather than the relative measurement of signal intensity.

Acknowledgements

M.S.S.A received a long-term fellowship from the Deutscher Akademischer Austausch Dienst (DAAD). Work in the laboratory was supported by the EU-funded projects ProteomeBinders and AffinityProteome.

References

Aebersold, R. and Mann, M. (2003) Mass spectrometry-based proteomics. *Nature* **422**: 198–207.

Afanassiev, V., Hanemann, V. and Wolfl, S. (2000) Preparation of DNA and protein micro arrays on glass slides coated with an agarose film. *Nucleic Acids Res* **28**: E66.

Ahn, E.H., Kang, D.K., Chang, S.I. *et al.* (2006) Profiling of differential protein expression in angiogenin-induced HUVECs using antibody-arrayed ProteoChip. *Proteomics* **6**: 1104–1109.

Alhamdani, M.S., Schröder, C. and Hoheisel, J.D. (2009) Oncoproteomic profiling with anti-body microarrays. *Genome Med* **1**: 68.

Alhamdani, M.S., Schröder, C., Werner, J. *et al.* (2010) Single-step procedure for the isolation of proteins at near-native conditions from mammalian tissue for proteomic analysis on antibody microarrays. *J Proteome Res* **9**: 963–971.

Anderson, K., Potter, A., Baban, D. and Davies, K.E. (2003) Protein expression changes in spinal muscular atrophy revealed with a novel antibody array technology. *Brain* **126**: 2052–2064.

Andersson, O., Kozlowski, M., Garachtchenko, T. *et al.* (2005) Determination of relative protein abundance by internally normalized ratio algorithm with antibody arrays. *J Proteome Res* **4**: 758–767.

Angenendt, P. (2005) Progress in protein and antibody microarray technology. *Drug Discov Today* **10**: 503–511.

Angenendt, P., Glokler, J., Murphy, D. *et al.* (2002) Toward optimized antibody micro-arrays: a comparison of current microarray support materials. *Anal Biochem* **309**: 253–260.

Angenendt, P., Glokler, J., Sobek, J. *et al.* (2003) Next generation of protein microarray support materials: evaluation for protein and antibody microarray applications. *J Chromatogr A* **1009**: 97–104.

Ansorge, W.J. (2009) Next-generation DNA sequencing techniques. *N Biotechnol* **25**: 195–203.

Arora, P., Hammes, G.G. and Oas, T.G. (2006) Folding mechanism of a multiple independently-folding domain protein: double B domain of protein A. *Biochemistry* **45**: 12312–12324.

Azizan, A., Sweat, J., Espino, C. *et al.* (2006) Differential proinflammatory and angiogenesis-specific cytokine production in human pulmonary endothelial cells, HPMEC-ST1.6R infected with dengue-2 and dengue-3 virus. *J Virol Methods* **138**: 211–217.

Backmann, N., Zahnd, C., Huber, F. *et al.* (2005) A label-free immunosensor array using single-chain antibody fragments. *Proc Natl Acad Sci USA* **102**: 14587–14592.

Bandyopadhyay, S., Romero, J.R. and Chattopadhyay, N. (2008) Kaempferol and quercetin stimulate granulocyte-macrophage colony-stimulating factor secretion in human prostate cancer cells. *Mol Cell Endocrinol* **287**: 57–64.

Barber, N., Gez, S., Belov, L. *et al.* (2009) Profiling CD antigens on leukaemias with an antibody microarray. *FEBS Lett* **583**: 1785–1791.

Barreiro, E., Schols, A.M., Polkey, M.I. *et al.* (2008) Cytokine profile in quadriceps muscles of patients with severe COPD. *Thorax* **63**: 100–107.

Barta, I., Kullmann, T., Csiszer, E. and Antus, B. (2010) Analysis of cytokine pattern in exhaled breath condensate of patients with squamous cell lung carcinoma. *Int J Biol Markers* **25**: 52–56.

Belov, L., Huang, P., Barber, N. *et al.* (2003) Identification of repertoires of surface antigens on leukemias using an antibody microarray. *Proteomics* **3**: 2147–2154.

Belov, L., Huang, P., Chrisp, J. S. *et al.* (2005) Screening microarrays of novel monoclonal antibodies for binding to T-, B- and myeloid leukaemia cells. *J Immunol Methods* **305**: 10–19.

Belov, L., Mulligan, S. P., Barber, N. *et al.* (2006) Analysis of human leukaemias and lym-phomas using extensive immunophenotypes from an antibody microarray. *Br J Haematol* **135**: 184–197.

Beranova-Giorgianni, S. (2003) Proteome analysis by two-dimensional gel electrophoresis and mass spectrometry: strengths and limitations. *Trend Anal Chem* **22**: 273–281.

Bereczki, E., Gonda, S., Csont, T. *et al.* (2007) Overexpression of biglycan in the heart of transgenic mice: an antibody microarray study. *J Proteome Res* **6**: 854–861.

Berglund, L., Bjorling, E., Oksvold, P. *et al.* (2008) A genecentric Human Protein Atlas for expression profiles based on antibodies. *Mol Cell Proteomics* **7**: 2019–2027.

Bergsma, D., Chen, S., Buchweitz, J. *et al.* (2010) Antibody-array interaction mapping, a new method to detect protein complexes applied to the discovery and study of serum amyloid P interactions with kininogen in human plasma. *Mol Cell Proteomics* **9**: 446–456.

Beste, G., Schmidt, F.S., Stibora, T. and Skerra, A. (1999) Small antibody-like proteins with prescribed ligand specificities derived from the lipocalin fold. *Proc Natl Acad Sci USA* **96**: 1898–1903.

Bianchi, E., Folgori, A., Wallace, A. *et al.* (1995) A conformationally homogeneous combinatorial peptide library. *J Mol Biol* **247**: 154–160.

Binz, H. K., Amstutz, P., Kohl, A. *et al.* (2004) High-affinity binders selected from designed ankyrin repeat protein libraries. *Nat Biotechnol* **22**: 575–582.

Birney, E., Stamatoyannopoulos, J.A., Dutta, A. *et al.* (2007) Identification and analysis of functional elements in 1% of the human genome by the ENCODE pilot project. *Nature* **447**: 799–816.

Bonroy, K., Frederix, F., Reekmans, G. *et al.* (2006) Comparison of random and oriented immobilisation of antibody fragments on mixed self-assembled monolayers. *J Immunol Methods* **312**: 167–181.

Borrebaeck, C.A. (2006) Antibody microarray-based oncoproteomics. *Expert Opin Biol Ther* **6**: 833–838.

Bunai, K. and Yamane, K. (2005) Effectiveness and limitation of two-dimensional gel electrophoresis in bacterial membrane protein proteomics and perspectives. *J Chromatogr B Analyt Technol Biomed Life Sci* **815**: 227–236.

Butler, J.E., Ni, L., Brown, W.R. *et al.* (1993) The immunochemistry of sandwich ELISAs – VI. Greater than 90% of monoclonal and 75% of polyclonal anti-fluorescyl capture antibodies (CAbs) are denatured by passive adsorption. *Mol Immunol* **30**: 1165–1175.

Butler, J.E., Ni, L., Nessler, R. *et al.* (1992) The physical and functional behavior of capture antibodies adsorbed on polystyrene. *J Immunol Methods* **150**: 77–90.

Cai, Z., Chen, Q., Chen, J. *et al.* (2009) Monocyte chemotactic protein 1 promotes lung cancer-induced bone resorptive lesions in vivo. *Neoplasia* **11**: 228–236.

Cao, X., Plasencia, C., Kanzaki, A. *et al.* (2009) Elucidation of the molecular mechanisms of a salicylhydrazide class of compounds by proteomic analysis. *Curr Cancer Drug Targets* **9**: 189–201.

Carlsson, A., Wingren, C., Ingvarsson, J. *et al.* (2008) Serum proteome profiling of metastatic breast cancer using recombinant antibody microarrays. *Eur J Cancer* **44**: 472–480.

Chaga, G.S. (2008) Antibody arrays for determination of relative protein abundances. *Methods Mol Biol* **441**: 129–151.

Chang, D.T., Jones, J.A., Meyerson, H. *et al.* (2008) Lymphocyte/macrophage interactions: biomaterial surface-dependent cytokine, chemokine, and matrix protein production. *J Biomed Mater Res A* **87**: 676–687.

Chatterjee, M., Wojciechowski, J. and Tainsky, M.A. (2009) Discovery of antibody biomarkers using protein microarrays of tumor antigens cloned in high throughput. *Methods Mol Biol* **520**: 21–38.

Chen, L.L., Ye, F., Lu, W.G. *et al.* (2009a) Evaluation of immune inhibitory cytokine profiles in epithelial ovarian carcinoma. *J Obstet Gynaecol Res* **35**: 212–218.

Chen, M.H., Wang, Q.F., Chen, L.G. *et al.* (2009b) The inhibitory effect of Gynostemma pentaphyllum on MCP-1 and type I procollagen expression in rat hepatic stellate cells. *J Ethnopharmacol* **126**: 42–49.

Chen, S., Laroche, T., Hamelinck, D. *et al.* (2007) Multiplexed analysis of glycan variation on native proteins captured by antibody microarrays. *Nat Methods* **4**: 437–444.

Cheran, L.E., Chacko, M., Zhang, M. and Thompson, M. (2004) Protein microarray scanning in label-free format by Kelvin nanoprobe. *Analyst* **129**: 161–168.

Cho, J.H., Cho, C.K., Shin, J.W. *et al.* (2009) Myelophil, an extract mix of Astragali Radix and Salviae Radix, ameliorates chronic fatigue: a randomised, double-blind, controlled pilot study. *Complement Ther Med* **17**: 141–146.

Chou, S.Y., Weng, J.Y., Lai, H.L. *et al.* (2008) Expanded-polyglutamine huntingtin protein suppresses the secretion and production of a chemokine (CCL5/RANTES) by astrocytes. *J Neurosci* **28**: 3277–3290.

Christopherson, R.I., Stoner, K., Barber, N. *et al.* (2006) Classification of AML using a monoclonal antibody microarray. *Methods Mol Med* **125**: 241–251.

Chung, A.S. and Chin, Y.E. (2009) Antibody array platform to monitor protein tyrosine phosphorylation in mammalian cells. *Methods Mol Biol* **527**: 247–255, ix.

Clamp, M., Fry, B., Kamal, M. *et al.* (2007) Distinguishing protein-coding and noncoding genes in the human genome. *Proc Natl Acad Sci USA* **104**: 19428–19433.

Cui, Y., Wei, Q., Park, H. and Lieber, C.M. (2001) Nanowire nanosensors for highly sensitive and selective detection of biological and chemical species. *Science* **293**: 1289–1292.

Danczyk, R., Krieder, B., North, A. *et al.* (2003) Comparison of antibody functionality using different immobilization methods. *Biotechnol Bioeng* **84**: 215–223.

Dewing, P., Christensen, A., Bondar, G. and Micevych, P. (2008) Protein kinase C signaling in the hypothalamic arcuate nucleus regulates sexual receptivity in female rats. *Endocrinology* **149**: 5934–5942.

Dhungana, S., Sharrack, B. and Woodroofe, N. (2009) Cytokines and chemokines in idiopathic intracranial hypertension. *Headache* **49**: 282–285.

Diamandis, E.P. and Van Der Merwe, D.E. (2005) Plasma protein profiling by mass spectrometry for cancer diagnosis: opportunities and limitations. *Clin Cancer Res* **11**: 963–965.

Doerr, A. (2005) Protein microarray velcro. *Nat Methods* **2**: 642–643.

Drouvalakis, K.A., Bangsaruntip, S., Hueber, W. *et al.* (2008) Peptide-coated nanotube-based biosensor for the detection of disease-specific autoantibodies in human serum. *Biosens Bioelectron* **23**: 1413–1421.

Eads, D., Hansen, R., Oyegunwa, A. *et al.* (2009) Terameprocol, a methylated derivative of nordihydroguaiaretic acid, inhibits production of prostaglandins and several key inflammatory cytokines and chemokines. *J Inflamm (Lond)* **6**: 2.

Ebihara, N., Chen, L., Tokura, T. *et al.* (2007) Distinct functions between toll-like receptors 3 and 9 in retinal pigment epithelial cells. *Ophthalmic Res* **39**: 155–163.

Ehrlich, J.R., Tang, L., Caiazzo, R.J., Jr *et al.* (2008) The "reverse capture" autoantibody microarray: an innovative approach to profiling the autoantibody response to tissue-derived native antigens. *Methods Mol Biol* **441**: 175–192.

Ekins, R.P. (1989) Multi-analyte immunoassay. *J Pharm Biomed Anal* **7**: 155–168.

Ekins, R.P. (1998) Ligand assays: from electrophoresis to miniaturized microarrays. *Clin Chem* **44**: 2015–2030.

Ekins, R.P. and Chu, F.W. (1991) Multianalyte microspot immunoassay –microanalytical "compact disk" of the future. *Clin Chem* **37**: 1955–1967.

Ellmark, P., Ingvarsson, J., Carlsson, A. *et al.* (2006) Identification of protein expression signatures associated with Helicobacter pylori infection and gastric adenocarcinoma using recombinant antibody microarrays. *Mol Cell Proteomics* **5**: 1638–1646.

Fei, Y.Y., Landry, J.P., Sun, Y.S. *et al.* (2008) A novel high-throughput scanning microscope for label-free detection of protein and small-molecule chemical microarrays. *Rev Sci Instrum* **79**: 013708.

Fenton, J.I., Nunez, N.P., Yakar, S. *et al.* (2009) Diet-induced adiposity alters the serum profile of inflammation in C57BL/6N mice as measured by antibody array. *Diabetes Obes Metab* **11**: 343–354.

Fischer, I., Schulze, S., Kuhn, C. *et al.* (2009) Inhibiton of RET and JAK2 signals and upregulation of VEGFR3 phosphorylation in vitro by galectin-1 in trophoblast tumor cells BeWo. *Placenta* **30**: 1078–1082.

Foster, R., Segers, I., Smart, D. *et al.* (2010) A differential cytokine expression profile is induced by highly purified human menopausal gonadotropin and recombinant follicle-stimulating hormone in a pre- and postovulatory mouse follicle culture model. *Fertil Steril* **93**: 1464–1476.

Fujita, K., Ewing, C.M., Sokoll, L.J. *et al.* (2008) Cytokine profiling of prostatic fluid from cancerous prostate glands identifies cytokines associated with extent of tumor and inflammation. *Prostate* **68**: 872–882.

Gao, W.M., Kuick, R., Orchekowski, R.P. *et al.* (2005) Distinctive serum protein profiles involving abundant proteins in lung cancer patients based upon antibody microarray analysis. *BMC Cancer* **5**: 110.

Gasparrini, M., Rivas, D., Elbaz, A. and Duque, G. (2009) Differential expression of cytokines in subcutaneous and marrow fat of aging C57BL/6J mice. *Exp Gerontol* **44**: 613–618.

Gembitsky, D.S., Lawlor, K., Jacovina, A. *et al.* (2004) A prototype antibody microarray platform to monitor changes in protein tyrosine phosphorylation. *Mol Cell Proteomics* **3**: 1102–1118.

Ghose, S., Hubbard, B. and Cramer, S.M. (2007) Binding capacity differences for antibodies and Fc-fusion proteins on protein A chromatographic materials. *Biotechnol Bioeng* **96**: 768–779.

Glokler, J. and Angenendt, P. (2003) Protein and antibody microarray technology. *J Chromatogr B Analyt Technol Biomed Life Sci* **797**: 229–240.

Grassel, S., Ahmed, N., Gottl, C. and Grifka, J. (2009) Gene and protein expression profile of naive and osteo-chondrogenically differentiated rat bone marrow-derived mesenchymal progenitor cells. *Int J Mol Med* **23**: 745–755.

Gruber, H.E., Hoelscher, G., Loeffler, B. *et al.* (2009) Prostaglandin E1 and misoprostol increase epidermal growth factor production in 3D-cultured human annulus cells. *Spine J* **9**: 760–766.

Guss, B., Eliasson, M., Olsson, A. *et al.* (1986) Structure of the IgG-binding regions of streptococcal protein G. *EMBO J* **5**: 1567–1575.

Haab, B.B. (2003) Methods and applications of antibody microarrays in cancer research. *Proteomics* **3**: 2116–2122.

Haab, B.B. (2005) Antibody arrays in cancer research. *Mol Cell Proteomics* **4**: 377–383.

Haddad, G. and Belosevic, M. (2009) Transferrin-derived synthetic peptide induces highly conserved pro-inflammatory responses of macrophages. *Mol Immunol* **46**: 576–586.

Hahn, C.D., Leitner, C., Weinbrenner, T. *et al.* (2007) Self-assembled monolayers with latent aldehydes for protein immobilization. *Bioconjug Chem* **18**: 247–253.

Hamelinck, D., Zhou, H., Li, L. *et al.* (2005) Optimized normalization for antibody microarrays and application to serum-protein profiling. *Mol Cell Proteomics* **4**: 773–784.

Han, M.K., Hong, M.Y., Lee, D. *et al.* (2006) Expression profiling of proteins in L-threonine biosynthetic pathway of Escherichia coli by using antibody microarray. *Proteomics* **6**: 5929–5940.

Han, M.K., Oh, Y.H., Kang, J. *et al.* (2009) Protein profiling in human sera for identification of potential lung cancer biomarkers using antibody microarray. *Proteomics* **9**: 5544–5552.

Han, X., Aslanian, A. and Yates, J.R., 3rd (2008) Mass spectrometry for proteomics. *Curr Opin Chem Biol* **12**: 483–490.

Hao, Q., Wang, L. and Tang, H. (2009) Vascular endothelial growth factor induces protein kinase D-dependent production of proinflammatory cytokines in endothelial cells. *Am J Physiol Cell Physiol* **296**: C821–C827.

Hao, Y., Yu, Y., Wang, L. *et al.* (2008) IPO-38 is identified as a novel serum biomarker of gastric cancer based on clinical proteomics technology. *J Proteome Res* **7**: 3668–3677.

Harper, J.I., Godwin, H., Green, A. *et al.* (2010) A study of matrix metalloproteinase expression and activity in atopic dermatitis using a novel skin wash sampling assay for functional biomarker analysis. *Br J Dermatol* **162**: 397–403.

He, M. and Taussig, M.J. (2005) Ribosome display of antibodies: expression, specificity and recovery in a eukaryotic system. *J Immunol Methods* **297**: 73–82.

Hergenroeder, G.W., Moore, A.N., Mccoy, J.P., Jr *et al.* (2010) Serum IL-6: a candidate biomarker for intracranial pressure elevation following isolated traumatic brain injury. *J Neuroinflammation* **7**: 19.

Hoheisel, J.D. (2006) Microarray technology: beyond transcript profiling and genotype analysis. *Nat Rev Genet* **7**: 200–210.

Hon, K.L., Ching, G.K., Wong, K.Y. *et al.* (2008) A pilot study to explore the usefulness of antibody array in childhood atopic dermatitis. *J Natl Med Assoc* **100**: 500–504.

Hu, H., Kwun, J., Aizenstein, B.D. and Knechtle, S.J. (2009) Noninvasive detection of acute and chronic injuries in human renal transplant by elevation of multiple cytokines/chemokines in urine. *Transplantation* **87**: 1814–1820.

Huang, R.P. (2007) An array of possibilities in cancer research using cytokine antibody arrays. *Expert Rev Proteomics* **4**: 299–308.

Huang, R.P., Huang, R., Fan, Y. and Lin, Y. (2001) Simultaneous detection of multiple cytokines from conditioned media and patient's sera by an antibody-based protein array system. *Anal Biochem* **294**: 55–62.

Hudelist, G., Pacher-Zavisin, M., Singer, C.F. *et al.* (2004) Use of high-throughput protein array for profiling of differentially expressed proteins in normal and malignant breast tissue. *Breast Cancer Res Treat* **86**: 281–291.

Huff, J.L., Lynch, M.P., Nettikadan, S. *et al.* (2004) Label-free protein and pathogen detection using the atomic force microscope. *J Biomol Screen* **9**: 491–497.

Iiizumi, M., Bandyopadhyay, S., Pai, S.K. *et al.* (2008) RhoC promotes metastasis via activation of the Pyk2 pathway in prostate cancer. *Cancer Res* **68**: 7613–7620.

Inai, K., Takagi, K., Takimoto, N. *et al.* (2008) Multiple inflammatory cytokine-productive ThyL-6 cell line established from a patient with thymic carcinoma. *Cancer Sci* **99**: 1778–1784.

Ingvarsson, J., Larsson, A., Sjoholm, A.G. *et al.* (2007) Design of recombinant antibody microarrays for serum protein profiling: targeting of complement proteins. *J Proteome Res* **6**: 3527–3536.

Ingvarsson, J., Wingren, C., Carlsson, A. *et al.* (2008) Detection of pancreatic cancer using antibody microarray-based serum protein profiling. *Proteomics* **8**: 2211–2219.

Jamesdaniel, S., Ding, D., Kermany, M.H. *et al.* (2008) Proteomic analysis of the balance between survival and cell death responses in cisplatin-mediated ototoxicity. *J Proteome Res* **7**: 3516–3524.

Jarvinen, K., Vuolteenaho, K., Nieminen, R. *et al.* (2008) Selective iNOS inhibitor 1400W enhances anti-catabolic IL-10 and reduces destructive MMP-10 in OA cartilage. Survey of the effects of 1400W on inflammatory mediators produced by OA cartilage as detected by protein antibody array. *Clin Exp Rheumatol* **26**: 275–282.

Jiang, X., Zhang, D., Shi, J. *et al.* (2009) Increased inflammatory response both in brain and in periphery in presenilin 1 and presenilin 2 conditional double knock-out mice. *J Alzheimers Dis* **18**: 515–523.

Jin, G., Zhao, Z.Y., Wang, Z.H. *et al.* (2004) The development of biosensor with imaging ellipsometry. *Conf Proc IEEE Eng Med Biol Soc* **3**: 1975–1978.

Jin, K.B., Choi, H.J., Kim, H.T. *et al.* (2008) Cytokine array after cyclosporine treatment in rats. *Transplant Proc* **40**: 2682–2684.

Jostock, T. and Dübel, S. (2005) Screening of molecular repertoires by microbial surface display. *Comb Chem High Throughput Screen* **8**: 127–133.

Jung, Y., Kang, H.J., Lee, J.M. *et al.* (2008) Controlled antibody immobilization onto immuno-analytical platforms by synthetic peptide. *Anal Biochem* **374**: 99–105.

Kader, H.A., Tchernev, V.T., Satyaraj, E. *et al.* (2005) Protein microarray analysis of disease activity in pediatric inflammatory bowel disease demonstrates elevated serum PLGF, IL-7, TGF-beta1, and IL-12p40 levels in Crohn's disease and ulcerative colitis patients in remission versus active disease. *Am J Gastroenterol* **100**: 414–423.

Karaca, B., Kucukzeybek, Y., Gorumlu, G. *et al.* (2008) Profiling of angiogenic cytokines produced by hormone- and drug-refractory prostate cancer cell lines, PC-3 and DU-145 before and after treatment with gossypol. *Eur Cytokine Netw* **19**: 176–184.

Kaukinen, A., Kuusniemi, A.M., Lautenschlager, I. and Jalanko, H. (2008) Glomerular endothelium in kidneys with congenital nephrotic syndrome of the Finnish type (NPHS1). *Nephrol Dial Transplant* **23**: 1224–1232.

Kim, H., Park, J.H., Cho, I.H. *et al.* (2009) Selective immobilization of proteins on gold dot arrays and characterization using chemical force microscopy. *J Colloid Interface Sci* **334**: 161–166.

Kline, M., Donovan, K., Wellik, L. *et al.* (2007) Cytokine and chemokine profiles in multiple myeloma; significance of stromal interaction and correlation of IL-8 production with disease progression. *Leuk Res* **31**: 591–598.

Knight, P.R., Sreekumar, A., Siddiqui, J. *et al.* (2004) Development of a sensitive microarray immunoassay and comparison with standard enzyme-linked immunoassay for cytokine analysis. *Shock* **21**: 26–30.

Ko, I.K., Kato, K. and Iwata, H. (2005) Parallel analysis of multiple surface markers expressed on rat neural stem cells using antibody microarrays. *Biomaterials* **26**: 4882–4891.

Konthur, Z., Hust, M. and Dübel, S. (2005) Perspectives for systematic in vitro antibody generation. *Gene* **364**: 19–29.

Koomen, J.M., Li, D., Xiao, L.C. *et al.* (2005) Direct tandem mass spectrometry reveals limitations in protein profiling experiments for plasma biomarker discovery. *J Proteome Res* **4**: 972–981.

Korf, U., Henjes, F., Schmidt, C. *et al.* (2008) Antibody microarrays as an experimental platform for the analysis of signal transduction networks. *Adv Biochem Eng Biotechnol* **110**: 153–175.

Kullmann, T., Barta, I., Csiszer, E. *et al.* (2008) Differential cytokine pattern in the exhaled breath of patients with lung cancer. *Pathol Oncol Res* **14**: 481–483.

Kusnezow, W., Banzon, V., Schröder, C. *et al.* (2007) Antibody microarray-based profiling of complex specimens: systematic evaluation of labeling strategies. *Proteomics* **7**: 1786–1799.

Kusnezow, W. and Hoheisel, J.D. (2002) Antibody microarrays: promises and problems. *Biotechniques* Suppl: 14–23.

Kusnezow, W. and Hoheisel, J.D. (2003) Solid supports for microarray immunoassays. *J Mol Recognit* **16**: 165–176.

Kusnezow, W., Jacob, A., Walijew, A. *et al.* (2003) Antibody microarrays: an evaluation of production parameters. *Proteomics* **3**: 254–264.

Kusnezow, W., Syagailo, Y.V., Goychuk, I. *et al.* (2006b) Antibody microarrays: the crucial impact of mass transport on assay kinetics and sensitivity. *Expert Rev Mol Diagn* **6**: 111–124.

Kusnezow, W., Syagailo, Y.V., Ruffer, S. *et al.* (2006a) Optimal design of microarray immunoassays to compensate for kinetic limitations: theory and experiment. *Mol Cell Proteomics* **5**: 1681–1696.

Kyo, M., Usui-Aoki, K. and Koga, H. (2005) Label-free detection of proteins in crude cell lysate with antibody arrays by a surface plasmon resonance imaging technique. *Anal Chem* **77**: 7115–7121.

Ladd, J., Taylor, A.D., Piliarik, M. *et al.* (2009) Label-free detection of cancer biomarker candidates using surface plasmon resonance imaging. *Anal Bioanal Chem* **393**: 1157–1563.

Lal, S., Brown, A., Nguyen, L. *et al.* (2009) Using antibody arrays to detect microparticles from acute coronary syndrome patients based on cluster of differentiation (CD) antigen expression. *Mol Cell Proteomics* **8**: 799–804.

Lal, S., Lui, R., Nguyen, L. *et al.* (2004) Increases in leukocyte cluster of differentiation antigen expression during cardiopulmonary bypass in patients undergoing heart transplantation. *Proteomics* **4**: 1918–1926.

Lausted, C., Hu, Z. and Hood, L. (2008) Quantitative serum proteomics from surface plasmon resonance imaging. *Mol Cell Proteomics* **7**: 2464–2474.

Lee, H.J., Nedelkov, D. and Corn, R.M. (2006) Surface plasmon resonance imaging measurements of antibody arrays for the multiplexed detection of low molecular weight protein biomarkers. *Anal Chem* **78**: 6504–6510.

Lee, M.H., Choi, B.Y., Kundu, J.K. *et al.* (2009) Resveratrol suppresses growth of human ovarian cancer cells in culture and in a murine xenograft model: eukaryotic elongation factor 1A2 as a potential target. *Cancer Res* **69**: 7449–7458.

Lee, Y.A., Cho, E.J. and Yokozawa, T. (2010) Oligomeric proanthocyanidins improve memory and enhance phosphorylation of vascular endothelial growth factor receptor-2 in senescence-accelerated mouse prone/8. *Br J Nutr* **103**: 479–489.

Lee, Y.C., Chiou, T.J., Tzeng, W.F. and Chu, S.T. (2008) Macrophage inflammatory protein-3alpha influences growth of K562 leukemia cells in co-culture with anticancer drug-pretreated HS-5 stromal cells. *Toxicology* **249**: 116–122.

Leonardi, A., Sathe, S., Bortolotti, M. *et al.* (2009) Cytokines, matrix metalloproteases, angiogenic and growth factors in tears of normal subjects and vernal keratoconjunctivitis patients. *Allergy* **64**: 710–717.

Letarte, M., Voulgaraki, D., Hatherley, D. *et al.* (2005) Analysis of leukocyte membrane protein interactions using protein microarrays. *BMC Biochem* **6**: 2.

Levy, R., Forsyth, C.M., Laporte, S.L. *et al.* (2007) Fine and domain-level epitope mapping of botulinum neurotoxin type A neutralizing antibodies by yeast surface display. *J Mol Biol* **365**: 196–210.

Li, C., Simeone, D.M., Brenner, D.E. *et al.* (2009) Pancreatic cancer serum detection using a lectin/glyco-antibody array method. *J Proteome Res* **8**: 483–492.

Li, S., Sack, R., Vijmasi, T. *et al.* (2008) Antibody protein array analysis of the tear film cytokines. *Optom Vis Sci* **85**: 653–660.

Li, X., Wang, H., Touma, E. *et al.* (2007) Genetic network and pathway analysis of differentially expressed proteins during critical cellular events in fracture repair. *J Cell Biochem* **100**: 527–543.

Lin, Y., Huang, R., Cao, X. *et al.* (2003) Detection of multiple cytokines by protein arrays from cell lysate and tissue lysate. *Clin Chem Lab Med* **41**: 139–145.

Lipovsek, D. and Plückthun, A. (2004) In-vitro protein evolution by ribosome display and mRNA display. *J Immunol Methods* **290**: 51–67.

Liu, B.C., Zhang, L., Lv, L.L. *et al.* (2006) Application of antibody array technology in the analysis of urinary cytokine profiles in patients with chronic kidney disease. *Am J Nephrol* **26**: 483–490.

Loch, C.M., Ramirez, A.B., Liu, Y. *et al.* (2007) Use of high density antibody arrays to validate and discover cancer serum biomarkers. *Mol Oncol* **1**: 313–320.

Lu, Y., Cai, Z., Xiao, G. *et al.* (2007) Monocyte chemotactic protein-1 mediates prostate cancer-induced bone resorption. *Cancer Res* **67**: 3646–3653.

Lu, Y., Wang, J., Xu, Y. *et al.* (2008) CXCL16 functions as a novel chemotactic factor for prostate cancer cells in vitro. *Mol Cancer Res* **6**: 546–554.

Lukesova, S., Kopecky, O., Vroblova, V. *et al.* (2008) Determination of angiogenic factors in serum by protein array in patients with renal cell carcinoma. *Folia Biol (Praha)* **54**: 134–140.

Lv, L.L. and Liu, B.C. (2007) High-throughput antibody microarrays for quantitative proteomic analysis. *Expert Rev Proteomics* **4**: 505–513.

Macbeath, G. and Schreiber, S.L. (2000) Printing proteins as microarrays for high-throughput function determination. *Science* **289**: 1760–1763.

Madoz-Gurpide, J., Canamero, M., Sanchez, L. *et al.* (2007) A proteomics analysis of cell signaling alterations in colorectal cancer. *Mol Cell Proteomics* **6**: 2150–2164.

Martin, D.B. and Nelson, P.S. (2001) From genomics to proteomics: techniques and applications in cancer research. *Trends Cell Biol* **11**: S60–S65.

Morais, S., Tamarit-Lopez, J., Carrascosa, J. *et al.* (2008) Analytical prospect of compact disk technology in immunosensing. *Anal Bioanal Chem* **391**: 2837–2844.

Moschos, S.J., Smith, A.P., Mandic, M. *et al.* (2007) SAGE and antibody array analysis of melanoma-infiltrated lymph nodes: identification of Ubc9 as an important molecule in advanced-stage melanomas. *Oncogene* **26**: 4216–4225.

Mychaleckyj, J.C. (2007) Genome mapping statistics and bioinformatics. *Methods Mol Biol* **404**: 461–488.

Nedelkov, D., Tubbs, K.A. and Nelson, R.W. (2006) Surface plasmon resonance-enabled mass spectrometry arrays. *Electrophoresis* **27**: 3671–3675.

Neuhoff, S., Moers, J., Rieks, M. *et al.* (2007) Proliferation, differentiation, and cytokine secretion of human umbilical cord blood-derived mononuclear cells in vitro. *Exp Hematol* **35**: 1119–1131.

Nilsson, P., Paavilainen, L., Larsson, K. *et al.* (2005) Towards a human proteome atlas: high-throughput generation of mono-specific antibodies for tissue profiling. *Proteomics* **5**: 4327–4337.

Nisnevitch, M., Kolog-Gulco, M., Trombka, D. *et al.* (2000) Immobilization of antibodies onto glass wool. *J Chromatogr B Biomed Sci Appl* **738**: 217–223.

Nonomura, T., Masaki, T., Morishita, A. *et al.* (2007) Identification of c-Yes expression in the nuclei of hepatocellular carcinoma cells: involvement in the early stages of hepatocarcinogenesis. *Int J Oncol* **30**: 105–111.

Nygren, P.A. (2008) Alternative binding proteins: affibody binding proteins developed from a small three-helix bundle scaffold. *FEBS J* **275**: 2668–2676.

Nygren, P.A. and Skerra, A. (2004) Binding proteins from alternative scaffolds. *J Immunol Methods* **290**: 3–28.

Nystrom, A., Shaik, Z.P., Gullberg, D. *et al.* (2009) Role of tyrosine phosphatase SHP-1 in the mechanism of endorepellin angiostatic activity. *Blood* **114**: 4897–4906.

Ohshima, M., Yamaguchi, Y., Micke, P. *et al.* (2008) In vitro characterization of the cytokine profile of the epithelial cell rests of Malassez. *J Periodontol* **79**: 912–919.

Okuno, J., Maehashi, K., Kerman, K. *et al.* (2007) Label-free immunosensor for prostate-specific antigen based on single-walled carbon nanotube array-modified microelectrodes. *Biosens Bioelectron* **22**: 2377–2381.

Olle, E.W., Messamore, J., Deogracias, M.P. *et al.* (2005) Comparison of antibody array substrates and the use of glycerol to normalize spot morphology. *Exp Mol Pathol* **79**: 206–209.

Orchekowski, R., Hamelinck, D., Li, L. *et al.* (2005) Antibody microarray profiling reveals individual and combined serum proteins associated with pancreatic cancer. *Cancer Res* **65**: 11193–11202.

Oscarsson, J., Karched, M., Thay, B. *et al.* (2008) Proinflammatory effect in whole blood by free soluble bacterial components released from planktonic and biofilm cells. *BMC Microbiol* **8**: 206.

Pavlickova, P., Schneider, E.M. and Hug, H. (2004) Advances in recombinant antibody microarrays. *Clin Chim Acta* **343**: 17–35.

Peczuh, M.W. and Hamilton, A.D. (2000) Peptide and protein recognition by designed molecules. *Chem Rev* **100**: 2479–2494.

Pelech, S., Jelinkova, L., Susor, A. *et al.* (2008) Antibody microarray analyses of signal transduction protein expression and phosphorylation during porcine oocyte maturation. *J Proteome Res* **7**: 2860–2871.

Peluso, P., Wilson, D.S., Do, D. *et al.* (2003) Optimizing antibody immobilization strategies for the construction of protein microarrays. *Anal Biochem* **312**: 113–124.

Perera, C.N., Spalding, H.S., Mohammed, S.I. and Camarillo, I.G. (2008) Identification of proteins secreted from leptin stimulated MCF-7 breast cancer cells: a dual proteomic approach. *Exp Biol Med (Maywood)* **233**: 708–720.

Plückthun, A. (2009) Alternative Scaffolds: Expanding the Options of Antibodies, in *Recombinant Antibodies for Immunotherapy* (ed. M. Little), Cambridge University Press, New York, pp. 243–271.

Podlaski, F.J. and Stern, A.S. (2000) Site-specific immobilization of antibodies to protein G-derivatized solid supports. *Methods Mol Biol* **147**: 41–48.

Prakesch, M., Denisov, A.Y., Naim, M. *et al.* (2008) The discovery of small molecule chemical probes of Bcl-X(L) and Mcl-1. *Bioorg Med Chem* **16**: 7443–7449.

Qi, W., Cooke, L.S., Stejskal, A. *et al.* (2009) MP470, a novel receptor tyrosine kinase inhibitor, in combination with Erlotinib inhibits the HER family/PI3K/Akt pathway and tumor growth in prostate cancer. *BMC Cancer* **9**: 142.

Rajala, R.V. (2008) Phospho-site-specific antibody microarray to study the state of protein phosphorylation in the retina. *J Proteomics Bioinform* **1**: 242.

Ray, R.B., Raychoudhuri, A., Steele, R. and Nerurkar, P. (2010) Bitter melon (Momordica charantia) extract inhibits breast cancer cell proliferation by modulating cell cycle regulatory genes and promotes apoptosis. *Cancer Res* **70**: 1925–1931.

Ray, S., Mehta, G. and Srivastava, S. (2010) Label-free detection techniques for protein microarrays: prospects, merits and challenges. *Proteomics* **10**: 731–748.

Rimini, R., Schwenk, J.M., Sundberg, M. *et al.* (2009) Validation of serum protein profiles by a dual antibody array approach. *J Proteomics* **73**: 252–266.

Rivas, L.A., Garcia-Villadangos, M., Moreno-Paz, M. *et al.* (2008) A 200-antibody microarray biochip for environmental monitoring: searching for universal microbial biomarkers through immunoprofiling. *Anal Chem* **80**: 7970–7979.

Roque, A.C., Taipa, M.A. and Lowe, C.R. (2004) A new method for the screening of solid-phase combinatorial libraries for affinity chromatography. *J Mol Recognit* **17**: 262–267.

Royce, T.E., Rozowsky, J.S., Luscombe, N.M. *et al.* (2006) Extrapolating traditional DNA microarray statistics to tiling and protein microarray technologies. *Methods Enzymol* **411**: 282–311.

Rubina, A.Y., Dementieva, E.I., Stomakhin, A.A. *et al.* (2003) Hydrogel-based protein microchips: manufacturing, properties, and applications. *Biotechniques* **34**: 1008–1014, 1016–1020, 1022.

Sanchez-Carbayo, M. (2006) Antibody arrays: technical considerations and clinical applications in cancer. *Clin Chem* **52**: 1651–1659.

Sanchez-Carbayo, M., Socci, N.D., Lozano, J.J. *et al.* (2006) Profiling bladder cancer using targeted antibody arrays. *Am J Pathol* **168**: 93–103.

Schmidt, M.W., Houseman, A., Ivanov, A.R. and Wolf, D.A. (2007) Comparative proteomic and transcriptomic profiling of the fission yeast Schizosaccharomyces pombe. *Mol Syst Biol* **3**: 79.

Schröder, C., Jacob, A., Tonack, S. *et al.* (2010) Dual-color proteomic profiling of complex samples with a microarray of 810 cancer-related antibodies. *Mol Cell Proteomics* **9**: 1271–1280.

Schuffenhauer, A., Ruedisser, S., Marzinzik, A.L. *et al.* (2005) Library design for fragment based screening. *Curr Top Med Chem* **5**: 751–762.

Scupoli, M.T., Donadelli, M., Cioffi, F. *et al.* (2008) Bone marrow stromal cells and the upregulation of interleukin-8 production in human T-cell acute lymphoblastic leukemia through the CXCL12/CXCR4 axis and the NF-kappaB and JNK/AP-1 pathways. *Haematologica* **93**: 524–532.

Seeber, J.W., Zorn-Kruppa, M., Lombardi-Borgia, S. *et al.* (2008) Characterisation of human corneal epithelial cell cultures maintained under serum-free conditions. *Altern Lab Anim* **36**: 569–583.

Seong, S.Y. (2002) Microimmunoassay using a protein chip: optimizing conditions for protein immobilization. *Clin Diagn Lab Immunol* **9**: 927–930.

Shafer, M.W., Mangold, L., Partin, A.W. and Haab, B.B. (2007) Antibody array profiling reveals serum TSP-1 as a marker to distinguish benign from malignant prostatic disease. *Prostate* **67**: 255–267.

Sinensky, A.K. and Belcher, A.M. (2007) Label-free and high-resolution protein/DNA nanoarray analysis using Kelvin probe force microscopy. *Nat Nanotechnol* **2**: 653–659.

Skalnikova, H., Vodicka, P., Pelech, S. *et al.* (2008) Protein signaling pathways in differentiation of neural stem cells. *Proteomics* **8**: 4547–4559.

Smith, L., Watson, M.B., O'Kane, S.L. *et al.* (2006) The analysis of doxorubicin resistance in human breast cancer cells using antibody microarrays. *Mol Cancer Ther* **5**: 2115–2120.

Soellner, M.B., Dickson, K.A., Nilsson, B.L. and Raines, R.T. (2003) Site-specific protein immobilization by Staudinger ligation. *J Am Chem Soc* **125**: 11790–11791.

Song, S., Li, B., Wang, L. *et al.* (2007) A cancer protein microarray platform using antibody fragments and its clinical applications. *Mol Biosyst* **3**: 151–158.

Sorribas, H., Padeste, C. and Tiefenauer, L. (2002) Photolithographic generation of protein micropatterns for neuron culture applications. *Biomaterials* **23**: 893–900.

Spisak, S., Galamb, B., Sipos, F. *et al.* (2010) Applicability of antibody and mRNA expression microarrays for identifying diagnostic and progression markers of early and late stage colorectal cancer. *Dis Markers* **28**: 1–14.

Sreekumar, A., Nyati, M.K., Varambally, S. *et al.* (2001) Profiling of cancer cells using protein microarrays: discovery of novel radiation-regulated proteins. *Cancer Res* **61**: 7585–7593.

Srivastava, M., Eidelman, O., Jozwik, C. *et al.* (2006) Serum proteomic signature for cystic fibrosis using an antibody microarray platform. *Mol Genet Metab* **87**: 303–310.

Stoevesandt, O. and Taussig, M.J. (2007) Affinity reagent resources for human proteome detection: initiatives and perspectives. *Proteomics* **7**: 2738–2750.

Stumpp, M.T. and Amstutz, P. (2007) DARPins: a true alternative to antibodies. *Curr Opin Drug Discov Devel* **10**: 153–159.

Sun, H., Chua, M.S., Yang, D. *et al.* (2008) Antibody arrays identify potential diagnostic markers of hepatocellular carcinoma. *Biomark Insights* **3**: 1–18.

Suraniti, E., Sollier, E., Calemczuk, R. *et al.* (2007) Real-time detection of lymphocytes binding on an antibody chip using SPR imaging. *Lab Chip* **7**: 1206–8.

Sze, S.K., De Kleijn, D.P., Lai, R.C. *et al.* (2007) Elucidating the secretion proteome of human embryonic stem cell-derived mesenchymal stem cells. *Mol Cell Proteomics* **6**: 1680–1689.

Takeda, N., Sumi, Y., Prefontaine, D. *et al.* (2009) Epithelium-derived chemokines induce airway smooth muscle cell migration. *Clin Exp Allergy* **39**: 1018–1026.

Tapias, A., Ciudad, C.J., Roninson, I.B. and Noe, V. (2008) Regulation of Sp1 by cell cycle related proteins. *Cell Cycle* **7**: 2856–2867.

Taussig, M.J., Stoevesandt, O., Borrebaeck, C.A. *et al.* (2007) ProteomeBinders: planning a European resource of affinity reagents for analysis of the human proteome. *Nat Methods* **4**: 13–17.

Templin, M.F., Stoll, D., Bachmann, J. and Joos, T. O. (2004) Protein microarrays and multiplexed sandwich immunoassays: what beats the beads? *Comb Chem High Throughput Screen* **7**: 223–229.

Thompson, M., Cheran, L.E., Zhang, M. *et al.* (2005) Label-free detection of nucleic acid and protein microarrays by scanning Kelvin nanoprobe. *Biosens Bioelectron* **20**: 1471–1481.

Toh, H.C., Wang, W.W., Chia, W.K. *et al.* (2009) Clinical benefit of allogeneic melanoma cell lysate-pulsed autologous dendritic cell vaccine in MAGE-positive colorectal cancer patients. *Clin Cancer Res* **15**: 7726–7736.

Tolmachev, V., Orlova, A., Nilsson, F.Y. *et al.* (2007) Affibody molecules: potential for in vivo imaging of molecular targets for cancer therapy. *Expert Opin Biol Ther* **7**: 555–568.

Tomchuck, S.L., Zwezdaryk, K.J., Coffelt, S.B. *et al.* (2008) Toll-like receptors on human mesenchymal stem cells drive their migration and immunomodulating responses. *Stem Cells* **26**: 99–107.

Tsai, M.C., Wei, C.P., Lee, D.Y. *et al.* (2008) Inflammatory mediators of cerebrospinal fluid from patients with spinal cord injury. *Surg Neurol* **70** Suppl 1: 19–24; discussion 24.

Tuerk, C. and Gold, L. (1990) Systematic evolution of ligands by exponential enrichment: RNA ligands to bacteriophage T4 DNA polymerase. *Science* **249**: 505–510.

Uemura, N., Nakanishi, Y., Kato, H. *et al.* (2009) Antibody-based proteomics for esophageal cancer: Identification of proteins in the nuclear factor-kappaB pathway and mitotic checkpoint. *Cancer Sci* **100**: 1612–1622.

Uhlen, M. (2008) Affinity as a tool in life science. *Biotechniques* **44**: 649–654.

Uhlen, M., Bjorling, E., Agaton, C. *et al.* (2005) A human protein atlas for normal and cancer tissues based on antibody proteomics. *Mol Cell Proteomics* **4**: 1920–1932.

Uhlen, M. and Hober, S. (2009) Generation and validation of affinity reagents on a proteome-wide level. *J Mol Recognit* **22**: 57–64.

Usui-Aoki, K., Shimada, K., Nagano, M. *et al.* (2005) A novel approach to protein expression profiling using antibody microarrays combined with surface plasmon resonance technology. *Proteomics* **5**: 2396–2401.

Valsesia, A., Colpo, P., Meziani, T. *et al.* (2006) Immobilization of antibodies on biosensing devices by nanoarrayed self-assembled monolayers. *Langmuir* **22**: 1763–1767.

Wang, X., Zhao, M. and Nolte, D.D. (2008) Area-scaling of interferometric and fluorescent detection of protein on antibody microarrays. *Biosens Bioelectron* **24**: 987–993.

Wang, Z.H. and Jin, G. (2003) A label-free multisensing immunosensor based on imaging ellipsometry. *Anal Chem* **75**: 6119–6123.

Wang, Z.H., Meng, Y.H., Ying, P.Q. *et al.* (2006) A label-free protein microfluidic array for parallel immunoassays. *Electrophoresis* **27**: 4078–4085.

Watanabe, M., Guo, W., Zou, S. *et al.* (2005) Antibody array analysis of peripheral and blood cytokine levels in rats after masseter inflammation. *Neurosci Lett* **382**: 128–133.

Wilkins, M.R., Appel, R.D., Van Eyk, J.E. *et al.* (2006) Guidelines for the next 10 years of proteomics. *Proteomics* **6**: 4–8.

Wilkins, M.R., Pasquali, C., Appel, R.D. *et al.* (1996) From proteins to proteomes: large scale protein identification by two-dimensional electrophoresis and amino acid analysis. *Biotechnology (NY)* **14**: 61–65.

Wilson, D.S., Keefe, A.D. and Szostak, J.W. (2001) The use of mRNA display to select high-affinity protein-binding peptides. *Proc Natl Acad Sci USA* **98**: 3750–3755.

Wingren, C., Ingvarsson, J., Dexlin, L. *et al.* (2007) Design of recombinant antibody microarrays for complex proteome analysis: choice of sample labeling-tag and solid support. *Proteomics* **7**: 3055–3065.

Winter, G., Griffiths, A.D., Hawkins, R.E. and Hoogenboom, H.R. (1994) Making antibodies by phage display technology. *Annu Rev Immunol* **12**: 433–455.

Wittmann-Liebold, B., Graack, H.R. and Pohl, T. (2006) Two-dimensional gel electrophoresis as tool for proteomics studies in combination with protein identification by mass spectrometry. *Proteomics* **6**: 4688–4703.

Wong, L.L., Chang, C.F., Koay, E.S. and Zhang, D. (2009) Tyrosine phosphorylation of PP2A is regulated by HER-2 signalling and correlates with breast cancer progression. *Int J Oncol* **34**: 1291–1301.

Woolfson, A., Stebbing, J., Tom, B.D. *et al.* (2005) Conservation of unique cell-surface CD antigen mosaics in HIV-1-infected individuals. *Blood* **106**: 1003–1007.

Wu, J.Q., Dyer, W.B., Chrisp, J. *et al.* (2008) Longitudinal microarray analysis of cell surface antigens on peripheral blood mononuclear cells from HIV+ individuals on highly active antiretroviral therapy. *Retrovirology* **5**: 24.

Wu, J.Q., Wang, B., Belov, L. *et al.* (2007) Antibody microarray analysis of cell surface antigens on CD4+ and CD8+ T cells from HIV+ individuals correlates with disease stages. *Retrovirology* **4**: 83.

Xi, L., Wang, S., Wang, C. *et al.* (2009) The pro-angiogenic factors stimulated by human papillomavirus type 16 E6 and E7 protein in C33A and human fibroblasts. *Oncol Rep* **21**: 25–31.

Xiao, X., Feng, Y., Vu, B.K. *et al.* (2009) A large library based on a novel (CH2) scaffold: identification of HIV-1 inhibitors. *Biochem Biophys Res Commun* **387**: 387–392.

Xu, L., Aha, P., Gu, K. *et al.* (2002) Directed evolution of high-affinity antibody mimics using mRNA display. *Chem Biol* **9**: 933–942.

Yang, H., Gurgel, P.V. and Carbonell, R.G. (2005) Hexamer peptide affinity resins that bind the Fc region of human immunoglobulin G. *J Pept Res* **66** Suppl 1: 120–137.

Yang, L., Nuraje, N., Bai, H. and Matsui, H. (2008) Crossbar assembly of antibody-functionalized peptide nanotubes via biomimetic molecular recognition. *J Pept Sci* **14**: 203–209.

Yeretssian, G., Lecocq, M., Lebon, G. *et al.* (2005) Competition on nitrocellulose-immobilized antibody arrays: from bacterial protein binding assay to protein profiling in breast cancer cells. *Mol Cell Proteomics* **4**: 605–617.

Yin, M., Zhang, L., Sun, X.M. *et al.* (2010) Lack of apoE causes alteration of cytokines expression in young mice liver. *Mol Biol Rep* **37**: 2049–2054.

Yuan, Y., He, H. and Lee, L.J. (2009) Protein A-based antibody immobilization onto polymeric microdevices for enhanced sensitivity of enzyme-linked immunosorbent assay. *Biotechnol Bioeng* **102**: 891–901.

Zander, K.A., Saavedra, M.T., West, J. *et al.* (2009) Protein microarray analysis of nasal polyps from aspirin-sensitive and aspirin-tolerant patients with chronic rhinosinusitis. *Am J Rhinol Allergy* **23**: 268–272.

Zeng, Z., Hincapie, M., Haab, B.B. *et al.* (2010) The development of an integrated platform to identify breast cancer glycoproteome changes in human serum. *J Chromatogr A* **1217**: 3307–3315. Epub Sep 2009.

Zheng, G., Patolsky, F., Cui, Y. *et al.* (2005) Multiplexed electrical detection of cancer markers with nanowire sensor arrays. *Nat Biotechnol* **23**: 1294–1301.

Zhong, D., Xiong, L., Liu, T. *et al.* (2009) The glycolytic inhibitor 2-deoxyglucose activates multiple prosurvival pathways through IGF1R. *J Biol Chem* **284**: 23225–23233.

Zhou, H., Bouwman, K., Schotanus, M. *et al.* (2004) Two-color, rolling-circle amplification on antibody microarrays for sensitive, multiplexed serum-protein measurements. *Genome Biol* **5**: R28.

Zhu, X., Landry, J.P., Sun, Y.S. *et al.* (2007) Oblique-incidence reflectivity difference microscope for label-free high-throughput detection of biochemical reactions in a microarray format. *Appl Opt* **46**: 1890–1895.

11

Biomarker Detection and Molecular Profiling by Multiplex Microbead Suspension Array Based Immunoproteomics

V. V. Krishhan, Imran H. Khan and Paul A. Luciw

Introduction

Recent developments in the microbead suspension array based multiplex immunoassays have the potential to evolve into efficient, relatively inexpensive and flexible tools for both researchers and clinicians (Fulton et al. 1997; Gordon and McDade 1997; Vignali 2000; Dunbar 2006; Nolan, Iannone and Lizard 2006; Hsu, Joos and Koga 2009; Krishhan, Khan and Luciw 2009). The full potential of this multiplex approach has only been realized recently. Small, cell-sized polystyrene/latex microbeads (~3–5 μm in diameter) have proved to be the ideal platform for this purpose. The first immunoassays performed by flow cytometry used microbeads with chemically modified surfaces (Horan and Kappler 1977). As the number of multiplexed analytes increased, hybrid flow cytometers were designed to include a special signal processing board (e.g. FACScan and FACS Calibur systems from Becton Dickinson Biosciences). Multiplexed microbead immunoassays, have been designed for up to 100-plexed detection (the 'Luminex 100' instrument, Austin, TX). More recently, Luminex has designed an instrument that will enable

Molecular Analysis and Genome Discovery, Second Edition. Edited by Ralph Rapley and Stuart Harbron.
© 2012 John Wiley & Sons, Ltd. Published 2012 by John Wiley & Sons, Ltd.

500-plex testing (DiIulio 2007). These specialized instruments for microbead-based multiplex detection are capable of digital signal processing and operate with two lasers: an inexpensive red diode to identify the microbeads and a green yttrium aluminum garnet (YAG) laser to measure the reporter fluorochrome. Current trends for faster and cheaper computers, improved digital processing and smaller and cheaper lasers will contribute to the eventual arrival of miniaturized field equipment that operates on battery power and allows five-plexed to ten-plexed assays at low cost (Mandy *et al.* 2008). Three companies offer microbead-based technology: Luminex (Austin TX, www.luminexcorp.com), Becton Dickinson Biosciences (San Jose, CA, www.bdfacs.com or www.pharmingen.com) and DiaSorin (Saluggia, Italy, www.diasorincopalis.com). Several vendors market the Luminex system, including Bio-Rad (Richmond, CA) and Qiagen (Germantown, MD). In this chapter, we will focus on the Luminex technology because this is the system with which we, and many other investigators, have had extensive experience and which we consider to be most robust for both basic research and clinical applications.

Principles of microbead-based multiplexing

The operation of multiplex microbead immunoassay can be divided into two basic parts: (1) bead-based immunoassay; and (2) a flow cytometer-based approach for multiplexing. Additionally, specialized software is used for instrument operation and analysis of data collected on numerous analytes in a single sample. Polystyrene latex microbeads have chemically modified surfaces for enhanced coupling to proteins (Langer and Tirrell 2004). Surfaces of the fluorescent microbeads constitute the platform for specific binding reactions and the presence of bead-bound analyte is detected with a conjugate coupled to a reporter fluorochrome. Sets of microbeads are embedded with precise ratios of red and infrared fluorescent dyes for classification of specific bead sets. Each bead can be thought of as an ELISA well, and therefore a sample containing 'N' number of beads is considered to be 'N' number of independent ELISA measurements. In one format, each bead set is coupled to a distinct antigen that binds the antibody to be measured. In another format, akin to a capture or sandwich ELISA, the capture antibody of each immunosorbent assay is coupled to one of a multiple of different microbead sets, each set containing a unique ratio of the red and orange (or infrared) fluorochrome labels (Figure 11.1). Once coupled with either a capture antibody or protein target, microbeads from different sets can be pooled together for the assay and individually identified during data acquisition in the flow cytometer by measuring the intensities of the two classifier fluorochrome dyes.

For immobilization, two types of beads can be used: chemically activated carboxylated beads or Penta-His beads, which have antibodies against His-tags on their surface. Recently, Verkaik *et al.* (2008) have compared carboxylated and Penta-His beads and found that for carboxylated beads, the non-specific background is lower,

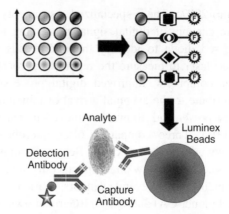

Figure 11.1 Basic principles of microbead sandwich immunoassay. Microbeads are covalently coupled with capture antibody. Proteins captured on an antibody-conjugated microbeads are detected by tagged detection antibodies. Inward and outward arrows represent the excitation and emission beams, respectively. Each bead class can be coated with a different capture antibody specific for a single antigen. After incubation with samples, the beads form an antibody–antigen complex. This step is followed by the addition of biotinylated detector antibodies. Incubation with streptavidin-linked fluorescent reporter, R-phycoerythrin (SA-PE), completes the 'antigen sandwich'. In the flow cytometer, beads are interrogated one at a time. This example consists of four different microbead sets conjugated to a distinct antigen as shown in the figure. In addition, this multiplex assay includes one bead set coated with biotin as a positive control for instrument operation (not shown). All five bead sets are mixed and assayed simultaneously in a single tube and subsequently analysed in the flow cytometer

the specific signal intensity is higher and not dependent on the configuration of the protein and beads are useful over longer periods.

In the flow cytometer, bead-bound analyte is detected with a conjugated antibody (e.g. biotin conjugate), which allows subsequent binding of a reporter fluorochrome (e.g. streptavidin-conjugated phycoerythrin – SA-PE). This immunoassay produces a variable amount of the green fluorescent reporter dye, which is proportional to the amount of analyte bound to the surface of each microbead. Data are acquired on a conventional flow cytometer connected to a data acquisition system. The instrument quantitates the green, orange (or infrared) and red fluorescence of each microbead using three independent detectors. A red laser excites the dye molecules inside the bead and classifies the bead to its unique bead set and a green laser quantifies the assay at the bead surface. Beads are interrogated one at a time in the flow cytometer. Lumimex's bead array has a similar operating procedure to that of a conventional flow cytometer, except for the choice of laser frequencies for excitation and detection. In a generic flow cytometer, the reporter fluorescence is generally green (such as FITC or Alexa 488), while the Luminex system uses the SA-PE complex, an orange (or infrared) florescence reporter that is measured by the YAG laser (excitation at 532 nm).

Software manipulates the near red florescence data (e.g. infrared and red) to separate the pool of microbeads into individual bead sets and presents the average green fluorescence intensity for each bead set. The concentration of analytes in the sample can be determined by extrapolation from an internal standard. Because the individual bead sets are separated by fluidics in the flow cytometer, many assays can be performed simultaneously to detect multiple analytes in a single sample. This approach has been extended to generate a 100-plex or 500-plex microbead array by Luminex.

Experimental aspects of the multiplex microbead assay

Our laboratory uses the instrument originally developed by Luminex and also the similar Bio-Plex (Bio-Rad) system, to develop assays that quantitate multiple protein analytes simultaneously (Khan *et al.* 2005, 2006a, 2006b, 2009, 2010; Feng *et al.* 2004). This experience is directly applicable to many future particle-based flow cytometric assay systems that might be developed by other vendors and has a unique combination of features: high throughput capacity, analyte quantification over a wide range of concentrations, small sample volume, high reproducibility, high sensitivity especially with efficient fluorochromes such as phycoerythrin (PE), a wide dynamic range and, most importantly, the capacity for multiplexing. Here we highlight a recent study of molecular profiling cytokines and chemokines in rheumatoid arthritis to demonstrate the robustness of multiplex microbead immunoassay and to the value of computational tools to establish protein profiles with potential for clinical diagnostics.

Problem definition: Rheumatoid arthritis (RA) is a systemic, potentially debilitating, chronic autoimmune disease characterized by inflammation and destruction of the joints. Rituximab, a chimeric anti-CD20 monoclonal antibody that is FDA approved for the treatment of non-Hodgkins lymphoma and RA, reduces antibody production by depleting B-cells. However, its effect on cytokines and chemokines is not understood. The purpose of this study was to evaluate changes that may occur over time in immunomodulator protein levels in RA patients undergoing rituximab therapy and develop a biomarker profile. Blood cytokine and chemokine levels are likely to be altered because of RA. Therefore, plasma levels of cytokines and chemokines may serve as useful biomarkers for patient stratification and for monitoring efficacy of therapy. Furthermore, it is practical to use multiplex technology with the capability to measure a large number of immunomodulators simultaneously for this and other clinical applications.

Experimental details: Patient and treatment details are given in the original references (Tuscano *et al.* 2005; Khan *et al.* 2009; Tuscano and Sands 2009). Briefly, the subjects were adult, male or female patients who had active, seropositive RA with at least two swollen and tender joints and functional class I, II or III, as defined by the American College of Rheumatology (ACR) revised criteria. A total of 17 patients were enrolled in the original study (Tuscano *et al.* 2005). Of the 17 patients in the original study, plasma samples from six patients were available at multiple time points (baseline and 3, 6 and 9 months). Patients continued with their baseline disease-modifying antirheumatic drugs (DMARDs) and other medications at a stable dose throughout the study period. Rituximab was administered as an intravenous infusion weekly for four consecutive weeks according to a dose escalation schedule (Tuscano *et al.* 2005). Blood samples were obtained at baseline and at follow-up visits for assessment of conventional clinical laboratory parameters including erythrocyte sedimentation rate (ESR), rheumatoid factor (RF), peripheral B and T cell counts, complement C3 and C4 levels and immunoglobulin isotype (IgA, IgG and IgM) and human anti-chimeric antibody (HACA) titres. Samples were processed and analysed using standard techniques (Tuscano *et al.* 2005). Control plasma samples were obtained randomly from 14 healthy individuals without regard to sex but they were of advanced age (median age of 58 years). The median age of patients was 66 years.

Multiplex kits. Multiplex panels for measuring cytokines and chemokines on the Luminex platform were the Bio-Plex kits from Bio-Rad (Hercules, CA). A total of 27 analytes are profiled; Eotaxin, FGF-basic, G-CSF, GM-CSF, IFNγ, IL-10, IL-12p70, IL-13, IL-15, IL-17, IL-1b, IL-1ra, IL-2, IL-4, IL-5, IL-6, IL-7, IL-8, IL-9, IP-10, MCP-1, MIP-1α, MIP-1B, PDGF-bb, RANTES, TNFα and VEGF. Detailed comparison of the performance of multiplex kits are available from the original work by us (Khan *et al.* 2009) in addition to detailed work by Djoba Siawaya *et al.* (2008). The kit and the plasma samples were diluted using the appropriate sample diluents provided in each kit in accordance with the manufacturer's instructions.

Data analysis: MFI (median florescent intensity), obtained as the raw data from the instrument, is converted to concentrations (pg/ml) of different analytes in each plasma sample using the standard curves generated in the multiplex assays following the standard procedures. The analyte concentrations were transformed to $\log^2$ scale for further analysis to accommodate the dynamic range in the concentration values. Investigation of reproducible differences between treatments was performed with the Bioconductor and R software package. Differential measurements in patient samples across the four time points, detected by an F test, and a separate F test was performed on each analyte and each panel. *p*-values for different analytes were transformed to compensate for multiple comparisons using the false discovery rate (FDR) adjustment for multiple comparisons using the Benjamini-Hochberg procedure. Cluster analysis was based on Euclidean distance on the fold changes without standardization to identify the natural grouping of analyte expression profiles (Dudoit, Popper-Shaffer and Boldrick 2002). All of data processing was performed using R statistical environment (Gentleman 1996; Gentleman *et al.* 2004). Plots were generated with a combination of Matlab (Mathworks Inc. 2005) and SigmaPlot (Systat Software 2005).

Biomarker profile: Computer modelling and hierarchical cluster analysis of the multiplex data allowed a comparison of the performance of multiplex assay kits and revealed profiles of immunomodulators in the RA patients. Multivariate analysis

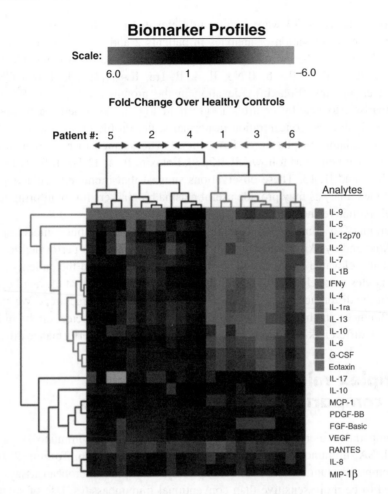

Figure 11.2 Cluster analysis of analytes measured by Bioplex. Fold changes of the analytes measured using Bioplex kit ($\log^2(\text{FC}) > 2$ and p-value < 0.05) are classified by hierarchical cluster analysis (Eucledian distance metric). The clustering approach groups the patients into two major groups denoted by bidirectional arrows: red (elevated analytes) and blue (healthy control levels) along the top horizontal direction); and the respective analytes: red bars along the right vertical axis denote significantly elevated analytes. The heat-map scale at the top indicates analyte levels in patient plasma as compared to healthy controls ($n = 14$) (Khan *et al*. 2009). *Reproduced with permission from Cytmetry B Clinical Cytometry*

was used to determine the number of analytes shared between the various kits at the statistical selection criteria of $\log^2$ (FC) > 2 and p-value < 0.05. To identify the patients as well as the immunomodulators that contributed maximally to the observed elevation in the levels of different analytes, a hierarchical cluster analysis was performed. Figure 11.2 cluster analysis of RA patient data was obtained by the Bio-Plex assay kit.

Out of 23 analytes, 14 were significantly elevated (p-value < 0.05) in patients 1, 3 and 6, whereas elevation of analytes in patients 2 and 4 was not significant. Out of the 14 analytes that were significantly elevated, 13 are cytokines (IL-9, IL-5, IL-12p70, IL-2, IL-7, IL-1b, IFNg, IL-4, IL-1ra, IL-13, IL-10, IL-6, GCSF) and one is a chemokine (Eotaxin). Elevation in the protein levels in the RA patients in comparison to healthy controls is given in $\log^2$ scale, where each unit corresponds to double the concentration in normal scale. Profile of significantly elevated immunomodulators in plasma of RA patients who seemed to have benefited from rituximab treatment is as follows: IL-12p70, Eotaxin, IL-4, TNFα, Il-9, IL-1β, IFNγ, IL-10, IL-6 and IL-13. These observations suggest that immunomodulator profiling by multiplex analysis may provide useful plasma biomarkers for monitoring response to B-cell reductive therapy in RA patients.

In summary, a combination of high quality data and computer modeling of the chemokine and cytokine concentration data from patients was performed to identify significantly elevated (p-value < 0.05) immunomodulators and the patients in which these analytes were elevated. These differences in the detection of elevated cytokines highlight the differences in multiplex kits from different manufacturers. We speculate that differences in the antibody pairs used in capture and detection of individual analytes in different kits are likely to be a key factor in assay performance differences.

Multiplex microbead assay design and comparison with other methods

The comparative merits of flow cytometry vs. conventional immunoassay formats (e.g. ELISA, immunoblot) and planer array (Kingsmore 2006; Tozzoli 2007) have been compared (Jani *et al*. 2002) (Table 11.1). Multiplex microbead assays can be designed to be more sensitive than conventional immunoassays. Use of small particles (e.g. 5 micrometer polystyrene beads) as the solid phase of the assay produces reaction kinetics approaching liquid-phase conditions. Such kinetics enable the design of ultra-sensitive assays, which, in theory, can measure as little as 100 fg/mL; conventional ELISAs measure about 50 pg/mL (Vignali 2000). Sensitivity is further increased by the use of the chromophore phycoerythrin (PE), an exceptionally bright reporter dye. Optimal elicitation of this label is achieved, as in the Luminex system, with a green YAG laser (Mandy *et al*. 2001; Nolan and Mandy 2001). Conventional and hybrid cytometers use a blue laser that elicits PE signal less effectively. This technical improvement in the dedicated cytometers remains, however, to be proven in clinical settings. The high sensitivity of these multiplex microbead assays may also be of important for dealing with specificity issues; samples can be greatly diluted for a reduction of matrix effects to improve signal-to-noise. Nevertheless, it remains to be determined if PE can eventually be substituted by more robust fluorochromes that are suitable for adverse assay conditions (e.g. quantum dots).

Table 11.1 Comparative merits of flow cytometry vs. conventional immunoassay formats and planer array

Assay	ELISA[a]	Rapid EIA[b]	Planar arrays[d]	Multiplex microbead array
Equipment	Spectrophotometer	...	Laser fluorescence	Flow cytometer
Trained personal	Yes	No	Yes	Yes
Throughput	High	Low	High	High
Sensitivity	$\sim 10^{-11}$ g/ml	ND[c]	$\sim 10^{-12}$ g/ml	$\sim 10^{-13}$ g/ml
Specificity	Depend on antibody	Depend on antibody	Depend on antibody	Depend on antibody
Result	Quantitative (some titration required)	Qualitative	Qualitative	Quantitative
Dynamic range	Narrow	Narrow	Wide	Wide
Multiplexing	Very limited	Very limited	Large	Up to 100-plex and expandable
Time	3–4 h	<30 min	>2 h	2–3 h

[a]ELISA: Enzyme-linked immunosorbent assay
[b]EIA: Enzyme immunoassay
[c]ND: Not determined
[d]See (Kingsmore 2006; Tozzoli 2007) for more information

Simultaneous detection of multiple analytes decreases both the costs related to hands-on time and consumables. As early as 1999, when the Luminex format was first made available, quantitative analysis of six or more cytokines in the flow cytometric assay was much cheaper than conventional ELISA (Carson and Vignali 1999). For high-throughput studies, liquid-array assays based on microbeads can be automated to deliver results rapidly. With high affinity antibodies, such assays can be completed in 3 hours or less to produce reliable and quantitative results.

Importantly, quantification of analytes is reliable in the multiplex microbead immunoassay format. The amount of reporter dye on the beads, measured as the MFI, is proportional to the amount of captured analyte. Calibration curves are constructed in a fashion similar to traditional immunoassays. Nevertheless, due to the wide dynamic range of multiplex microbead assays, fewer dilutions are required; for many analytes it is possible to obtain quantitative estimates for a test sample analysed in a single dilution (Scillian et al. 1989). The ELISA format normally requires frequent sample dilutions to avoid misleading results. Multiplex assays are highly reproducible, with low coefficients of variation. Within a shared sample environment, each microbead behaves as an independent immunoassay. Because usually 100 microbeads of each type are collected for the analysis of a single sample, the final reading provides the median of 100 assays (Mandy et al. 2001; Nolan and Mandy 2001). Finally, multiplex assays can be readily customized to perform optimally with different tissue fluids in various clinical or epidemiological settings. All of these characteristics make this new multiplex microbead technology attractive for use in a wide range of settings, particularly for the surveillance of infectious diseases in developing countries.

Recently, several review articles in the literature have compared the performance of Luminex based assays with other methods; Richens *et al.* (2010) compared Cytokine Bead Array (CBA) and Luminex kits, and compared concentration measurements to those obtained using ELISA. Results from Luminex kits were found to be highly reproducible and reliable in general. Fu, Zhu and Van Eyk (2010) compared the assay platforms of MULTI-ARRAY (Meso Scale Discovery), Bio-Plex (Bio-Rad Laboratories), A2 (Beckman Coulter), FAST Quant (Whatman Schleicher & Schuell BioScience), and FlowCytomix (Bender MedSystems) and they concluded that MULTI-ARRAY and Bio-Plex multiplex immunoassay systems were the most suitable for biomarker analysis or quantification.

Applications of the multiplex microbead assay system for biomedical research and clinical studies

Efficient clinical management and epidemiological surveillance of infectious diseases depend upon accurate and rapid diagnosis (Hinman 1992; Shears 2000). In most diagnostic laboratories, this process relies on serological techniques aimed at the detection of pathogen-specific antibodies. These antibodies are frequently measured by ELISA, which is popular because of its high sensitivity and specificity, good reproducibility and a relatively high throughput capacity (Nkengasong *et al.* 1999). However, ELISAs require trained personnel and some specific equipment and are rarely suitable for field conditions in resource-poor settings (Spielberg *et al.* 1989). Also, ELISA is cost prohibitive for the analysis of multiple analytes and requires relatively large sample volumes. With the advent of monoclonal antibodies, microbead-based flow cytometric assays were touted as a viable alternative to the popular ELISA. Despite a number of successful applications, this approach has not yet impacted the dominance of microtitre plate ELISA. The first bead immunoassay was developed in 1982 to quantify human IgG (Lisi *et al.* 1982). In many ways, this study was ahead of its time, because the assay was also performed without any wash steps. With this first generation microbead diagnostic system, a number of assays have been successfully established to measure antibodies against *Helicobacter pylori* (Best *et al.* 1992), hepatitis C virus (McHugh *et al.* 1997), α-gliadin in coeliac disease (Presani, Perticarari and Mangiarotti 1989), immunoglobulin and immune complexes (McHugh *et al.* 1986; Syrjala *et al.* 1991; Labus and Petersen 1992) and phospholipids in autoimmune diseases (Stewart *et al.* 1993; Laakel, Bouchard and Lagace 1996; Drouvalakis, Neeson and Buchanan 1999). In some instances, these assays were more sensitive than the conventional ELISA and could resolve indeterminate clinical samples (McHugh *et al.* 1997). Particle-based flow cytometric assays have also been used to measure other soluble analytes, such as von Willebrand factor (vWf) multimers (Kempfer *et al.* 1999). Likewise, such assays have also been established to detect RNA and DNA genomes for a number of viruses (Mehrpouyan *et al.* 1997; P.L. Smith *et al.* 1998; Van Cleve *et al.* 1998).

More recently, microbeads labelled with precise quantities of fluorescent dye, or conjugated with anti-mouse IgG antibody, have been used to accurately quantify cell surface expression of a variety of molecules (Brockhoff, Hofstaedter and Knuechel 1994; Zagursky *et al.* 1995; Davis *et al.* 1998; Gratama *et al.* 1998; Iyer *et al.* 1998; Pannu, Joe and Iyer 2001). In one of the first studies, this approach was used to quantify the expression of EGF receptor on bladder carcinoma cells (Brockhoff, Hofstaedter and Knuechel 1994); the data obtained by the microbead method compared favourably with that derived from conventional Scatchard analysis. It has also been suggested that this approach can facilitate inter-laboratory standardization, which is essential for multicentre clinical studies (Lenkei *et al.* 1998). A number of companies currently offer high-quality fluorescent calibration beads for such standardization within and between laboratories: QuantiBRITE, Becton Dickinson, San Jose, CA (www.bdfacs.com); QSC, Flow Cytometry Standards.

In 2002, during the initial developmental period of the multiplex microbead technology, the journal *Cytometry* published a report that summarized a workshop, sponsored by the Centers for Disease Control and Prevention, to examine issues surrounding these assays and the Luminex instrument. This workshop included topics on instrumentation, assay design, sample matrix and volume, quality control and development of commercial applications (Earley *et al.* 2002). More recently, the same journal took a lead in summarizing the recent advancement of multiplex microbead systems by compiling a wide range of articles (Nolan, Iannone and Lizard 2006). In this emerging area of diagnostics, a list of detection techniques may become obsolete in a short time. Many of the major vendors of antibodies have products specifically developed for a unique application of the multiplex microbead technology. A recent article by Schwenk *et al.* (2006), has reached a notable milestone; the authors present a protocol for an increased coupling throughput for immobilization of proteins to beads that enables the specificity of 96 antibodies vs. 100 different antigens to be determined in 2 hours. Advances *via* high-throughput analysis will potentially make a substantial impact on the use of microbead detection methods for proteomics research.

Multiplex microbead-based assays have been extended to include detection of nucleic acids (Dunbar 2006). These assays are generally based on the conjugation of synthetic oligonucleotides to specific sets of fluorochrome-labelled microbeads. In addition, fragments of DNA and RNA can also be attached to beads. In a general method, a test sample of nucleic acid is amplified and labelled with biotin by PCR, and then the product is hybridized with beads conjugated with specific oligonucleotides. After hybridization, the streptavidin-PE hybridization probe is added for detection of sequences bound to a specific bead set. Thus, essentially the same system is used for scoring the multiplex assay, i.e. identification of the specific microbead and detection of the reporter, as for protein analysis. Large amounts of genomic sequence information are being generated for a wide variety of organisms in all phyla. Accordingly, the multiplex microbead format will provide a rapid, cost-effective, high-throughput detection system for many investigational issues that require detection and measurement of specific DNA and RNA sequences. Applications include analysis of single nucleotide polymorphism (SNP) genotyping and genetic disease screening (Strom et al. 2006), gene expression profiling (Flagella et al. 2006), HLA DNA typing (Itoh et al. 2005), microbial pathogen detection (Brunstein and Thomas 2006; Han et al. 2006; Tang et al. 2007; Hindson et al. 2008) and analysis of small interfering RNA (siRNA) (Zhang et al. 2005) and microRNA (Lu et al. 2005).

With the advantages of inherent sensitivity and easy experimental setup for the multiplex microbead format, another potential area of application is the development of handheld, portable assays. In particular, the multiplex immunoassay format could make a significant impact on infectious diseases by enabling simultaneous detection of antibodies to and/or antigens of multiplex infectious pathogens, such as human immunodeficiency virus (HIV), the hepatitis A, B and C viruses, *Mycobacterium tuberculosis,* as well as many other viral, bacterial and parasitic pathogens (Girosi et al. 2006; Urdea et al. 2006). For example, for HIV and AIDS the need for field-testing devices in developing countries is well illustrated by efforts to implement simple and rapid diagnostic kits, such as agglutination tests, flow-through cassettes, solid phase (dipstick) tests and immunochromatographic strips (http://www.rapid-diagnostics.org/rti-hiv-diag.htm) (Roberts, Grusky and Swanson 2007). With these tests, which are done by technicians with basic training and interpreted visually, it has been possible to extend the diagnosis of AIDS to less well-equipped laboratories outside major medical centres. Moreover, with results available on the same day, these assays now constitute an important component of volunteer-testing clinics (Kassler et al. 1998; Roberts, Grusky and Swanson 2007). Nevertheless, even these simple assays are not ideal; their weaknesses include a lower sensitivity compared with ELISA (Kuun, Brashaw and Heyns 1997; Meda et al. 1999), difficulties in interpreting weakly positive sera (Malone et al. 1993) and limitations to sample size and detection of one or a very small number of analytes. Although the need for such instrumentation is more critical now, a truly affordable, portable multiplex microbead assay is not here yet. Hopeful signs include the development of innovative and practical technical components that could make feasible point-of-care testing (Mandy et al. 2008). Among the features desirable for an ideal assay, some are compatible and others tend to be mutually exclusive. On one hand, the most obvious dilemma is that a sophisticated assay is unlikely to be cost effective and/or suited for field conditions. Nonetheless, the multiplex microbead immunoassay presents significant advantages for diagnostics through its inherent flexibility by selection of microbead sets targeted for specific analytes and its capability for detecting multiple analytes in a small sample in a quantitative fashion (Z. Lukacs et al. 2005). For example, these advantages could significantly improve detection of HIV infection in very challenging situations such as in infants (Creek *et al.* 2007) and in individuals co-infected with *M. tuberculosis* (Cohen 2007).

Selected investigational fields for multiplex analysis and examples of applications

To provide examples of the use of the multiplex microbead suspension array technology, we have selected several representative areas of biomedical research and clinical application: (i) immunology and transplantation, (ii) infectious diseases, (iii) cancer, (iv) neurological diseases and (v) paediatric medicine. Many other fields of medicine, requiring detection of multiple proteins or nucleic acids, have also seen the application of this technology. The references in this section are intended to highlight the range and diversity of investigational and clinical issues that can be robustly addressed by the multiplex technology. Table 11.2 lists the examples of detection of both protein analytes and nucleic acids.

Highlighting the recent impact of the multiplex microbead immunoassay capability, this detection system is being offered as a commercial service for the analysis of biomarkers associated with various diseases (Rules-Based Medicine, Austin TX). In addition, academic centres and research institutions are beginning to offer multiplex assays as core services for a wide variety of basic and clinical applications.

Challenges and current limitations

To fully realize the potential of flow cytometry to meet the challenges of high throughput and content, informatics systems need to be integrated with the collection and analysis of data. This challenge is augmented by multiplexing, where each cluster of events might represent 10 or more assay data points. In addition, the data must be accurately tracked to the contents of the sample, with each sample potentially containing information on multiple analytes. The data need to be converted to a format compatible with biostatistical and informatics analysis of the assay result, such as the activity of the analyte. These computational issues will continue to impact both the development and application of high-throughput flow cytometry systems and multiplex methodologies.

A major problem of antibody assays is the need to acquire a large collection of specific antibodies for a wide range of analytes. Although there are many monoclonal and polyclonal antibodies commercially available, it is not feasible to standardize and calibrate several hundreds of these reagents to produce a reliable assay in a multiplex format while fulfilling the required level of sensitivity and specificity. The same is true for protein microarray technology in which the production of a large set of proteins remains a challenging task. To solve this problem for antibody microarrays,

Table 11.2 Selected applications of using multiplex analysis

Application	References
Immunology and transplantation	
Cytokines and chemokines secreted by human lymphocytes stimulated in vitro	Christiansen *et al*. 2006; Laugel *et al*. 2007; Meyer *et al*. 2007
Population-based study of immunomodulators in serum of different age groups	Shurin *et al*. 2007; Shurin and Smolkin 2007
Simultaneous analysis of binding specificities to a large number of different antibodies	Jia *et al*. 2004; Schwenk *et al*. 2007
Cytokine production by respiratory and peripheral blood dendritic cells	McDonald *et al*. 2005
Transplantation: HLA and histocompatability antigens and antibodies.	Colombo *et al*. 2007; Panigrahi *et al*. 2007; Suarez-Alvarez *et al*. 2007
Transplantation: genotyping	Pietz *et al*. 2005; Dalva and Beksac 2007; Iihara *et al*. 2007.
Infectious diseases	
Detection of serum antibodies to viral and bacterial pathogens in humans.	Clavijo *et al*. 2006; Laher *et al*. 2006; Morrow *et al*. 2006
Animals	Khan *et al*. 2006b, 2008
Antibody analysis in blood spots	N.W. Lukacs, Hogaboam and Kunkel 2005
Serotyping of viral and bacterial pathogens.	Lal *et al*. 2004, 2005; Dias *et al*. 2005; Lin *et al*. 2007
Population-based study exploring viral infection and autoimmune disease	Barzilai, Ram and Shoenfeld 2007; Barzilai *et al*. 2007.
Immunomodulators and cell activation pathways induced by infectious agents.	Kurkjian *et al*. 2006; Lindemans *et al*. 2006a, 2006b; Schmeck *et al*. 2006; Lee *et al*. 2007
Detection of nucleic acids of infectious pathogens	Page and Kurtzman 2005; Brunstein and Thomas 2006; McNamara *et al*. 2006; Page *et al*. 2006; Oh *et al*. 2007; Tang, Yuan and Chai 2007
Cancer	
Analysis of cell signaling pathways in tumor cell lines	Khan *et al*. 2006a; Johnson *et al*. 2007
Immunomodulators and biomarkers for cancer diagnosis and prognosis	D.A. Deans *et al*. 2006; C. Deans *et al*. 2007
Immunomodulators and biomarkers in cancer therapy	Allen *et al*. 2007; Schwenk *et al*. 2007; Shurin and Smolkin 2007; Yurkovetsky *et al*. 2007
Immunomodulators for vaccine evaluation in cancer patients.	Ullenhag *et al*. 2006; Dehqanzada *et al*. 2007; Kyte *et al*. 2007; Zhong *et al*. 2007
Assessment of anti-tumor antibodies in cancer patients	Tanaka *et al*. 2006
Cancer biology studies in cell culture and animal models.	Keyes *et al*. 2003; Zartman *et al*. 2004; Allen *et al*. 2007; Lewczuk *et al*. 2007; Raskovalova *et al*. 2007; Wolf *et al*. 2007
MicroRNA analysis of cancer.	Goff *et al*. 2005; Lu *et al*. 2005

Table 11.2 (*continued*)

Application	References
Neurological diseases	
Alzheimer's disease	Hansson *et al*. 2006; Nielsen *et al*. 2007
Neurodevelopmental diseases	Nelson *et al*. 2006
Cognitive and other neurological conditions	Ichiyama *et al*. 2006; Nagafuchi, *et al*. 2006; Lewczuk *et al*. 2007; Li *et al*. 2007; Millward *et al*. 2007
Pediatric medicine	
Screening for mutations	Strom *et al*. 2006; Pyatt, Mihal and Prior 2007; Yeom *et al*. 2008
Analysis of protein analytes in dried blood spots	Bellisario, Colinas and Pass 2001; Skogstrand *et al*. 2005
Immunomodulators and pediatric disease	An *et al*. 2004; Hodge *et al*. 2004; Langouche *et al*. 2005; Skogstrand *et al*. 2005; Carey *et al*. 2007; Ekelund *et al*. 2008; Strunk *et al*. 2007

several attempts have been made to replace expensive antibody products (Hanes *et al*. 2000; Hallborn and Carlsson 2002); this includes the application of aptamers from nucleic acid libraries to serve as ligands for the specific detection of proteins (D. Smith *et al*. 2003; Porschewski *et al*. 2006).

Previous studies have shown that antibodies display widely varying performance on microarrays, with no activity, decreased specificity or a lowered affinity (Arenkov *et al*. 2000; Haab, Dunham and Brown 2001). Although antibody performance can be increased by optimizing the surface binding and applying indirect immobilization strategies (Kusnezow and Hoheisel 2003; Kusnezow *et al*. 2003; Peluso *et al*. 2003), it would be advantageous for the suppliers of antibodies to include information regarding the suitability of their product for immobilization on microarrays. This could be done in a similar manner as currently practiced for application in immunoblotting, ELISA, or immunohistochemistry.

Typically, bead-based assays use the mean or median value of a population of measurements to evaluate ligand binding or other activity that produces a change in fluorescence intensity. The final representative signal, measured at the end of an experiment, is a population parameter, i.e. a statistically integrated distribution of individual bead measurements. The sensitivity of these measurements, therefore, is estimated by parametric or nonparametric statistical methods, as well as regression analysis over a limited range of titration (concentration vs. response profile). Although this approach does not pose any difficulty for analytes that are present at high concentration in the sample, quantification may be problematic for proteins of low abundance. The origins of such variations have been recently discussed by Jacobson *et al*. (2006). Performing a heterogeneous assay with a wash step produces an assay with very little free fluorescence. Thus, with the inclusion of a background subtraction algorithm, photon-counting error (Poisson influence) causes this low free fluorescence measurement to carry with it a high variance. At low analyte concentrations that are near the limit of detection of an assay, microbead-associated fluorescence is low and a high variance results from photon counting error. Therefore, at low analyte concentrations, when the free fluorescence measurement is subtracted from the microbead-associated fluorescence measurements, an abnormally high variance results. Accordingly, this abnormally high variance necessitates the collection of a large number of individual bead values from each population to ensure accuracy.

Detailed statistical analysis (Jacobson *et al.* 2006) demonstrates that the 95% confidence intervals of mean values, differences between two means and regression analyses are highly dependent upon the number of measurements performed. Using the standard Luminex 100 conditions (user manual) and considering a large number of measurements, groups or samples can be distinguished that differ by only a few tenths of an MFI (mean fluorescent intensity). Under standard conditions, one MFI unit on the Luminex 100 instrument is equivalent to the binding of $\sim$17 PE molecules to a bead (Jacobson *et al.* 2006). This analysis suggests that as few as three PE molecules (on average) need to bind to each bead to distinguish two populations when 1200 values of each population are collected. As anticipated, the sensitivity, in terms of confidence limits, improved markedly as the number of individual bead measurements was increased. This observation leads to the obvious conclusion that for precise determinations with bead-based assays, the number of measurements considered is directly related to the surety of the determined values.

This is not a limitation for Luminex, it is an advantage – labelling of proteins in a test sample is still a bottleneck of antibody microarray technology, in spite of the many techniques for the direct and indirect labelling of protein mixtures available. Reasons for this are not so much the absence of applicable coupling chemistries, but the diversity and differing quantities of available amino acids on protein surfaces as targets of these coupling chemistries. This diversity precludes direct coupling methods for homogeneous labelling of complex protein mixtures in defined stoichiometry. Therefore, absolute quantification of proteins in a complex is precluded and generally measurement of recombinant proteins that can be engineered to allow labelling by affinity tags or fusion with fluorescent reporter molecules are measured (Kukar *et al.* 2002). However, both of these approaches are expected to alter the structure of the protein and their binding properties as well. Therefore, the development of label-free detection methods that are amenable to high-throughput protein analysis is essential.

Protein microarrays have not developed at the same pace as that of nucleic acid microarrays due to the reasons described above. More importantly, equipment and procedures – such as the microscope slide format and its surface chemistry, fluorescent detection, the spotting devices and scanners – will need to be optimized to meet the different requirements of antibody and protein microarrays. These requirements include the development of a surface that avoids denaturing the proteins upon contact with the surface; linkers or activated layers may circumvent this problem. Another important feature is the provision of an environment that prevents dehydration of immobilized proteins and antibodies.

Summary and future directions

Completion of sequencing of the human genome and a wide range of organisms has accelerated the pace of gene discovery and subsequent functional analysis. It is clear that knowledge of genomes alone does not provide enough information to understand the complex cell networks that characterize and regulate cell functions. Genetics provides little or no information about protein localization, structure, modifications, interactions, activities and, ultimately, function. Although gene–protein dynamics have been analysed for many tissues (Kukar *et al.* 2002), there is still no reliable correlation between gene activity (i.e. transcript levels) and protein abundance or activity. Additionally, abundances of proteins and their entirety, i.e. the proteome, are highly dynamic and therefore require computational tools that are amenable for describing several variables simultaneously.

Although protein and antibody microarray technology is still in its early stages, rapid progress has been made in the past 5 years. Proof-of-principle experiments, as well as large-scale investigations, have applied microarrays with a diversity ranging from a few to several 10 000 s of different binders. Nevertheless, current progress in biochemistry and related fields supports the potential of protein and antibody microarray technology for basic research and diverse clinical applications. Thus, current progress in microarray assays systems, including the multiplex microbead array method, calls for a close watch on future developments.

Flow cytometry is rapidly evolving as a platform for automated high-throughput, high-content measurement of compound bioeffects in cell and bead-based assay systems. In a well-established laboratory setting with dedicated expertise, accurate analysis of endpoint assays can be typically performed at rates of 40–60 samples/min, on-line mixing assays at up to 10 samples/minute and mixing assays in conjunction with cell sorting (10 000 cells screened per assay) at 3 samples/minute (Edwards *et al.* 2004). Multiplexing of these assays offers the promise of increased throughput by a similar factor (or more). Because flow cytometers are widely available in nearly all research institutions, the instrument platforms reviewed here should be readily accessible to a broad spectrum of investigators.

Based on our experience and the recent published articles in this field, future research and technology development for multiplex flow cytometry are likely to focus on four main areas.

Technology development: It is likely that significant improvements will be made in bead chemistry, instrumentation and software. Various companies may develop their own versions of this technology, particularly companies that are already in the flow cytometry market. Acoustic radiation force has been used to improve the efficiency of bead-based immunoassays (Wiklund 2008). Various other advancements on different features of the multiplex microbead system could produce a significant increase in the range of analytes as well as the sensitivity of detection, particularly for proteins of low abundance.

Improved antibodies: Antibodies possess the remarkable qualities of recognizing native proteins and certain post-translational modifications with high specificity and sensitivity. These features are unlikely to be challenged by any other reagent. However, we believe that it would be a significant limitation to rely on currently available antibodies, for all detection systems. Multiplexed assays tend to be a direction for future developments; these assays pose unique challenges for the use of antibodies. As we and others have found, a number of antibodies cross-react with serum proteins or assay immunoglobulins (Andersson *et al.* 1989; Boscato, Egan and Stuart 1989; Balsari and Caruso 1997; Caruso *et al.* 1997). The aptamer approach offers an interesting alternative (Porschewski *et al.* 2006).

Rapid assays: Several investigators have developed rapid, 'no wash' multiplex assays that can be performed in 1–2 hours (Oliver, Kettman and Fulton 1998; Tripp *et al.* 2000). This approach is clearly advantageous and is likely to be the starting point for the development of novel multiplex microbead assays. However, in our experience, the sensitivity of the assays can drop 10- to 100-fold if washing is not included and if the sample incubation step is shortened.

Automation: Particularly for diagnostic laboratories and for proteomics research, automation is also likely to be an area for future development. All stages of the multiplex microbead assay system, from sample preparation to instrument operation, are highly compatible with robotics and automation. To this end, Bio-Rad has developed a multiplex instrument that addresses the needs for high-throughput analysis of clinical samples.

Particle (microbead)-based flow cytometric assays have great potential for both basic research and clinical applications. Despite the emergence a wide-range of assay technologies for protein analysis, methodologies that quantitate the native protein analyte will be an important part of medical diagnostics. Future improvements in flow cytometry assay systems will inevitably expand to include quantitation of multiple soluble factors to make a more informed disease diagnosis. Accordingly, flow cytometry, coupled with multiplex detection, will enable investigators to address complex biomedical research questions. Furthermore, it is likely that over the next decade a significant number of new markers, proteins and RNAs that are specifically related to a physiological state will be linked to normal and abnormal biological processes. Such discoveries of unique candidate biomarkers will undoubtedly increase the number of markers that need to be analysed and quantified in both experimental studies and clinical settings. This includes analysis of post-translational modifications that regulate the functions of many proteins. Therefore, pressure will continue to grow for the development of sensitive, specific, rapid and high-throughput assays with multiplex capabilities that can fulfil our expectations and demands for basic research and clinical applications.

Acknowledgements

Multiplex biomarker research for P.A.L. and I.H.K. is supported in part by NIH grants (Pacific Southwest Center for Biodefense and Emerging Infectious Diseases U54-AI065359, R24-RR022907 and the Base Grant to the California National Primate Research Center RR00169), a USAID grant (PGA-7251-07-001) and by funds from the Department of Pathology and Laboratory Medicine at U.C. Davis. P.A.L. was principal investigator of a research contract between U.C. Davis and the Millipore Corporation (Billerica, MA). V.V.K. is supported by an NIH grant: Research Infrastructure for Minority Institutions P20MD002732.

References

Allen, C., Duffy, S., Teknos, T. *et al.* (2007) Nuclear factor-kappaB-related serum factors as longitudinal biomarkers of response and survival in advanced oropharyngeal carcinoma. *Clin Cancer Res* **13**: 3182–3190.

An, H., Nishimaki, S., Ohyama, M. *et al.* (2004) Interleukin-6, interleukin-8, and soluble tumor necrosis factor receptor-I in the cord blood as predictors of chronic lung disease in premature infants. *Am J Obstet Gynecol* **191**: 1649–1654.

Andersson, G., Ekre, H.P., Alm, G. and Perlmann, P. (1989) Monoclonal antibody two-site ELISA for human IFN-gamma. Adaptation for determinations in human serum or plasma. *J Immunol Methods* **125**: 89–96.

Arenkov, P., Kukhtin, A., Gemmell, A. *et al.* (2000) Protein microchips: use for immunoassay and enzymatic reactions. *Anal Biochem* **278**: 123–131.

Balsari, A. and Caruso, A. (1997) Natural antibodies to IL-2. *Biotherapy* **10**: 25–28.

Barzilai, O., Ram, M. and Shoenfeld, Y. (2007) Viral infection can induce the production of autoantibodies. *Curr Opin Rheumatol* **19**: 636–643.

Barzilai, O., Sherer, Y., Ram, M. *et al.* (2007) Epstein-Barr virus and cytomegalovirus in autoimmune diseases: are they truly notorious? A preliminary report. *Ann N Y Acad Sci* **1108**: 567–577.

Bellisario, R., Colinas, R.J. and Pass, K.A. (2001) Simultaneous measurement of antibodies to three HIV-1 antigens in newborn dried blood-spot specimens using a multiplexed microsphere-based immunoassay. *Early Hum Dev* **64**: 21–25.

Best, L.M., Veldhuyzen van Zanten, S.J., Bezanson, G.S. *et al.* (1992) Serological detection of Helicobacter pylori by a flow microsphere immunofluorescence assay. *J Clin Microbiol* **30**: 2311–2317.

Boscato, L.M., Egan, G.M. and Stuart, M.C. (1989) Specificity of two-site immunoassays. *J Immunol Methods* **117**: 221–229.

Brockhoff, G., Hofstaedter, F. and Knuechel, R. (1994) Flow cytometric detection and quantitation of the epidermal growth factor receptor in comparison to Scatchard analysis in human bladder carcinoma cell lines. *Cytometry* **17**: 75–83.

Brunstein, J. and Thomas, E. (2006) Direct screening of clinical specimens for multiple respiratory pathogens using the Genaco Respiratory Panels 1 and 2. *Diagn Mol Pathol* **15**: 169–173.

Carey, W.A., Talley, L.I., Sehring, S.A. *et al.* (2007) Outcomes of dialysis initiated during the neonatal period for treatment of end-stage renal disease: a North American Pediatric Renal Trials and Collaborative Studies special analysis. *Pediatrics* **119**: e468–473.

Carson, R.T. and Vignali, D.A. (1999) Simultaneous quantitation of 15 cytokines using a multiplexed flow cytometric assay. *J Immunol Methods* **227**: 41–52.

Caruso, A., Licenziati, S., Corulli, M. *et al.* (1997) Flow cytometric analysis of activation markers on stimulated T cells and their correlation with cell proliferation. *Cytometry* **27**: 71–76.

Christiansen, J., Farm, G., Eid-Forest, R. *et al.* (2006) Interferon-gamma secreted from peripheral blood mononuclear cells as a possible diagnostic marker for allergic contact dermatitis to gold. *Contact Dermatitis* **55**: 101–112.

Clavijo, A., Hole, K., Li, M. and Collignon, B. (2006) Simultaneous detection of antibodies to foot-and-mouth disease non-structural proteins 3ABC, 3D, 3A and 3B by a multiplexed Luminex assay to differentiate infected from vaccinated cattle. *Vaccine* **24**: 1693–1704.

Cohen, G.M. (2007) Access to diagnostics in support of HIV/AIDS and tuberculosis treatment in developing countries. *AIDS* **21** Suppl 4: S81–87.

Colombo, M.B., Haworth S.E., Poli, F. *et al.* (2007) Luminex technology for anti-HLA antibody screening: evaluation of performance and of impact on laboratory routine. *Cytometry B Clin Cytom* **72**: 465–471.

Creek, T. L., Sherman, G.G., Nkengasong, J. *et al.* (2007) Infant human immunodeficiency virus diagnosis in resource-limited settings: issues, technologies, and country experiences. *Am J Obstet Gynecol* **197**(3 Suppl): S64–71.

Dalva, K. and Beksac, M. (2007) HLA typing with sequence-specific oligonucleotide primed PCR (PCR-SSO)and use of the Luminex technology. *Methods Mol Med* **134**: 61–69.

Davis, K.A., Abrams, B., Iyer, S.B. *et al.* (1998) Determination of CD4 antigen density on cells: role of antibody valency, avidity, clones, and conjugation. *Cytometry* **33**: 197–205.

Deans, C., Rose-Zerilli, M., Wigmore, S. *et al.* (2007) Host cytokine genotype is related to adverse prognosis and systemic inflammation in gastro-oesophageal cancer. *Ann Surg Oncol* **14**: 329–339.

Deans, D.A., Wigmore, S.J., Gilmour, H. *et al.* (2006) Elevated tumour interleukin-1beta is associated with systemic inflammation: A marker of reduced survival in gastro-oesophageal cancer. *Br J Cancer* **95**: 1568–1575.

Dehqanzada, Z.A., Storrer, C.E., Hueman, M.T. *et al.* (2007) Assessing serum cytokine profiles in breast cancer patients receiving a HER2/neu vaccine using Luminex technology. *Oncol Rep* **17**: 687–694.

Dias, D., Van Doren, J., Schlottmann, S. *et al.* (2005) Optimization and validation of a multiplexed luminex assay to quantify antibodies to neutralizing epitopes on human papillomaviruses 6, 11, 16, and 18. *Clin Diagn Lab Immunol* **12**: 959–969.

DiIulio, R. (2007) Multiplexing and Microassays. Clinical Lab Products. Retrieved 14 March 2008, from http://www.clpmag.com/issues/articles/2007–2010_06.asp.

Djoba Siawaya, J.F., Roberts, T., Babb, C. *et al.* (2008) An evaluation of commercial fluorescent bead-based luminex cytokine assays. *PLoS One* **3**: e2535.

Drouvalakis, K.A., Neeson, P.J. and Buchanan, R.R. (1999) Detection of anti-phosphatidylethanolamine antibodies using flow cytometry. *Cytometry* **36**: 46–51.

Dudoit, S., Popper-Shaffer, J. and Boldrick, J.C. (2002) Multiple hypothesis testing in microarray experiments. *U.C. Berkeley Division of Biostatistics Working Paper Series*

Working Paper 110. www.bepress.com/ucbbiostat/paper110 (Published in *Stat Sci* **18**: 71–103).

Dunbar, S.A. (2006) Applications of Luminex xMAP technology for rapid, high-throughput multiplexed nucleic acid detection. *Clin Chim Acta* **363**: 71–82.

Earley, M.C., Vogt, R.F., Jr, Shapiro, H.M. *et al.* (2002) Report from a workshop on multianalyte microsphere assays. *Cytometry* **50**: 239–242.

Edwards, B.S., Oprea, T., Prossnitz, E.R. and Sklar, L.A. (2004) Flow cytometry for high-throughput, high-content screening. *Curr Opin Chem Biol* **8**: 392–398.

Ekelund, C.K., Vogel, I., Skogstrand, K. *et al.* (2008) Interleukin-18 and interleukin-12 in maternal serum and spontaneous preterm delivery. *J Reprod Immunol.* **77**: 179–185. Epub 2007 Sep 11.

Feng, S., Kendall, L.V., Hodzic, E. *et al.* (2004) Recombinant Helicobacter bilis protein P167 for mouse serodiagnosis in a multiplex microbead assay. *Clin Diagn Lab Immunol* **11**: 1094–1099.

Flagella, M., Bui, S., Zheng, Z. *et al.* (2006) A multiplex branched DNA assay for parallel quantitative gene expression profiling. *Anal Biochem* **352**: 50–60.

Fu, Q., Zhu, J. and Van Eyk, J.E. (2010) Comparison of multiplex immunoassay platforms. *Clin Chem* **56**: 314–318.

Fulton, R.J., McDade, R.L., Smith, P.L. *et al.* (1997) Advanced multiplexed analysis with the FlowMetrix system. *Clin Chem* **43**: 1749–1756.

Gentleman, R.C. (1996) A language for data analysis and graphics. *Journal of Computational Graphics* **5**: 299–314.

Gentleman, R.C., Carey, V.J., Bates, D.M. *et al.* (2004) Bioconductor: open software development for computational biology and bioinformatics. *Genome Biol* **5**: R80.

Girosi, F., Olmsted, S.S., Keeler, E. *et al.* (2006) Developing and interpreting models to improve diagnostics in developing countries. *Nature* **444** Suppl 1: 3–8.

Goff, L.A., Yang, M., Bowers, J. *et al.* (2005) Rational probe optimization and enhanced detection strategy for microRNAs using microarrays. *RNA Biol* **2**: 93–100.

Gordon, R.F. and McDade, R.L. (1997) Multiplexed quantification of human IgG, IgA, and IgM with the FlowMetrix system. *Clin Chem* **43**: 1799–1801.

Gratama, J.W., D'Hautcourt, J.L., Mandy, F. *et al.* (1998) Flow cytometric quantitation of immunofluorescence intensity: problems and perspectives. European Working Group on Clinical Cell Analysis. *Cytometry* **33**: 166–178.

Haab, B.B., Dunham, M.J. and Brown, P.O. (2001) Protein microarrays for highly parallel detection and quantitation of specific proteins and antibodies in complex solutions. *Genome Biol* **2**: RESEARCH0004.

Hallborn, J. and Carlsson, R. (2002) Automated screening procedure for high-throughput generation of antibody fragments. *Biotechniques* Suppl: 30–37.

Han, J., Swan, D.C., Smith, S.J. *et al.* (2006) Simultaneous amplification and identification of 25 human papillomavirus types with Templex technology. *J Clin Microbiol* **44**: 4157–4162.

Hanes, J., Schaffitzel, C., Knappik, A. and Pluckthun, A. (2000) Picomolar affinity antibodies from a fully synthetic naive library selected and evolved by ribosome display. *Nat Biotechnol* **18**: 1287–1292.

Hansson, O., Zetterberg, H., Buchhave, P. *et al.* (2006) Association between CSF biomarkers and incipient Alzheimer's disease in patients with mild cognitive impairment: a follow-up study. *Lancet Neurol* **5**: 228–234.

Hindson, B.J., Reid, S.M., Baker, B.R. *et al.* (2008) Diagnostic evaluation of multiplexed reverse transcription-PCR microsphere array assay for detection of foot-and-mouth and look-alike disease viruses. *J Clin Microbiol* **46**: 1081–1089.

Hinman, A.R. (1992) The laboratory's role in prevention and control of infectious diseases. *Clin Chem* **38** (8B Pt 2): 1532–1538.

Hodge, G., Hodge, S., Haslam, R. *et al.* (2004) Rapid simultaneous measurement of multiple cytokines using 100 microl sample volumes – association with neonatal sepsis. *Clin Exp Immunol* **137**: 402–407.

Horan, P.K. and Kappler, J.W. (1977) Automated fluorescent analysis for cytotoxicity assays. *J Immunol Methods* **18**: 309–316.

Hsu, H.Y., Joos, T.O. and Koga, H. (2009) Multiplex microsphere-based flow cytometric platforms for protein analysis and their application in clinical proteomics – from assays to results. *Electrophoresis* **30**: 4008–4019.

Ichiyama, T., Siba, P., Suarkia, D. *et al.* (2006) Analysis of serum and cerebrospinal fluid cytokine levels in subacute sclerosing panencephalitis in Papua New Guinea. *Cytokine* **33**: 17–20.

Iihara, H., Niwa, T., Shah, M.M. *et al.* (2007) Rapid multiplex immunofluorescent assay to detect antibodies against Burkholderia pseudomallei and taxonomically closely related nonfermenters. *Jpn J Infect Dis* **60**: 230–234.

Itoh, Y., Mizuki, N., Shimada, T. *et al.* (2005) High-throughput DNA typing of HLA-A, -B, -C, and -DRB1 loci by a PCR-SSOP-Luminex method in the Japanese population. *Immunogenetics* **57**: 717–729.

Iyer, S.B., Hultin, L.E., Zawadzki, J.A. *et al.* (1998) Quantitation of CD38 expression using QuantiBRITE beads. *Cytometry* **33**: 206–212.

Jacobson, J.W., Oliver, K.G., Weiss, C. and Kettman, J. (2006) Analysis of individual data from bead-based assays (bead arrays) *Cytometry A* **69**: 384–390.

Jani, I.V., Janossy, G., Brown, D.W. and Mandy, F. (2002) Multiplexed immunoassays by flow cytometry for diagnosis and surveillance of infectious diseases in resource-poor settings. *Lancet Infect Dis* **2**: 243–250.

Jia, X.C., Raya, R., Zhang, L. *et al.* (2004) A novel method of multiplexed competitive antibody binning for the characterization of monoclonal antibodies. *J Immunol Methods* **288**: 91–98.

Johnson, T.J., Wannemuehler, Y.M., Johnson, S.J. *et al.* (2007) Plasmid replicon typing of commensal and pathogenic Escherichia coli isolates. *Appl Environ Microbiol* **73**: 1976–1983.

Kassler, W.J., Alwano-Edyegu, M.G., Marum, E. *et al.* (1998) Rapid HIV testing with same-day results: a field trial in Uganda. *Int J STD AIDS* **9**: 134–138.

Kempfer, A.C., Silaf, M.R., Farias, C.E. *et al.* (1999) Binding of von Willebrand factor to collagen by flow cytometry. *Am J Clin Pathol* **111**: 418–423.

Keyes, K.A., Mann, L., Cox, K. *et al.* (2003) Circulating angiogenic growth factor levels in mice bearing human tumors using Luminex Multiplex technology. *Cancer Chemother Pharmacol* **51**: 321–327.

Khan, I.H., Kendall, L.V., Ziman, M. *et al.* (2005) Simultaneous serodetection of 10 highly prevalent mouse infectious pathogens in a single reaction by multiplex analysis. *Clin Diagn Lab Immunol* **12**: 513–519.

Khan, I.H., Krishnan, V.V., Ziman, M. *et al.* (2009) A comparison of multiplex suspension array large-panel kits for profiling cytokines and chemokines in rheumatoid arthritis patients. *Cytometry B Clin Cytom* **76**: 159–168.

Khan, I.H., Mendoza, S., Rhyne, P. *et al.* (2006a) Multiplex analysis of intracellular signaling pathways in lymphoid cells by microbead suspension arrays. *Mol Cell Proteomics* **5**: 758–768.

Khan, I.H., Mendoza, S., Yee, J. *et al.* (2006b) Simultaneous detection of antibodies to six nonhuman-primate viruses by multiplex microbead immunoassay. *Clin Vaccine Immunol* **13**: 45–52.

Khan, I.H., Ravindran, R., Yee, J. *et al.* (2008) Profiling antibodies to *Mycobacterium tuberculosis* by multiplex microbead suspension arrays for serodiagnosis of tuberculosis. *Clin Vaccine Immunol.* **15**: 433–438.

Khan, I.H., Zhao, J., Ghosh, P. *et al.* (2010) Microbead arrays for the analysis of ErbB receptor tyrosine kinase activation and dimerization in breast cancer cells. *Assay Drug Dev Technol* **8**: 27–36.

Kingsmore, S.F. (2006) Multiplexed protein measurement: technologies and applications of protein and antibody arrays. *Nat Rev Drug Discov* **5**: 310–320.

Krishhan, V.V., Khan, I.H. and Luciw, P.A. (2009) Multiplexed microbead immunoassays by flow cytometry for molecular profiling: basic concepts and proteomics applications. *Crit Rev Biotechnol* **29**: 29–43.

Kukar, T., Eckenrode, S., Gu, Y. *et al.* (2002) Protein microarrays to detect protein–protein interactions using red and green fluorescent proteins. *Anal Biochem* **306**: 50–54.

Kurkjian, K.M., Mahmutovic, A.J., Kellar, K.L. *et al.* (2006) Multiplex analysis of circulating cytokines in the sera of patients with different clinical forms of visceral leishmaniasis. *Cytometry A* **69**: 353–358.

Kusnezow, W. and Hoheisel, J.D. (2003) Solid supports for microarray immunoassays. *J Mol Recognit* **16**: 165–176.

Kusnezow, W., Jacob, A., Walijew, A. *et al.* (2003) Antibody microarrays: an evaluation of production parameters. *Proteomics* **3**: 254–264.

Kuun, E., Brashaw, M. and Heyns, A.D. (1997) Sensitivity and specificity of standard and rapid HIV-antibody tests evaluated by seroconversion and non-seroconversion low-titre panels. *Vox Sang* **72**: 11–15.

Kyte, J.A., Kvalheim, G., Lislerud, K. *et al.* (2007) T cell responses in melanoma patients after vaccination with tumor-mRNA transfected dendritic cells. *Cancer Immunol Immunother* **56**: 659–675.

Laakel, M., Bouchard, M. and Lagace, J. (1996) Measurement of mouse anti-phospholipid antibodies to solid-phase microspheres by both flow cytofluorometry and Alcian blue-pretreated microtitre plates in an ELISA. *J Immunol Methods* **190**: 267–273.

Labus, J.M. and Petersen, B.H. (1992) Quantitation of human anti-mouse antibody in serum by flow cytometry. *Cytometry* **13**: 275–281.

Laher, G., Balmer, P., Gray, S.J. *et al.* (2006) Development and evaluation of a rapid multianalyte particle-based flow cytometric assay for the quantification of meningococcal serogroup B-specific IgM antibodies in sera for nonculture case confirmation. *FEMS Immunol Med Microbiol* **48**: 34–43.

Lal, G., Balmer, P., Joseph, H. *et al.* (2004) Development and evaluation of a tetraplex flow cytometric assay for quantitation of serum antibodies to Neisseria meningitidis serogroups A, C, Y, and W-135. *Clin Diagn Lab Immunol* **11**: 272–279.

Lal, G., Balmer, P., Stanford, E. *et al.* (2005) Development and validation of a nonaplex assay for the simultaneous quantitation of antibodies to nine Streptococcus pneumoniae serotypes. *J Immunol Methods* **296**: 135–147.

Langer, R. and Tirrell, D.A. (2004) Designing materials for biology and medicine. *Nature* **428**: 487–492.

Langouche, L., Vanhorebeek, I., Vlasselaers, D. *et al.* (2005) Intensive insulin therapy protects the endothelium of critically ill patients. *J Clin Invest* **115**: 2277–2286.

Laugel, B., Price, D.A., Milicic, A. and Sewell, A.K. (2007) CD8 exerts differential effects on the deployment of cytotoxic T lymphocyte effector functions. *Eur J Immunol* **37**: 905–913.

Lee, W.M., Grindle, K., Pappas, T. *et al.* (2007) High-throughput, sensitive, and accurate multiplex PCR-microsphere flow cytometry system for large-scale comprehensive detection of respiratory viruses. *J Clin Microbiol* **45**: 2626–2634.

Lenkei, R., Gratama, J.W., Rothe, G. *et al.* (1998) Performance of calibration standards for antigen quantitation with flow cytometry. *Cytometry* **33**: 188–196.

Lewczuk, P., Kornhuber, J., Vanderstichele, H. *et al.* (2007) Multiplexed quantification of dementia biomarkers in the CSF of patients with early dementias and MCI: a multicenter study. *Neurobiol Aging* **29**: 812–818.

Li, G., Sokal, I., Quinn, J.F. *et al.* (2007) CSF tau/Abeta42 ratio for increased risk of mild cognitive impairment: a follow-up study. *Neurology* **69**: 631–639.

Lin, X., Flint, J.A., Azaro, M. *et al.* (2007) Microtransponder-based multiplex assay for genotyping cystic fibrosis. *Clin Chem* **53**: 1372–1376.

Lindemans, C.A., Coffer, P.J., Schellens, I.M. *et al.* (2006a) Respiratory syncytial virus inhibits granulocyte apoptosis through a phosphatidylinositol 3-kinase and NF-kappaB-dependent mechanism. *J Immunol* **176**: 5529–5537.

Lindemans, C.A., Kimpen, J.L., Luijk, B. *et al.* (2006b) Systemic eosinophil response induced by respiratory syncytial virus. *Clin Exp Immunol* **144**: 409–417.

Lisi, P.J., Huang, C.W., Hoffman, R.A. and Teipel, J.W. (1982) A fluorescence immunoassay for soluble antigens employing flow cytometric detection. *Clin Chim Acta* **120**: 171–179.

Lu, J., Getz, G., Miska, E.A. *et al.* (2005) MicroRNA expression profiles classify human cancers. *Nature* **435**: 834–838.

Lukacs, N.W., Hogaboam, C.M. and Kunkel, S.L. (2005) Chemokines and their receptors in chronic pulmonary disease. *Curr Drug Targets Inflamm Allergy* **4**: 313–317.

Lukacs, Z., Dietrich, A., Ganschow, R. *et al.* (2005) Simultaneous determination of HIV antibodies, hepatitis C antibodies, and hepatitis B antigens in dried blood spots – a feasibility study using a multi-analyte immunoassay. *Clin Chem Lab Med* **43**: 141–145.

Malone, J.D., Smith, E.S., Sheffield, J. *et al.* (1993) Comparative evaluation of six rapid serological tests for HIV-1 antibody. *J Acquir Immune Defic Syndr* **6**: 115–119.

Mandy, F., Janossy, G., Bergeron, M. *et al.* (2008) Affordable CD4 T-cell enumeration for resource-limited regions: a status report for 2008. *Cytometry B Clin Cytom* **74** Suppl 1: 527–539.

Mandy, F.F., Nakamura, T., Bergeron, M. and Sekiguchi, K. (2001) Overview and application of suspension array technology. *Clin Lab Med* **21**: 713–729, vii.

Mathworks Inc. (2005) Matlab. Natick, MA 01760, USA, Mathworks Inc.

McDonald, C.G., Dailey, V.K., Bergstrom, H.C. *et al.* (2005) Periadolescent nicotine administration produces enduring changes in dendritic morphology of medium spiny neurons from nucleus accumbens. *Neurosci Lett* **385**: 163–167.

McHugh, T.M., Stites, D.P., Casavant, C.H. and Fulwyler, M.J. (1986) Flow cytometric detection and quantitation of immune complexes using human C1q-coated microspheres. *J Immunol Methods* **95**: 57–61.

McHugh, T.M., Viele, M.K., Chase, E.S. and Recktenwald, D.J. (1997) The sensitive detection and quantitation of antibody to HCV by using a microsphere-based immunoassay and flow cytometry. *Cytometry* **29**: 106–112.

McNamara, D.T., Kasehagen, L.J., Grimberg, B.T. *et al.* (2006) Diagnosing infection levels of four human malaria parasite species by a polymerase chain reaction/ligase detection reaction fluorescent microsphere-based assay. *Am J Trop Med Hyg* **74**: 413–421.

Meda, N., Gautier-Charpentier, L., Soudre, R.B. *et al.* (1999) Serological diagnosis of human immuno-deficiency virus in Burkina Faso: reliable, practical strategies using less expensive commercial test kits. *Bull World Health Org* **77**: 731–739.

Mehrpouyan, M., Bishop, J.E., Ostrerova, N. *et al.* (1997) A rapid and sensitive method for non-isotopic quantitation of HIV-1 RNA using thermophilic SDA and flow cytometry. *Mol Cell Probes* **11**: 337–347.

Meyer, E.H., Wurbel, M.A., Staton, T.L. *et al.* (2007) iNKT cells require CCR4 to localize to the airways and to induce airway hyperreactivity. *J Immunol* **179**: 4661–4671.

Millward, J.M., Caruso, M., Campbell, I.L. *et al.* (2007) IFN-{gamma}-induced chemokines synergize with pertussis toxin to promote t cell entry to the central nervous system. *J Immunol* **178**: 8175–8182.

Morrow, D.A., de Lemos, J.A., Sabatine, M.S. *et al.* (2006) Clinical relevance of C-reactive protein during follow-up of patients with acute coronary syndromes in the Aggrastat-to-Zocor Trial. *Circulation* **114**: 281–288.

Nagafuchi, M., Nagafuchi, Y., Sato, R. *et al.* (2006) Adult meningism and viral meningitis, 1997–2004: clinical data and cerebrospinal fluid cytokines. *Intern Med* **45**: 1209–1212.

Nelson, P.G., Kuddo, T., Song, E.Y. *et al.* (2006) Selected neurotrophins, neuropeptides, and cytokines: developmental trajectory and concentrations in neonatal blood of children with autism or Down syndrome. *Int J Dev Neurosci* **24**: 73–80.

Nielsen, H.M., Minthon, L., Londos, E. *et al.* (2007) Plasma and CSF serpins in Alzheimer disease and dementia with Lewy bodies. *Neurology* **69**: 1569–1579.

Nkengasong, J.N., Maurice, C., Koblavi, S. *et al.* (1999) Evaluation of HIV serial and parallel serologic testing algorithms in Abidjan, Cote d'Ivoire. *Aids* **13**: 109–117.

Nolan, J.P., Iannone, M. and Lizard, G.E. (2006) Special issue: multiplexed and microsphere-based analysis. *Cytometry A* **69**: 317–476.

Nolan, J.P. and Mandy, F.F. (2001) Suspension array technology: new tools for gene and protein analysis. *Cell Mol Biol (Noisy-le-grand)* **47**: 1241–1256.

Oh, Y., Bae, S.M., Kim, Y.W. *et al.* (2007) Polymerase chain reaction-based fluorescent Luminex assay to detect the presence of human papillomavirus types. *Cancer Sci* **98**: 549–554.

Oliver, K.G., Kettman, J.R. and Fulton, R.J. (1998) Multiplexed analysis of human cytokines by use of the FlowMetrix system. *Clin Chem* **44**: 2057–2060.

Page, B.T. and Kurtzman, C.P. (2005) Rapid identification of Candida species and other clinically important yeast species by flow cytometry. *J Clin Microbiol* **43**: 4507–4514.

Page, B.T., Shields, C.E., Merz, W.G. and Kurtzman, C.P. (2006) Rapid identification of ascomycetous yeasts from clinical specimens by a molecular method based on flow cytometry and comparison with identifications from phenotypic assays. *J Clin Microbiol* **44**: 3167–3171.

Panigrahi, A., Gupta, N., Siddiqui, J.A. *et al.* (2007) Post transplant development of MICA and anti-HLA antibodies is associated with acute rejection episodes and renal allograft loss. *Hum Immunol* **68**: 362–367.

Pannu, K.K., Joe, E.T. and Iyer, S.B. (2001) Performance evaluation of QuantiBRITE phycoerythrin beads. *Cytometry* **45**: 250–258.

Peluso, P., Wilson, D.S., Do, D. *et al.* (2003) Optimizing antibody immobilization strategies for the construction of protein microarrays. *Anal Biochem* **312**: 113–124.

Pietz, B.C., Warden, M.B., DuChateau, B.K. and Ellis, T.M. (2005) Multiplex genotyping of human minor histocompatibility antigens. *Hum Immunol* **66**: 1174–82.

Porschewski, P., Grattinger, M.A., Klenzke, K. *et al.* (2006) Using aptamers as capture reagents in bead-based assay systems for diagnostics and hit identification. *J Biomol Screen* **11**: 773–781.

Presani, G., Perticarari, S. and Mangiarotti, M.A. (1989) Flow cytometric detection of anti-gliadin antibodies. *J Immunol Methods* **119**: 197–202.

Pyatt, R.E., Mihal, D.C. and Prior, T.W. (2007) Assessment of liquid microbead arrays for the screening of newborns for spinal muscular atrophy. *Clin Chem* **53**: 1879–1885.

Raskovalova, T., Lokshin, A., Huang, X. *et al.* (2007) Inhibition of cytokine production and cytotoxic activity of human antimelanoma specific CD8+ and CD4+ T lymphocytes by adenosine-protein kinase A type I signaling. *Cancer Res* **67**: 5949–5956.

Richens, J.L., Urbanowicz, R.A., Metcalf, R. *et al.* (2010) Quantitative Validation and Comparison of Multiplex Cytokine Kits. *J Biomol Screen.* **5**: 562–568.

Roberts, K.J., Grusky, O. and Swanson, A.N. (2007) Outcomes of blood and oral fluid rapid HIV testing: a literature review, 2000–2006. *AIDS Patient Care STDS* **21**: 621–637.

Schmeck, B., Moog, K., Zahlten, J. *et al.* (2006) Streptococcus pneumoniae induced c-Jun-N-terminal kinase- and AP-1 -dependent IL-8 release by lung epithelial BEAS-2B cells. *Respir Res* **7**: 98.

Schwenk, J.M., Lindberg, J., Sundberg, M. *et al.* (2006) Determination of binding specificities in highly multiplexed bead-based assays for antibody proteomics. *Mol Cell Proteomics* **6**: 125–132.

Schwenk, J.M., Lindberg, J., Sundberg, M. *et al.* (2007) Determination of binding specificities in highly multiplexed bead-based assays for antibody proteomics. *Mol Cell Proteomics* **6**: 125–132.

Scillian, J.J., McHugh, T.M., Busch, M.P. *et al.* (1989) Early detection of antibodies against rDNA-produced HIV proteins with a flow cytometric assay. *Blood* **73**: 2041–2048.

Shears, P. (2000) Emerging and reemerging infections in Africa: the need for improved laboratory services and disease surveillance. *Microbes Infect* **2**: 489–495.

Shurin, G.V., Yurkovetsky, Z.R., Chatta, G.S. *et al.* (2007) Dynamic alteration of soluble serum biomarkers in healthy aging. *Cytokine* **39**: 123–129.

Shurin, M.R. and Smolkin, Y.S. (2007) Immune-mediated diseases: where do we stand? *Adv Exp Med Biol* **601**: 3–12.

Skogstrand, K., Thorsen, P., Norgaard-Pedersen, B. *et al.* (2005) Simultaneous measurement of 25 inflammatory markers and neurotrophins in neonatal dried blood spots by immunoassay with xMAP technology. *Clin Chem* **51**: 1854–1866.

Smith, D., Collins, B.D., Heil, J. and Koch, T.H. (2003) Sensitivity and specificity of photoaptamer probes. *Mol Cell Proteomics* **2**: 11–18.

Smith, P.L., WalkerPeach, C.R., Fulton, R.J. and DuBois, D.B. (1998) A rapid, sensitive, multiplexed assay for detection of viral nucleic acids using the FlowMetrix system. *Clin Chem* **44**: 2054–2056.

Spielberg, F., Kabeya, C.M., Ryder, R.W. *et al.* (1989) Field testing and comparative evaluation of rapid, visually read screening assays for antibody to human immunodeficiency virus. *Lancet* **1**: 580–584.

Stewart, M.W., Etches, W.S., Russell, A.S. *et al.* (1993) Detection of antiphospholipid antibodies by flow cytometry: rapid detection of antibody isotype and phospholipid specificity. *Thromb Haemost* **70**: 603–607.

Strom, C.M., Janeszco, R., Quan, F. *et al.* (2006) Technical validation of a TM Biosciences Luminex-based multiplex assay for detecting the American College of Medical Genetics recommended cystic fibrosis mutation panel. *J Mol Diagn* **8**: 371–375.

Strunk, T., Hartel, C., Temming, P. *et al.* (2007) Erythropoietin inhibits cytokine production of neonatal and adult leukocytes. *Acta Paediatr.* **97**: 16–20.

Suarez-Alvarez, B., Lopez-Vazquez, A., Gonzalez, M.Z. *et al.* (2007) The relationship of anti-MICA antibodies and MICA expression with heart allograft rejection. *Am J Transplant* **7**: 1842–1848.

Syrjala, M.T., Tolo, H., Koistinen, J. and Krusius, T. (1991) Determination of anti-IgA antibodies with a flow cytometer-based microbead immunoassay (MIA) *J Immunol Methods* **139**: 265–270.

Systat Software, I. (2005) SigmaPlot. San Jose, CA, USA, Systat Software, Inc.

Tanaka, M., Komatsu, N., Yanagimoto, Y. *et al.* (2006) Development of a new diagnostic tool for pancreatic cancer: simultaneous measurement of antibodies against peptides recognized by cytotoxic T lymphocytes. *Kurume Med J* **53** (3–4): 63–70.

Tang, D., Yuan, R. and Chai, Y. (2007) Magnetic control of an electrochemical microfluidic device with an arrayed immunosensor for simultaneous multiple immunoassays. *Clin Chem* **53**: 1323–1329.

Tang, Y.W., Kilic, A., Yang, Q. *et al.* (2007) StaphPlex system for rapid and simultaneous identification of antibiotic resistance determinants and Panton-Valentine leukocidin detection of staphylococci from positive blood cultures. *J Clin Microbiol* **45**: 1867–1873.

Tozzoli, R. (2007) Recent advances in diagnostic technologies and their impact in autoimmune diseases. *Autoimmun Rev* **6**: 334–340.

Tripp, R.A., Jones, L., Anderson, L.J. and Brown, M.P. (2000) CD40 ligand (CD154) enhances the Th1 and antibody responses to respiratory syncytial virus in the BALB/c mouse. *J Immunol* **164**: 5913–5921.

Tuscano, J.M., Martin, S., Song, K. and Wun, T. (2005) B cell reductive therapy in the treatment of autoimmune diseases: a focus on monoclonal antibody treatment of rheumatoid arthritis. *Hematology* **10**: 521–527.

Tuscano, J.M. and Sands, J. (2009) B cell reductive therapy with rituximab in the treatment of rheumatoid arthritis. *Biologics* **3**: 225–232.

Ullenhag, G.J., Spendlove, I., Watson, N.F. *et al.* (2006) A neoadjuvant/adjuvant randomized trial of colorectal cancer patients vaccinated with an anti-idiotypic antibody, 105AD7, mimicking CD55. *Clin Cancer Res* **12**: 7389–7396.

Urdea, M., Penny, L.A., Olmsted, S.S. *et al.* (2006) Requirements for high impact diagnostics in the developing world. *Nature* **444** Suppl 1: 73–79.

Van Cleve, M., Ostrerova, N., Tietgen, K. *et al.* (1998) Direct quantitation of HIV by flow cytometry using branched DNA signal amplification. *Mol Cell Probes* **12**: 243–247.

Verkaik, N., Brouwer, E., Hooijkaas, H. *et al.* (2008) Comparison of carboxylated and Penta-His microspheres for semi-quantitative measurement of antibody responses to His-tagged proteins. *J Immunol Methods* **335**: 121–125.

Vignali, D.A.A. (2000) Multiplexed particle-based flow cytometric assays. *J Immunol Methods* **243**: 243–255.

Wiklund, M. (2008) Enhancement of bead-based immunoassays by the use of acoustic radiation force. *J Acoust Soc Am* **123**: 3794.

Wolf, J.S., Li, G., Varadhachary, A. *et al.* (2007) Oral lactoferrin results in T cell-dependent tumor inhibition of head and neck squamous cell carcinoma in vivo. *Clin Cancer Res* **13**: 1601–1610.

Yeom, H.J., Her, Y.S., Oh, M.J. *et al.* (2008) Application of multiplex bead array assay for Yq microdeletion analysis in infertile males. *Mol Cell Probes* **22**: 76–82. Epub 2007 Jun 30.

Yurkovetsky, Z.R., Kirkwood, J.M., Edington, H.D. *et al.* (2007) Multiplex analysis of serum cytokines in melanoma patients treated with interferon-alpha2b. *Clin Cancer Res* **13**: 2422–2428.

Zagursky, R.J., Sharp, D., Solomon, K.A. and Schwartz, A. (1995) Quantitation of cellular receptors by a new immunocytochemical flow cytometry technique. *Biotechniques* **18**: 504–509.

Zartman, J.K., Foreman, N.K., Donson, A.M. and Fleitz, J.M. (2004) Measurement of tamoxifen-induced apoptosis in glioblastoma by cytometric bead analysis of active caspase-3. *J Neurooncol* **67**: 3–7.

Zhang, A., Pastor, L., Nguyen, Q. *et al.* (2005) Small interfering RNA and gene expression analysis using a multiplex branched DNA assay without RNA purification. *J Biomol Screen* **10**: 549–556.

Zhong, H., Han, B., Tourkova, I.L. *et al.* (2007) Low-dose paclitaxel prior to intratumoral dendritic cell vaccine modulates intratumoral cytokine network and lung cancer growth. *Clin Cancer Res* **13**(18 Pt 1): 5455–5462.

12

Mass Spectrometry in Metabolomics

William J. Griffiths and Yuqin Wang

Introduction

The metabolome represents the natural compliment to genome and proteome. In its most basic form the metabolome comprises the entire complement of small molecules (typically <1000 Da) in a cell, tissue, biofluid or organism. Inclusive components may be endogenous molecules, but also exogenous compounds derived from food and pharmaceutical sources. At present it is not possible to characterize the entire metabolome; however, it is possible to get a snap-shot of the metabolome by profiling the more abundant and readily detected molecules. Metabolite detection depends on the analytical technology used and at present the dominating technologies are nuclear magnetic resonance (NMR) and mass spectrometry (MS), although other spectroscopes are also used, e.g. Fourier Transform infrared (FT-IR), Raman (Viant 2008; Scherling 2010). In this chapter we will restrict our attention to MS methods which are currently dominating the field.

Metabolomics can be approached from two quite differing starting points leading to targeted and untargeted endpoints. In an untargeted (global) approach the concept is to profile the metabolome in an unbiased fashion. In practise this is of course impossible as all analytical methods are themselves biased. However, within the confines of analytical bias, untargeted metabolomics can be used as a hypothesis-generating exercise, by acquiring data and subsequently generating a hypothesis. Targeted metabolomics on the other hand uses analytical bias to its advantage allowing the analytical technology to focus on the class of metabolite of interest. Targeted

Molecular Analysis and Genome Discovery, Second Edition. Edited by Ralph Rapley and Stuart Harbron.
© 2012 John Wiley & Sons, Ltd. Published 2012 by John Wiley & Sons, Ltd.

metabolomics is hypothesis driven as the metabolites of interest are predicted to be of importance in the study.

There are numerous reasons for the current interest in metabolomics. Many endogenous metabolites represent the end-products of enzyme-catalysed reactions and can thus be considered to reflect the phenotype more closely than either expressed proteins (proteome) or mRNA (transcriptomics). This leads to the idea of biomarker discovery, where a single metabolite or profile of metabolites is characteristic of a disease. This concept is not new and has been exploited for many years in clinical chemistry (Chace 2005), and for decades MS-generated metabolite profiles in urine and plasma have been used to diagnose inborn errors of metabolism in children (Sjövall 2004; Clayton 2003; Setchell and Heubi 2006). While inborn errors of metabolism are often a consequence of a single enzymatic defect (e.g. in 7-dehydrocholesterol reductase, the enzyme responsible for one of the final steps in cholesterol biosynthesis resulting in Smith Lemli Opitz syndrome; Porter and Herman 2011), and are often simply identified by an elevated level of enzyme substrate in plasma (Griffiths *et al.* 2008b), most diseases are multifactorial with less clearly defined markers. Much effort is currently being invested in metabolomics for biomarker discovery for diseases ranging from Alzheimer's disease (Greenberg *et al.* 2009) to nonalcoholic fatty liver disease (Puri *et al.* 2009) to cancer (Qui *et al.* 2010). Metabolomics is also important in the study of fundamental biochemical processes and those of the system (Weckwerth 2010), where only by analysing multiple analytes can the biology of the system be adequately described.

In the following sections we will discuss the various facets of the generic metabolomic pipeline, starting with sample collection and finishing with metabolite identification (Figure 12.1). In a final section we will discuss some applications of untargeted and targeted metabolomics.

Sample collection and preparation

These are probably the most important steps in any metabolomic study, whether its goal is biomarker discovery or an investigation of fundamental biochemistry. Distortion of the sample at the collection/storage and processing phase will inevitably have downstream consequences when data is analysed. Depending on the biological matrix, samples should be stored at $-80\,°C$ or $-20\,°C$ after collection, and freeze–thaw cycles minimized. It is advisable wherever possible to perform stability studies to evaluate optimal storage conditions.

Sample preparation depends greatly on whether subsequent MS analysis is by 'shotgun' – also called 'direct infusion' – electrospray ionization (ESI)-MS, liquid chromatography (LC)-MS, or gas chromatography (GC)-MS, and whether a targeted or global metabolic approach is adopted. Sample preparation itself introduces a targeted aspect to all metabolomic studies as the solvents used in each study dictate the metabolites which can be analysed. For example, in many metabolomic analyses

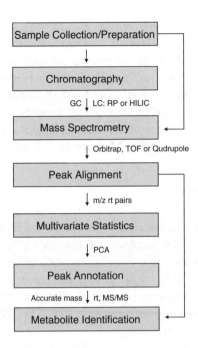

Figure 12.1 The generic metabolomics pipeline

of plasma or serum samples, preparation involves protein precipitation in organic solvent, centrifugation and reconstitution of soluble metabolites in highly aqueous solvents (Dunn *et al.* 2008a; Zelena *et al.* 2009; Michopoulos *et al.* 2009). This excludes the analysis of many lipids for the simple reason that they do not re-dissolve in the highly aqueous solvent. This is not a problem in itself as long as the analyst is aware that they are analysing only a section of the metabolome. Metabolite solubility is more of an issue in studies of plasma and serum than of urine, as many lipids are transported in blood attached to proteins as they are not soluble in the direct aqueous environment of blood. This issue of solubility dictates the separate analysis of many lipids and the development of lipidomics (Dennis 2009; Shevchenko and Simons 2010; Quehenberger *et al.* 2010).

Reproducibility of sample preparation is another important issue in all metabolomic studies. One way to monitor the reproducibility of sample preparation is by the addition of internal standards at an early step in the sample preparation process. A wide range of internal standards should be used, covering the major metabolite groups to be analysed (Mohamed *et al.* 2009). If internal standards are not added, abundant metabolites common to all samples may be used as a reference to which other metabolites can be normalized (Griffiths *et al.* 2010). When a targeted approach is used, the classical MS method of isotope dilution MS can be used. Here stable isotope analogues of representative analytes are added during extraction. Losses of analyte during sample processing and handling will be identical for endogenous molecules and their stable isotope surrogates, allowing correction for losses to be made. In the absence of isotope-labelled standards, structural analogues can be used.

For monitoring the reproducibility of subsequent analytical steps it is advisable to generate a quality control (QC) sample. This usually is made up, following sample processing, from a combination of aliquots from all samples to be analysed. In this way the QC sample will be representative of all samples to be analysed, and allows the monitoring of the performance of the analytical steps (Dunn *et al.* 2008a; Zelena *et al.* 2009; Michopoulos *et al.* 2009; Mohamed *et al.* 2009).

Data acquisition

Data acquisition in MS-based metabolomics can be subdivided into four processes: sample introduction, ionization, analysis and detection (Figure 12.2). There are currently three basic formats used in metabolomics: GC-MS, direct-infusion or shotgun ESI-MS and LC-MS (Griffiths *et al.* 2010), and each places their own specific requirements on the MS setup to be used. Unlike GC-MS and LC-MS, shotgun ESI-MS does not include an on-line separation step. This reduces the dimensionality of the data but simplifies the analysis.

Chromatography

GC achieves separation of analytes by partition between a mobile gas phase and a liquid-bound stationary phase. A prerequisite for GC is the presence of analytes in the vapour phase. GC is thus suitable for volatile and thermally stable compounds or those made so by derivatization. In fact GC and GC-MS were first used in metabolomic studies as far back as the 1960s, although at this time 'omics' was yet to enter the scientific vocabulary (Horning *et al.* 1963; Pauling *et al.* 1971). Currently, GC-MS is widely used by the plant metabolomics community (Scherling 2010; Weckwerth 2010; Galindo *et al.* 2009), but the requirements of volatility and thermal stability constrain its use in metabolomic studies of animal samples. Nevertheless,

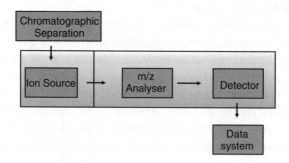

Figure 12.2 Steps in MS data acquisition

simple derivatization protocols are available that enhance the analyte volatility and thermal stability of many biomolecules, thereby allowing their analysis by GC-MS (Blau and Halket 1993). Almost by definition, the application of derivatization chemistry introduces a targeted aspect to GC-MS based metabolomics. However, the unsurpassed resolution provided by capillary GC columns and the reproducibility of the retention index makes GC-MS an important methodology. GC-MS protocols and studies have been published for the metabolomic analysis of biofluids including plasma and serum (Fiehn and Kind 2007; O'Hagan *et al.* 2005; Begley *et al.* 2009; Zhang *et al.* 2007).

It was not until the introduction of ESI by John Fenn (Nobel Prize in Chemistry 2002) and colleagues that LC could be routinely coupled with MS. ESI works well with solvents based on methanol or acetonitrile and water and is thus most compatible with reversed-phase (RP)-LC. In RP chromatography the stationary phase has a non-polar coating (e.g. C_{18}) while the mobile phase is polar, usually water/acetonitrile or water/methanol with organic acid (e.g. 0.1% formic acid) or buffer (e.g. 10 mM ammonium acetate) added. Non-polar molecules are most strongly retained by the column with polar ones being eluted first. Gradient elution is performed with the content of organic solvent in the mobile phase increasing with time. For the metabolomic analysis of biofluids, which are rich in polar molecules, RP-LC is not ideal as there may be insufficient retention of polar metabolites. An alternative to RP-LC is normal phase (NP)-LC, where the more non-polar molecules are eluted first, but mobile phases typically exclude water thereby limiting analyte solubility. A second alternative to RP-LC is hydrophilic interaction (HILIC)-LC, where the separation principle is similar to NP chromatography, retention being based on polar interactions between analytes and the polar stationary phase (Cubbon *et al.* 2010), but with a mobile phase made up with water and water-miscible organic solvents such as methanol and acetonitrile, and being fully compatible with ESI-MS. A gradient is used where the water content of the mobile phase increase with time (Cubbon *et al.* 2010). As RP-LC and HILIC-LC offer complementary modes of retention, there are advantages in analysing samples by both RP-LC-ESI-MS and HILIC-LC-ESI-MS.

Both RP-LC and HILIC-LC columns come in many dimension and with differing sizes of packing materials. In recent years there has been a shift towards smaller particle size, and now many studies use columns packed with <2 μm particles in ultra-performance liquid chromatography (UPLC) or ultra (U)HPLC (Michopoulos *et al.* 2009; Nordström *et al.* 2006). Columns packed with smaller particles offer enhancement in resolution or speed of separation, although high back pressures are generated. As ESI is essentially a concentration-dependent process, there are obvious advantages in the sharper peaks provided by UPLC (<1 s half height), but very sharp peaks can challenge the scanning speed of the mass spectrometer, where older instruments may have scan times of about 1 s.

The proteomics community have widely adopted nanoflow-LC. This provides a sensitivity advantage by exploiting high-concentration, low volume chromatographic peaks and also more efficient ionization and ion sampling. While flow rates for conventional 2 mm i.d. columns may be 250 µl/min, when the column dimension is scaled down to 100 µm, i.d. flow rates are reduced to 250 nl/min. Nano-LC has not been widely adopted by the metabolomics community to date, but has been used for specialist applications where maximum sensitivity is required and sample quantities are limited (Liu *et al.* 2003). A penalty of nano-LC is long run times and subsequent reduced throughput. An added disadvantage is the problem of carryover.

It has been noticed by a number of groups that it is necessary to pre-condition the LC system (whether the system is a nano- or conventional-LC, HPLC or UPLC) with repeated injection of a representative biological extract (e.g. QC sample) to be able to subsequently achieve reproducible results (Zelena *et al.* 2009; Michopoulos *et al.* 2009; Mohamed *et al.* 2009). Importantly, iterative injections of standards fail to prime the system adequately. To avoid these issues, some groups omit on-line chromatography altogether, and analyse samples by direct infusion ESI-MS (Beckman *et al.* 2008; Boernsen *et al.* 2005). Chip-based-ESI – when linked to high-resolution, high-mass accuracy MS instruments – can provide a detailed fingerprint of the metabolome, but it is susceptible to ion-suppression effects which limit dynamic range and the absence of chromatography removes the resolution of isomers and of overlapping isotopic peaks from different metabolites. Ion-mobility (IM) offers an alternative strategy for separating ions. Ion-mobility separates ions according to their mobility in a buffer gas and can be linked to MS. Metabolomic studies have been performed on plasma exploiting ESI-IM-MS (Dwivedi *et al.* 2010).

Ionization techniques

When using GC-MS, electron ionization (EI) or chemical ionization (CI) are mostly used. In the positive-ion mode, EI relies on high-energy electrons colliding with analyte molecules and ionizing them by ejection of an electron. EI tends to generate low-abundance molecular ions $[M]^+$ and more abundant fragment-ions. This can be advantageous as libraries of EI spectra have been compiled that can be used for compound identification. The NIST library contains over 100 000 spectra. A disadvantage of EI can be the low abundance or absence of the molecular ion, making identification of unknowns difficult. CI is a less energetic form of ionization and is achieved by ionization of a reagent gas and subsequent ionization of the analyte in an ion-molecule reaction. If methane is the reagent gas, ionization will be achieved by protonation generating $[M+H]^+$ ions. Few fragment-ions are usually observed. Negative-ion CI can be used for specialist applications where the analytes are derivatized with electron capturing functionalities.

ESI is the dominant ionization mode in metabolomics; its compatibility with RP solvents allows its routine interface with RP-LC. ESI is mostly used in the positive-ion mode where ionization is via protonation or addition of some other cation.

In the absence of sample desalting on a LC column, analytes are often observed ionized with different adducts, e.g. $[M+H]^+$, $[M+Na]^+$, $[M+K]^+$. This is usually seen as a disadvantage as the analyte signal becomes diluted over several ions and the spectrum made more complex. The multiplexing of ions will also lead to difficulties when quantitative comparisons are made. However, advantages can be taken of ion multiplexing and a software MZedDB – taking advantage of the propensity of different analytes to give adduct ions – has been developed to enhance ion annotation and identification (Draper *et al.* 2009). Analytes that are basic are most readily ionized in the positive-ion mode, while acidic analytes are best analysed in the negative-ion mode. However, derivatization strategies exist to improve the ionization characteristics of poorly ionized analytes making them accessible to ESI (O'Maille *et al.* 2008). There are considerable advantages to performing ESI-MS studies in both the positive- and negative-ion modes. Depending on the mass spectrometer used, this may involve duplicate injections, but the two ionization modes will favour the ionization of complementary functional groups, that is acids or bases (Nordström *et al.* 2008). Other atmospheric pressure ionization (API) methods available are APCI and AP-photoionization. These ionization modes may be used to compliment ESI.

A relatively new API technique is desorption electrospray ionization (DESI). This requires minimal sample preparation and is achieved by electrospraying droplets at the target, analytes become ionized and vaporized and can be sampled by the source of the mass spectrometer (Ifa *et al.* 2010). Lipidomics is one area where DESI is being exploited (Manicke *et al.* 2009). Matrix-assisted laser desorption/ionization (MALDI) is also used in some metabolomic studies but has the problem of low-mass matrix ions complicating the metabolite spectra.

Mass (m/z) analysers

Current metabolomic studies use a wide range of mass analysers, extending from low-cost, small footprint ion-traps and single quadrupoles, to high-cost, large footprint Fourier transform (FT) ion cyclotron resonance (ICR) instruments. Whatever the mass analyser, its function is to determine the mass to charge ratio (m/z) of ions generated in the ion source and pass them on to the ion-detector for ion-counting (Figure 12.2).

Quadrupole mass filters are generally used for low-resolution (unit mass) studies, where high-mass accuracy (<0.1 Da) is not required. Quadrupole mass filters can be arranged in series separated by a quadrupole (or other multipole) ion guide that also acts as a collision cell in tandem mass spectrometers (MS/MS). MS/MS instruments can be operated in a number of ways. Just one analyser can be scanned in which case a basic mass spectrum is generated (MS^1). Alternatively, the first mass analyser (MS_1) can be set to transmit an ion of interest, which then collides with gas introduced into the collision cell, and the final mass analyser (MS_2) is scanned to record the resultant fragment-ion masses, i.e. MS/MS or MS^2. This is called a product-ion scan.

Other scans such as precursor-ion and neutral-loss scans can also be performed on tandem quadrupoles. In a precursor-ion scan, MS_2 is set to transmit a fragment-ion of interest and MS_1 scanned to identify precursor-ions that generate this fragment-ion. In a neutral-loss scan, MS_1 and MS_2 are scanned in parallel, but at an offset equivalent to the neutral-loss of interest. While product-ion scans are important for the structural determination of ions, precursor-ion and neutral-loss scans are used to identify ions with a particular structural feature. Precursor-ion and neutral-loss scans are widely used by the lipidomic community particularly in a methodology called multidimensional mass spectrometry (MDMS) (Han and Gross 2003, 2005).

Quadrupole mass filters separate ions of different m/z according to their behaviour in an electric field. The quadruple was invented by Wolfgang Paul (Nobel Prize in Physics 1989) who also invented the quadrupole ion-trap that similarly separates ions of different m/z according to their behaviour in an electric field. Like quadrupole mass filters quadrupole ion-traps tend to be used in applications where high resolution and high mass accuracy are not essential. However, as with quadrupole mass filters quadrupole ion-traps can be used in special circumstances to give high mass resolution or high mass accuracy (March 2009). The greatest advantage of the ion-trap is its ability to perform not only MS/MS (MS^2) but also MS/MS/MS (MS^3) and further stages of fragmentation (i.e. MS^n). This can generate considerable structural information, but is limited by the initial ion-current. A limitation of the ion-trap is the one-third rule, whereby the bottom one-third of the MS^n spectrum is lost.

In recent years, quadrupole ion-traps have been combined with high-resolution, high-mass accuracy analysers, e.g. in the LTQ-FT and LTQ-Orbitrap. The ion-trap is used to control the number of ions entering the FTICR cell or Orbitrap during the recording of high-resolution (up to 1 000 000, full width at half maximum, FWHM), accurate mass (<5 ppm) spectra, or where fragment-ions generated in the MS^n process in the ion-trap are mass (m/z) analysed in the FTICR or Orbitrap. Such instruments are now widely exploited by the metabolomics community (Dunn *et al.* 2008a; Beckman *et al.* 2008; Draper *et al.* 2009; Southam *et al.* 2007). As an alternative to an ion-trap preceding the FTICR cell, a quadrupole filter can be similarly used.

Another MS format popular in metabolomics is the combination of ESI with time-of-flight (TOF) mass analysers in the ESI-TOF and ESI-Q-TOF type instruments Zelena *et al.* 2009; Michopoulos *et al.* 2009; Mohamed *et al.* 2009; Nordström *et al.* 2006, 2008; Bruce *et al.* 2008; Want *et al.* 2010a,b). The ESI-TOF can generate mass spectral data at moderately high resolution (10 000–20 000, FWHM) and mass accuracies of <5ppm, while the Q-TOF type instruments have the added capabilities of being able to record MS/MS spectra. Newer instruments have fast scan times (≤ 0.2 s) making them compatible with the sharp chromatographic peaks of UPLC. A recent development in LC-ESI-TOF technology is the application of so called MS^E where mass spectra are alternately recorded at a very low collision voltage (<10 V) and then at elevated collision voltage (>20 eV) without mass selection. By using fast scan times and reconstructed ion chromatograms (RIC) over a chromatographic peak it is possible to match fragment-ions in the elevated collision voltage spectra

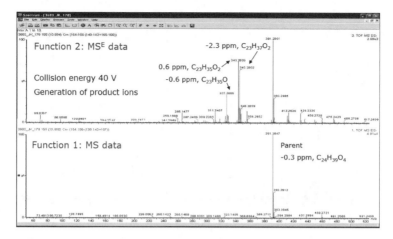

Figure 12.3 MS^E analysis of chenodeoxycholic acid, a metabolite of cholesterol found in urine, plasma and brain

to precursor-ions in the low collision energy spectra (Want *et al.* 2010a, 2010b) (Figure 12.3).

GC-MS analysis often uses single or tandem quadrupoles for mass analysis, although GC-TOF and GC-ion-traps are also commonly used in metabolomic work (Fiehn and Kind 2007; O'Hagan *et al.* 2005; Begley *et al.* 2009; Dunn *et al.* 2008b). The recent combination of GC with APCI now allows GC-MS and LC-MS to be performed on the same mass spectrometer (Figure 12.4).

Data analysis

All metabolomic studies generate a vast amount of data which require statistical processing. When chromatography is linked to MS, data consist of ion abundance at a given *m/z* and retention time (rt). Comparison of data sets is complicated by drift in both retention time and mass measurement accuracy. Comparison of data is further complicated by variation in instrument sensitivity with time. These drifts can be 'more' or 'less' corrected by instrument vendor software or freely downloadable software developed by independent metabolomic research groups (Smith *et al.* 2006; Tautenhahn *et al.* 2011).

Data treatment

The mass spectrometer comes with its own software that constructs lists of ion abundance against *m/z*. To these lists is added a time dimension to generated 3D ion intensity, *m/z,* rt plots. Following chromatographic alignment between runs it

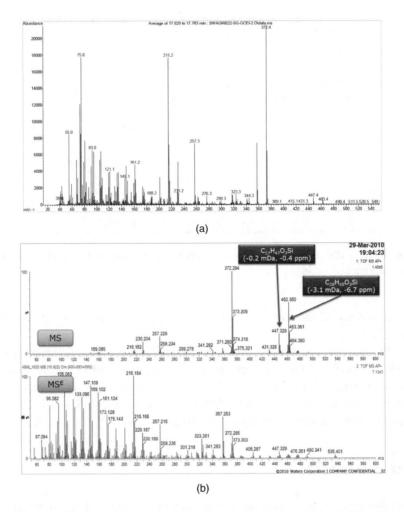

Figure 12.4 GC-EI-MS and GC-APCI-MS spectra of lithocholic acid

is necessary to generate a 'peak' or 'feature' list and use multivariate statistics to discover differences between samples that may be otherwise hidden. Peak picking software is provided by instrument manufacturers but is also available from different research groups in a freely downloadable form (Smith *et al.* 2006; Tautenhahn *et al.* 2011; Katajamaa and Oresic 2005; Broeckling *et al.* 2006). Metabolomic experiments are performed on multiple samples, thus multivariate statistics and pattern recognition methods are required to handle the data. Principle component analysis (PCA) is an unsupervised method often used in metabolomics to reduce data dimensionality, investigate clustering tendency and detect outliers (Figure 12.5). PCA is often followed by an analysis technique such as O-PLS (orthogonal projection latent structures), which is a supervised method where a training set is used to build a

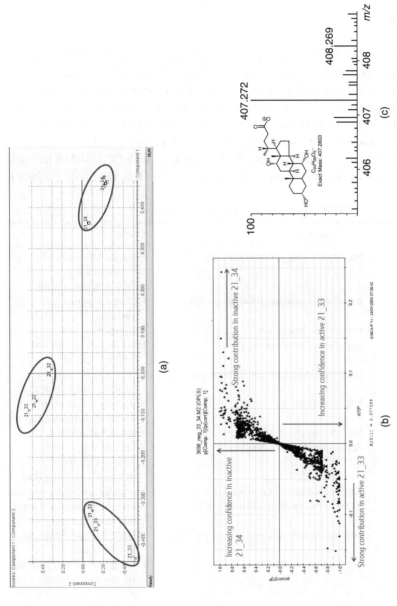

Figure 12.5 (A) PCA scores plot showing the variability between three different HPLC fractions 21_32, 21_33 and 21_34. Each fraction was analysed by LC-ESI-MS on a Q-TOF instrument in triplicate. (B) O-PLS S-plot comparing HPLC fractions 21_33 and 21_34. The ions contributing 'strongly' to variation between the fractions are indicated. (C) ESI-MS of one of the most statistically significant ions found in the O-PLS between fractions 21_33 and 21_34. Fraction 21_33 has a biological activity which is not found in fractions 21_32 or 21_34

model and estimate necessary parameters. Multivariate statistical analysis will point to peaks in the LC-ESI-MS data set that show difference in signal intensity between samples and potentially correspond to biomarkers.

Databases

Once a list of ions of interest (m/z, rt pairs) has been generated the next step is peak annotation and chemical identification. There are a number of databases in existence that can be searched to match mass values against chemical structure. These databases contain different amounts of information ranging from predicted mass and InCHI identifier to tissue location and metabolic pathway. A list of some of the available databases with web addresses is given in Table 12.1.

A list of contaminants potentially found when working with LC-ESI-MS has also been published and can be very useful when analysing data lists (Keller *et al.* 2008).

Table 12.1   Some metabolomic databases

Database	Web Address	Comment
NIST	http://www.nist.gov/srd/analy.cfm	Includes MS data, 191 436 EI spectra, 14 802 MS/MS spectra
KEGG	http://www.genome.jp/kegg/ligand.html	Includes enzymes and pathways
BioCyc	http://www.biocyc.org	Includes enzymes
ARM	http://www.metabolome.jp	Includes m/z search and MS/MS spectra
METLIN	http://metlin.scripps.edu/ions/neutral-loss	Search by *m/z*, also by MS/MS and fragment
HMDB	http://hmdb.ca/	Search by *m/z*, also by MS/MS, includes enzymes
Tumor Metabolome	http://metabolic-database.com/	Tumour metabolites
ChemSpider	http://www.chemspider.com/	Search by exact mass, includes chemical properties and suppliers
PubChem	http://pubchem.ncbi.nlm.nih.gov/	Search by molecular formula
EBI	http://www.ebi.ac.uk/chebi/init.do	Search by molecular formula and average mass
AAFS	http://www.ualberta.ca/~gjones/mslib.htm	Mini MS library of drugs and metabolites
DrugBank	http://www.drugbank.ca/	FDA-approved drugs, search by chemical formula
Lipid Bank	http://www.lipidbank.jp/index.html	Lipid data, includes EI-MS information
Lipid Search	http://lipidsearch.jp/LipidNavigator.htm	Lipid data
Lipid Maps	http://www.lipidmaps.org/	Lipid data, includes MS/MS spectra and m/z search

A number of groups have now built in-house databases made up of various combinations of the databases listed in Table 12.1 (Mohamed *et al.* 2009; Draper *et al.* 2009; Brown *et al.* 2009). For example, the Manchester Metabolomics Database (MMD) is built up from HMDB, KEGG, Lipid Maps, BioCyc and DrugBank and is available from the constructors (Brown *et al.* 2009). The Manchester group have also constructed UPLC-MS and GC-MS libraries of over 1000 commercially available standards. Similarly, MZedDB http://maltese.dbs.aber.ac.uk:8888/hrmet/index.html is a comprehensive archive generated from publically accessible databases (Draper *et al.* 2009). MZedDB, like many of the databases in Table 12.1 can be searched by *m/z* data using tools enabled on the web. Other databases (e.g. METLIN, HMDB and Lipid Maps) also allow database searches of MS/MS spectra.

Metabolite Identification

Once metabolites of interest have been annotated by database searching the next step is 'true' identification. This can be achieved by comparison of exact mass, retention time and MS/MS spectra of the metabolites of interest, to those of authentic standards. Alternatively, sufficient material may be collected for structural determination by NMR. In many cases authentic standards are not available and there is insufficient material for identification by NMR. In this case the researcher must hypothesize a structure based on existing exact mass, MS/MS spectra and retention time data and make a 'presumptive identification'. The application of microchemical reactions can be useful for confirming the presence of functional groups, but for a definitive identification, an authentic standard should be purchased or synthesized.

Applications

Metabolomic studies are numerous. Considerable efforts have been made by the plant community (Scherling *et al.* 2010), and intense endeavours have been made by pharmaceutical companies (Zelena *et al.* 2009; Michopoulos *et al.* 2009; Want *et al.* 2010b). It is beyond the scope of this chapter to consider more than a couple of applications of metabolomics and we will restrict our discussion to the metabolomic and lipidomic profiling of plasma/serum and CSF by LC-MS.

Plasma/serum and CSF profiling

The fact that metabolites found in plasma/serum are either water soluble or are bound to circulating proteins indicates that LC-MS is the preferred method of analysis. Nevertheless there have been a number of studies where GC-MS has been used following derivatization (Fiehn *et al.* 2007; Begley *et al.* 2009; Dunn *et al.* 2008b;

Jiye *et al.* 2005). However, the extent of the metabolome observed is defined more by the method of metabolite extraction and subsequent sample handling than by the ultimate detection method used. Table 12.2 summarizes a number of metabolomic studies performed on plasma/serum and CSF. All of the global metabolomic studies listed used metabolite extraction and protein precipitation in solutions of methanol. Then, following centrifugation, the supernatant is blow-down or lyophilized and reconstituted in the LC-MS solvent or, for GC-MS, a solvent appropriate for derivatization. Despite their presence in blood, many hydrophobic lipids are not soluble in highly aqueous methanol solutions, and thus will be excluded from further analysis. Additionally, the use of highly aqueous LC-mobile phases also discriminates against many lipids. It seems that many practitioners of 'global' metabolomics ignore these facts. Despite this, many potential metabolites can be observed in plasma/serum following a generic protocol based on extraction into methanol.

The metabolomics groups at the Scripps Institute (Nordström *et al.* 2006, 2008; O'Maille *et al.* 2008; Smith *et al.* 2006; Tautenhahn *et al.* 2011; Want *et al.* 2006, 2007; Crews *et al.* 2009) and at Manchester University (Dunn *et al.* 2008a, 2008b; Zelena *et al.* 2009; O'Hagan *et al.* 2005; Begley *et al.* 2009; Brown *et al.* 2009) have been particularly active in the development of methods for the metabolomic profiling of plasma/serum. Following metabolite extraction in methanol, centrifugation and reconstitution into LC-mobile phase LC-MS is performed. After data acquisition both groups use the software XCMS (Smith *et al.* 2006) to align data and generate feature tables. A feature being defined as an *m/z* peak with a signal to noise ratio of 10:1. The number of features observed depends on the chromatography used, and the scanning speed, resolution, sensitivity and dynamic range of the MS used. Data from different sample groups are then compared using multivariate statistics. From Table 12.2 it can be seen that the number of features observed in plasma/serum ranges from about 500 to 5000. The reproducibility of the observed features also varies and there seems to be a growing consensus that relative standard deviation (RSD) should be below 30% and preferably below 20% for replicate injections of the same sample. The Manchester group use multiple injections of a QC sample, made up of aliquots of all samples in a batch, dispersed throughout the batch. This allows monitoring of the precision of data acquisition throughout a batch of samples. To attain high reproducibility of data, many groups noted the requirement to condition the LC column before use. This can best be achieved by multiple injections of the QC sample prior to analysis of a batch of samples.

To date most published metabolomic studies have been more directed to method development and feature discrimination rather than compound identification. Compound identification is the current bottleneck in metabolomics, largely because of a lack of authentic standards and the sheer number of features listed.

In contrast to plasma/serum there have been few metabolomic studies of CSF. Those performed, however, have been more concerned with compound identification (Wishart *et al.* 2008; Stoop *et al.* 2010).

Table 12.2 Metabolomic and lipidomic studies of plasma/serum and CSF

Human Study	Extraction	GC- or LC-MS	Processing/ No. features	No. ID/ Annotation	Comment	Reference
Metabolomics						
Plasma profiling	1:9 (serum:90% MeOH), centrifugation (19 600 g × 10 min), derivatized to MO-TMS[1]	GC-TOF-MS	MATLAB[2]/>500 features[3] EI+, RSD <30%, 5 replicates	80 ID (LOD 0.1 −1 μM)	Pooled plasma, development of extraction and derivatization methods	Jiye et al. 2005
Serum profiling	3:1:9 (serum:H_2O:70%MeOH 30% CH_3CN), centrifugation 13 487g × 15 min), derivatized to MO-TMS[1]	GC-TOF-MS	ChromTof[2]/>200 features[3] EI, CV <20%, 3 replicates		Pooled serum, stability study	Dunn et al. 2008b
Serum profiling	1:3 (serum:MeOH), centrifugation (15 800g × 15 min), derivatize to MO-TMS[1]	GC-TOF-MS	ChromTof[2]/250 features[3] EI+, median RSD 18%, 18 replicates		Pooled serum, method development	Begley et al. 2009
Serum profiling	1:2 (serum:MeOH), centrifugation (13 200g × 10 min), reconstitute in 5% CH_3CN	HPLC-MS[4]	XCMS[2]/>2000 features[5] ESI+, ave RSD 25%, 6 replicates	4 ID[6]	Pooled serum, optimization of extraction	Want et al. 2006, 2007
Serum profiling	1:2 (serum:MeOH), centrifugation (13 200g × 10 min), reconstitute in MeOH/ iPrOH/acetone/H_2O (2.5:1.25:1.25:95)	HPLC-MS[4]	XCMS[2]/512 features[5] ESI+, 503 features[5] ESI-		Pooled serum, evaluation of multiple ionization methods	Nordström et al. 2008

(continued overleaf)

Table 12.2 (*continued*)

Human Study	Extraction	GC- or LC-MS	Processing/ No. features	No. ID/ Annotation	Comment	Reference
Plasma profiling	1:4 (plasma:MeOH), centrifugation (2053g × 10 min), reconstitute in 80% MeOH	UPLC-TOF-MS	MarkerLynx or MATLAB2/ >700 features[5] ESI+	10 ID[6]	Pooled serum, method development	Bruce *et al.* 2008
Serum profiling	1:3 (serum:MeOH), centrifugation (13 487g × 15 min), reconstitute in H$_2$O	UPLC-Orbitrap-MS	XCMS[2]/>1000 features[5] ESI-, RSD<20%, 14 injections		Pooled serum, instrument evaluation, also study of pre-enclampsia	Dunn *et al.* 2008a
Serum profiling	1:3 (serum:MeOH), centrifugation (15 871g × 15 min), reconstitute in H$_2$O	UPLC-TOF-MS	XCMS[2]/>1000 features[5] ESI+, RSD<20%, 60 injections, >1100 features[5] ESI-		Pooled serum, method development, importance of QC injection	Zelna *et al.* 2009
Serum profiling	1:3 (serum:MeOH), centrifugation (13 487g × 15 min), reconstitute in H$_2$O	UPLC-Orbitrap-MS	XCMS[2]/2079 features[5] ESI+, 4513 features[5] ESI-	ESI- 40% ESI+ 42% feature annotation[7,8]	Pooled serum, database development (MMD 42 687 entries)[7]	Brown *et al.* 2009

Plasma and CSF profiling	1:4 [plasma (CSF):MeOH (^{00}C)], centrifugation (13 000 rpm × 15 min), reconstitute in 5% CH_3CN	HPLC-TOF-MS	XCMS[2]/5300 features[5] ESI+ plasma and CSF, median CV 16% plasma, 15% CSF, 4 replicates	12 ID[6] CSF and plasma	Pooled plasma and CSF, also biological variation, median CV 46% plasma, 35% CSF	Crews *et al.* 2009
CSF profiling	Direct injection	UPLC-FTMS	200 features[5] ESI+/-	17 ID	Pooled CSF, search against HMDB[9]	Wishart *et al.* 2008
CSF profiling	1:4 (CSF:90% MeOH), centrifugation (5000 rpm × 5 min), derivatize to MO-TMS[1]	GC-MS[4]	AMDIS[2]/43 features[3] EI+	41 ID (1 < −400 μM)	Pooled CSF, ID and quantification	Wishat *et al.* 2008
CSF profiling	1:4 (CSF:MeOH), centrifugation (10000*g* × 10 min), derivatize to TMS[1]	GC-MS[4]	target lists/93 features[3] EI+, RSD <20%, 6 replicates	89 ID	Pooled CSF, also biological variation RSD 15–85%, in-house database	Stoop *et al.* 2010
Lipidomics Plasma lipid profiling	Folch-like ($CHCl_3$:MeOH:serum:H_2O, 66:132:10:200), centrifuge (10000*g*), lower phase, dilute 1:1 with iPrOH	UPLC-TOF-MS	Mass Hunter, Mass Profiler Professional2/ 1322 features[5] ESI+, 840 features[5] ESI-, RSD<20%, triplet extraction	55 ID	Method development, LPC (5), LPE (2), PC (22), PE (3), PI (1), SM (6), CE (5), DAG (4), TAG (7)	Sandra *et al.* 2010

(continued overleaf)

Table 12.2 (*continued*)

Human Study	Extraction	GC- or LC-MS	Processing/ No. features	No. ID/ Annotation	Comment	Reference
Plasma lipidomics	1:50:12.5 (plasma:MTBE:H$_2$O), centrifugation (4000 rpm), upper phase diluted 1:10 in CHCl$_3$:MeOH:iPrOH (1:2:4)	nanoflow-Orbirap-MS	LipidX^2/	95 ID	Screening of controls and hypertension patients, LPC (5), PC (13), PC-O (13), PE (8), PE-O (5), SM (10), CE (5), C, DAG (4), TAG (31),	Graessler *et al.* 2009
Plasma lipidomics	category specific	LC-MS/MS[10]		588 ID	NIST human plasma standard reference material (SRM 1950), LPC (12), PC (31), LPE (7), PE (38), PS (20), PG (16), PA (15), PI (19), N-acyl-PS (2), SP (204), ST (36), PL (8) FA (107),	Quehenberger *et al.* 2010

Steroidomics

Plasma lipidomics (steroidomics)	Modified Bligh/Dyer (plasma/CHCl3[3]/MeOH), alkaline hydrolysis (KOH in MeOH), modified Bligh/Dyer, NP-SPE	LC-MS/MS[10]	36 ID (1 nM–3.76 mM)	NIST human plasma standard reference material (SRM 1950), total and free sterol quantification[11], CE (22), free sterols 14	Quehenberger et al. 2010
Plasma sterol profiling	Alkaline hydrolysis, 1:10 (plasma:0.35M KOH in EtOH), Folch-like extraction (EtOH,CHCl3,0.15M NaCl), NP-SPE, derivatize to TMS[1]	GC-MS[4] with SIM	9 ID (4–400 nM)	Total oxysterol quantification[11]	Dzeletovic et al. 1995
Plasma sterol profiling	1:10 (plasma:EtOH), RP-SPE, with and without derivatization with GP reagent, RP-SPE	LC-Orbitrap-MS and nanoflow-Orbitrap-MS	14[12] ID, 38[12] + 5[13] annotated, (1 nM–3 μM)	Quantification of sterols/steroids	Griffiths et al. 2008; Ogundare et al. 2010
CSF oxysterol profiling	Alkaline hydrolysis, 1: 20 (CSF:0.35M KOH in EtOH), Folch-like extraction (EtOH,CHCl3,0.9% NaCl), NP-SPE, derivatize to TMS[1]	GC-MS[4] with SIM	7 ID (2 nM)	Total oxysterol quantification[11], control and multiple sclerosis patients	Leoni et al. 2005

(continued overleaf)

Table 12.2 (*continued*)

Human Study	Extraction	Processing/ No. features GC- or LC-MS	No. ID/ Annotation	Comment	Reference
CSF oxysterol profiling	1:10 (CSF:EtOH), RP-SPE, with derivatization with GP reagent, RP-SPE	LC-Orbitrap-MS	12[12] ID, 26[12] annotated (0.03–20 nM)	Quantification of sterols/steroids	Ogundare et al. 2010

Abbreviations: MeOH, methanol; MO, methyloxime; TMS, trimethylsilyl; LOD, limit of detection; CV, coefficient of variation; iPrOH, propan-2-ol; rpm, revolutions per min; MTBE, methyl-*tert*-butylether; SIM, selected ion monitoring; EtOH, ethanol.

Compound class: LPC, lysoglycerophosphocholine; PC, glycerophosphocholine; PC-O alk(en)yl ether linked PC; LPE, lysoglycerophosphoethanolamine; PE, glycerophosphoethanoloamine; PE-O alk(en)yl ether linked PE; PI, glycerophosphoinositol; PS, glycerophosphoserine; PG, glycerophosphoglycerol; PA, glycerophosphate; SP, sphingolipid; SM, sphingomyelin; ST, sterol lipid; CE, cholesterol ester; C, cholesterol; DAG, diacylglycerol; TAG, triacylglycerol; PL, prenol lipid; FA, fatty acyls.

[1]Oxo groups converted to methyloximes, alcohol groups converted to TMS ethers.

[2]MATLAB software 6.5 (Mathworks, Natick, MA); ChromaTof v2.12 or v3.25 (Leco, St Joseph, MO); XCMS http://metlin.scripps.edu/xcms/; MarkerLynx (Waters); AMDIS GC/MS software http://www.amdis.net/; MassHunter and Mass Profiler Professional software (Agilent Technologies); LipdX in-house software Max Plank Institute, Dresden.

[3]In GC-MS studies number of features relates to the number of chemical entities.

[4]Single quadrupole analyser.

[5]In LC-MS a feature is defined as a *m/z* peak usually with a s/n of >10:1.

[6]Compound ID was not the aim of the study.

[7]MMD (Manchester Metabolomics Database) available from developers http://dbkgroup.org/MMD/.

[8]Annotated by molecular formula derived from exact mass ±5 ppm.

[9]Search against http://www.csfmetabolome.ca/ part of HMDB http://hmdb.ca/.

[10]Mostly LC-MS/MS with multiple reaction monitoring.

[11]Total sterol concentrations are levels after alkaline hydrolysis of esters.

[12]Compound annotated by exact mass, retention time and MS[3], and identified by reference to authentic standards.

[13]Compounds annotated by exact mass and MS[3].

Targeted metabolomics (lipidomics)

For several decades the lipid community have adopted a targeted approach to metabolomics, although using the term metabolite profiling rather lipidomics (Sjövall *et al*. 2004; Dennis 2009). This is particularly true for sterol/steroid analysis where profiles have been generated in many body fluids including plasma and CSF (Sjövall *et al*. 2004; Dzeletovic *et al*. 1995; Leoni *et al*. 2005; Griffiths and Wang 2009). Studies targeting on the global plasma lipidome have been performed using LC-MS (Sandra *et al*. 2010) and also direct infusion MS, the latter exploiting the high resolution capability of the LTQ-Orbitrap (Graessler *et al*. 2009). However, the deepest study of the plasma lipidome comes from the Lipid Maps consortium (Dennis 2009; Quehenberger *et al*. 2010). They used different extraction methods and LC-MS protocols for the differing classes of lipids and were able to identify and quantify over 500 distinct molecular species from a pooled plasma sample representative of the population of the USA. Considering the sterol lipids, in total they were able to identify 36 molecular species, out of which 14 different sterols were released after alkaline hydrolysis of fatty acid esters. Levels ranged from 1.45 mg/mL (3.76 mM) for esterified cholesterol to 0.36 ng/mL (1 nM) for 25-hydroxycholesterol. Deuterated surrogate standards were used for quantification and precision was excellent (RSD <5%).

We have also investigated the plasma steroidome. We used LC-MS^n and positive-ion ESI on an LTQ-Orbitrap instrument following a 'charge-tagging' step designed to improve the sensitivity of sterol/steroid analysis. Charge-tagging as the name implies involves derivatising the target analyte with a charge-carrying group. Using this method we are able to uncover a greater portion of the steroidome than had previously been possible (Figure 12.6). Not all sterols and steroids are suitable for charge-tagging due to the absence of a suitable functional group for derivatisation. However, as many sterols/steroids are present in plasma as conjugates of sulphuric and/or glucuronic acid or as bile acids, application of negative-ion ESI can reveal these metabolites. Thus, in parallel to our derivatisation strategy we also use 'shotgun' lipidomics, where following a RP- solid phase extraction (SPE) step the steroid fraction is analysed by direct-infusion negative-ion ESI on an LTQ-Orbitrap, and compound annotation is based on exact mass and MS^3 spectra (Figure 12.7). It should be noted that in the absence of a chromatographic step isomers are not differentiated.

There have been far fewer metabolomic studies of CSF (Wishart *et al*. 2008; Stoop *et al*. 2010) than of plasma, this is also true with respect to lipidomics (O'Maille *et al*. 2008; Ogundare *et al*. 2010a,b) and steroidomics (Leoni *et al*. 2005), although a number of a number of studies targeted at specific metabolites have been performed (Kim *et al*. 2000). Using the charge-tagging strategy illustrated in Figure 12.6 we have been able to characterize 45 steroids and sterols in human CSF (Ogundare *et al*. 2010a,b).

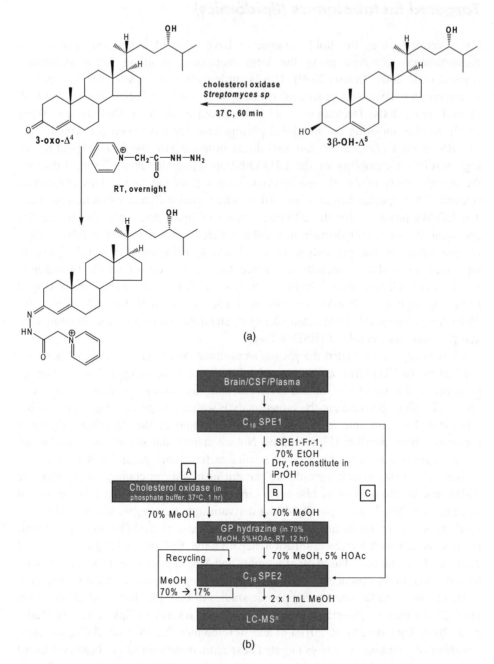

Figure 12.6 (A) Principal of charge-tagging with Girard P (GP) reagent. (B) Sample preparation. (C) Sterol analysis by LC-MS^n. (D) Spectra of GP-tagged 24S-hydroxycholesterol

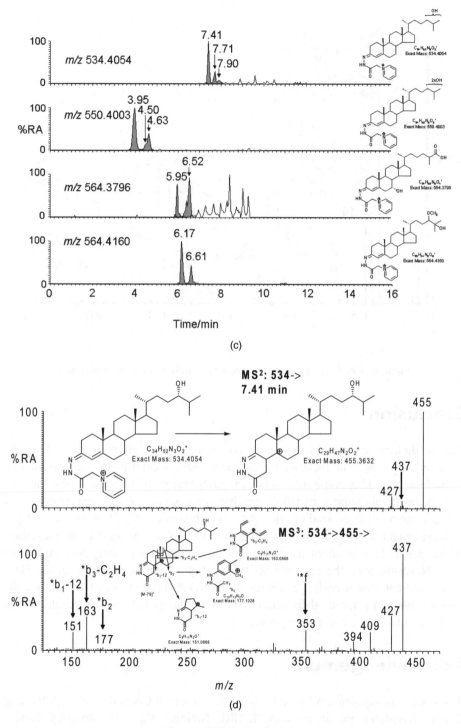

(c)

Time/min

(d)

Figure 12.6 (*continued*)

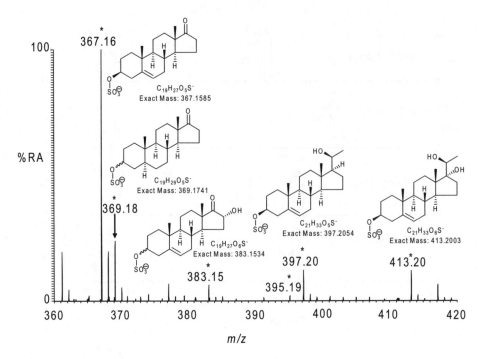

Figure 12.7   (A) Shotgun steroidomic analysis of human plasma

Conclusion

The metabolome constitutes the entire compliment of small molecules in a cell, tissue or body fluid. The metabolome can be thought to offer a closer representation of the biological phenotype than either the proteome or transcriptome and thus the study of the metabolome or metabolomics has great potential for biomarker discovery and in systems biology studies. However, metabolomics is still in its early stages of development. Currently, most metabolomic studies are biased to either to water soluble metabolites or alternatively towards lipids, and few investigators study both areas. Never-the-less, the example set by the Lipid Maps consortium in the USA where different groups each specialising in a different lipid class of lipid analyse the same sample and combine their data to determine the global lipidome could be one for the metabolomics community to follow.

Acknowledgement

This work was supported by funding from the Research Councils UK, BBSRC and EPSRC. Assistance provided by the EPSRC National Mass Spectrometry Service Centre is warmly acknowledged.

References

Beckmann, M., Parker, D., Enot, D.P. *et al.* (2008) High-throughput, nontargeted metabolite fingerprinting using nominal mass flow injection electrospray mass spectrometry. *Nat Protoc* **3**: 486–504.

Begley, P., Francis-McIntyre, S., Dunn, W.B. *et al.* (2009) Development and performance of a gas chromatography-time-of-flight mass spectrometry analysis for large-scale nontargeted metabolomic studies of human serum. *Anal Chem* Aug 15; **81**: 7038–7046.

Blau, K. and Halket, J.M. (1993) *Handbook of Derivatives for Chromatography*, 2nd edn. John Wiley & Sons, Ltd, Chichester.

Boernsen, K.O., Gatzek, S. and Imbert, G. (2005) Controlled protein precipitation in combination with chip-based nanospray infusion mass spectrometry. An approach for metabolomics profiling of plasma. *Anal Chem* Nov 15; **77**: 7255–7264.

Broeckling, C.D., Reddy, I.R., Duran, A.L. *et al.* (2006) MET-IDEA: data extraction tool for mass spectrometry-based metabolomics. *Anal Chem* Jul 1; **78**: 4334–4341.

Brown, M., Dunn, W.B., Dobson, P. *et al.* (2009) Mass spectrometry tools and metabolite-specific databases for molecular identification in metabolomics. *Analyst* **134**: 1322–1332.

Bruce, S.J., Jonsson, P., Antti, H., Cloarec, O., Trygg, J., Marklund, S.L. and Moritz, T. (2008) Evaluation of a protocol for metabolic profiling studies on human blood plasma by combined ultra-performance liquid chromatography/mass spectrometry: From extraction to data analysis. *Anal Biochem* **372**: 237–249.

Chace, D.H. and Kalas, T.A. (2005) A biochemical perspective on the use of tandem mass spectrometry for newborn screening and clinical testing. *Clin Biochem* **38**: 296–309.

Clayton, P.T. (2003) Diagnosis of inherited disorders of liver metabolism. *J Inherit Metab Dis* **26**: 135–146.

Crews, B., Wikoff, W.R., Patti, G.J. *et al.* (2009) Variability analysis of human plasma and cerebral spinal fluid reveals statistical significance of changes in mass spectrometry-based metabolomics data. *Anal Chem* **81**: 8538–8544.

Cubbon, S., Antonio, C., Wilson, J. and Thomas-Oates, J. (2010) Metabolomic applications of HILIC-LC-MS. *Mass Spectrom Rev* **29**: 671–684.

Dennis, E.A. (2009) Lipidomics joins the omics evolution. *Proc Natl Acad Sci USA* **106**: 2089–2090.

Draper, J., Enot, D.P., Parker, D. *et al.* (2009) Metabolite signal identification in accurate mass metabolomics data with MZedDB, an interactive m/z annotation tool utilising predicted ionization behaviour 'rules'. *BMC Bioinformatics* **10**: 227.

Dunn, W.B., Broadhurst, D., Brown, M. *et al.* (2008a) Metabolic profiling of serum using ultra performance liquid chromatography and the LTQ-Orbitrap mass spectrometry system. *J Chromatogr B Analyt Technol Biomed Life Sci* **871**: 288–298.

Dunn, W.B., Broadhurst, D., Ellis, D.I. *et al.* (2008b) A GC-TOF-MS study of the stability of serum and urine metabolomes during the UK Biobank sample collection and preparation protocols. *Int J Epidemiol* **37** Suppl 1:i23–30.

Dwivedi, P., Schultz, A.J. and Hill, H.H. (2010) Metabolic Profiling of human blood by high resolution Ion Mobility Mass Spectrometry (IM-MS). *Int J Mass Spectrom* **298**:7 8–90.

Dzeletovic, S., Breuer, O., Lund, E. and Diczfalusy, U. (1995) Determination of cholesterol oxidation products in human plasma by isotope dilution-mass spectrometry. *Anal Biochem* **225**: 73–80.

Fiehn, O. and Kind, T. (2007) Metabolite profiling in blood plasma. In *Metabolomics Methods and Protocols* (ed. Weckwerth, W.) *Methods in Molecular Biology*, Volume 358, Part I, Humana Press Inc., New Jersey, pp. 3–17.

Galindo, F.G., Dejmek, P., Lundgren, K. *et al.* (2009) Metabolomic evaluation of pulsed electric field-induced stress on potato tissue. *Planta* **230**: 469–479.

Graessler, J., Schwudke, D., Schwarz, P.E. *et al.* (2009) Top-down lipidomics reveals ether lipid deficiency in blood plasma of hypertensive patients. *PLoS One* **4**: e6261.

Greenberg, N., Grassano, A., Thambisetty, M. *et al.* (2009) A proposed metabolic strategy for monitoring disease progression in Alzheimer's disease. *Electrophoresis* **30**: 1235–1239.

Griffiths, W.J., Hornshaw, M., Woffendin, G. *et al.* (2008a) Discovering oxysterols in plasma: a window on the metabolome. *J Proteome Res* **7**: 3602–3612.

Griffiths, W.J., Koal, T., Wang, Y. *et al.* (2010) Targeted metabolomics for biomarker discovery. *Angew Chem Int Edit Engl* **49**: 5426–5445.

Griffiths, W.J. and Wang, Y. (2009) Analysis of neurosterols by GC-MS and LC-MS/MS. *J Chromatogr B Analyt Technol Biomed Life Sci* **877**: 2778–2805.

Griffiths, W.J., Wang, Y., Karu, K. *et al.* (2008b) Potential of sterol analysis by liquid chromatography-tandem mass spectrometry for the prenatal diagnosis of Smith-Lemli-Opitz syndrome. *Clin Chem* **54**: 1317–1324.

Han, X. and Gross, R.W. (2003) Global analyses of cellular lipidomes directly from crude extracts of biological samples by ESI mass spectrometry: a bridge to lipidomics. *J Lipid Res* **44**: 1071–1079.

Han, X. and Gross, R.W. (2005) Shotgun lipidomics: electrospray ionization mass spectrometric analysis and quantitation of cellular lipidomes directly from crude extracts of biological samples. *Mass Spectrom Rev* **24**: 367–412.

Horning, E.C., Vandenheuvel, W.J., Creech, B.G. (1963) Separation and determination of steroids by gas chromatography. *Methods Biochem Anal* **11**: 69–147.

Ifa, D.R., Wu, C., Ouyang, Z. and Cooks, R.G. (2010) Desorption electrospray ionization and other ambient ionization methods: current progress and preview. *Analyst* **135**: 669–681.

Jiye, A., Trygg, J., Gullberg, J. *et al.* (2005) Extraction and GC/MS analysis of the human blood plasma metabolome. *Anal Chem* **77**: 8086–8094.

Katajamaa, M. and Oresic, M. (2005) Processing methods for differential analysis of LC/MS profile data. *BMC Bioinformatics* **6**: 179.

Keller, B.O., Sui, J., Young, A.B. and Whittal, R.M. (2008) Interferences and contaminants encountered in modern mass spectrometry. *Anal Chim Acta* **627**: 71–81.

Kim, Y.S., Zhang, H. and Kim, H.Y. (2000) Profiling neurosteroids in cerebrospinal fluids and plasma by gas chromatography/electron capture negative chemical ionization mass spectrometry. *Anal Biochem* **277**: 187–195.

Leoni, V., Lütjohann, D. and Masterman, T. (2005) Levels of 7-oxocholesterol in cerebrospinal fluid are more than one thousand times lower than reported in multiple sclerosis. *J Lipid Res* **46**: 191–195.

Liu, S., Sjövall, J. and Griffiths, W.J. (2003) Neurosteroids in rat brain: extraction, isolation, and analysis by nanoscale liquid chromatography-electrospray mass spectrometry. *Anal Chem* **75**: 5835–5846.

McDonald, J.G., Thompson, B.M., McCrum, E.C. and Russell, D.W. (2007) Extraction and analysis of sterols in biological matrices by high performance liquid chromatography electrospray ionization mass spectrometry. *Methods Enzymol* **432**: 145–170.

Manicke, N.E., Nefliu, M., Wu, C. *et al.* (2009) Imaging of lipids in atheroma by desorption electrospray ionization mass spectrometry. *Anal Chem* **81**: 8702–8707.

March, R.E. (2009) Quadrupole ion traps. *Mass Spectrom Rev* **28**: 961–989.

Michopoulos, F., Lai, L., Gika, H. *et al.* (2009) UPLC-MS-based analysis of human plasma for metabonomics using solvent precipitation or solid phase extraction. *J Proteome Res* **8**: 2114–2121.

Mohamed, R., Varesio, E., Ivosev, G. *et al.* (2009) Comprehensive analytical strategy for biomarker identification based on liquid chromatography coupled to mass spectrometry and new candidate confirmation tools. *Anal Chem* **81**: 7677–7694.

Nordström, A., O'Maille, G., Qin, C. and Siuzdak, G. (2006) Nonlinear data alignment for UPLC-MS and HPLC-MS based metabolomics: quantitative analysis of endogenous and exogenous metabolites in human serum. *Anal Chem* **78**: 3289–3295.

Nordström, A., Want, E., Northen, T. *et al.* (2008) Multiple ionization mass spectrometry strategy used to reveal the complexity of metabolomics. *Anal Chem* **80**: 421–429.

O'Hagan, S., Dunn, W.B., Brown, M. *et al.* (2005) Closed-loop, multiobjective optimization of analytical instrumentation: gas chromatography/time-of-flight mass spectrometry of the metabolomes of human serum and of yeast fermentations. *Anal Chem* **77**: 290–303.

O'Maille, G., Go, E.P., Hoang, L. *et al.* (2008) Metabolomics relative quantitation with mass spectrometry using chemical derivatization and isotope labeling. *Spectroscopy* **22**: 327–343.

Ogundare, M., Griffiths, W.J., Lockhart, A. *et al.* (2010) Analysis of oxysterols and cholestenoic acids in plasma and CSF by charge-tagging and LC-MS[n]. Book of Abstracts, 31st Annual Meeting of the British Mass Spectrometry Society, Cardiff, September 5–8 (2010).

Ogundare, M., Theofilopoulos, S., Lockhart, A. *et al.* (2010) Cerebrospinal fluid steroidomics: are bioactive bile acids present in brain? *J Biol Chem* **285**: 4666–4679.

Pauling, L., Robinson, A.B., Teranishi, R. and Cary, P. (1971) Quantitative analysis of urine vapor and breath by gas-liquid partition chromatography. *Proc Natl Acad Sci USA* **68**: 2374–2376.

Porter, F.D. and Herman, G.E. (2011) Malformation syndromes caused by disorders of cholesterol synthesis. *J Lipid Res* **52**: 6–34.

Puri, P., Wiest, M.M., Cheung, O. *et al.* (2009) The plasma lipidomic signature of nonalcoholic steatohepatitis. *Hepatology* **50**: 1827–1838.

Qiu, Y., Cai, G., Su, M. *et al.* (2010) Urinary metabonomic study on colorectal cancer. *J Proteome Res* **9**: 1627–1634.

Quehenberger, O., Armando, A.M., Brown, A.H. *et al.* (2010) Lipidomics reveals a remarkable diversity of lipids in human plasma. *J Lipid Res* **51**: 3299–3305.

Sandra, K., Pereira Ados, S., Vanhoenacker, G. *et al.* (2010) Comprehensive blood plasma lipidomics by liquid chromatography/quadrupole time-of-flight mass spectrometry. *J Chromatogr A* **1217**: 4087–4099.

Scherling, C., Roscher, C., Giavalisco, P. *et al.* (2010) Metabolomics unravel contrasting effects of biodiversity on the performance of individual plant species. *PLoS One* **5**: e12569.

Setchell, K.D. and Heubi, J.E. (2006) Defects in bile acid biosynthesis – diagnosis and treatment. *J Pediatr Gastroenterol Nutr* **43** Suppl 1:S17–22.

Shevchenko, A. and Simons, K. (2010) Lipidomics: coming to grips with lipid diversity. *Nat Rev Mol Cell Biol* **11**: 593–598.

Sjövall, J. (2004) Fifty years with bile acids and steroids in health and disease. *Lipids* **39**: 703–722.

Smith, C.A., Want, E.J, O'Maille, G. *et al.* (2006) XCMS: processing mass spectrometry data for metabolite profiling using nonlinear peak alignment, matching, and identification. *Anal Chem* **78**: 779–787.

Southam, A.D., Payne, T.G., Cooper, H.J. *et al.* (2007) Dynamic range and mass accuracy of wide-scan direct infusion nanoelectrospray fourier transform ion cyclotron resonance mass spectrometry-based metabolomics increased by the spectral stitching method. *Anal Chem* **79**: 4595–4602.

Stoop, M.P., Coulier, L., Rosenling, T. *et al.* (2010) Quantitative proteomics and metabolomics analysis of normal human cerebrospinal fluid samples. *Mol Cell Proteomics* **9**: 2063–2075.

Tautenhahn, R., Patti, G.J., Kalisiak, E. *et al.* (2011) metaXCMS: second-order analysis of untargeted metabolomics data. *Anal Chem* **83**: 696–700.

Want, E.J., Coen, M., Masson, P. *et al.* (2010a) Ultra performance liquid chromatography-mass spectrometry profiling of bile acid metabolites in biofluids: application to experimental toxicology studies. *Anal Chem* **82**: 5282–5289.

Want, E.J., Nordström, A., Morita, H. and Siuzdak, G. (2007) From exogenous to endogenous: the inevitable imprint of mass spectrometry in metabolomics. *J Proteome Res* **6**: 459–468.

Want, E.J., O'Maille, G., Smith, C.A. *et al.* (2006) Solvent-dependent metabolite distribution, clustering, and protein extraction for serum profiling with mass spectrometry. *Anal Chem* **78**: 743–752.

Want, E.J., Wilson, I.D., Gika, H. *et al.* (2010b) Global metabolic profiling procedures for urine using UPLC-MS. *Nat Protoc* **5**: 1005–1018.

Weckwerth, W. (2010) Metabolomics: an integral technique in systems biology. *Bioanalysis* **2**: 829–836.

Wishart, D.S., Lewis, M.J., Morrissey, J.A. *et al.* (2008) The human cerebrospinal fluid metabolome. *J Chromatogr B Analyt Technol Biomed Life Sci* **871**: 164–173.

Viant, M.R. (2008) Recent developments in environmental metabolomics. *Mol Biosyst* **4**: 980–986.

Zelena, E., Dunn, W.B., Broadhurst, D. *et al.* (2009) Development of a robust and repeatable UPLC-MS method for the long-term metabolomic study of human serum. *Anal Chem* **81**: 1357–1364.

Zhang, Q., Wang, G., Du, Y. *et al.* (2007) GC/MS analysis of the rat urine for metabonomic research. *J Chromatogr B Analyt Technol Biomed Life Sci* **854**: 20–25.

Index

Molecular Analysis and Genome Discovery, Second Edition. Edited by Ralph Rapley and Stuart Harbron.
© 2012 John Wiley & Sons, Ltd. Published 2012 by John Wiley & Sons, Ltd.